William Snow Harris

A Treatise on Frictional Electricity

in theory and practice

William Snow Harris

A Treatise on Frictional Electricity
in theory and practice

ISBN/EAN: 9783337405731

Printed in Europe, USA, Canada, Australia, Japan

Cover: Foto ©berggeist007 / pixelio.de

More available books at **www.hansebooks.com**

A TREATISE

ON

FRICTIONAL ELECTRICITY,

In Theory and Practice.

BY

SIR WILLIAM SNOW HARRIS, F.R.S.

EDITED, WITH A MEMOIR OF THE AUTHOR,

BY CHARLES TOMLINSON, F.R.S.

LONDON:

VIRTUE AND CO., 26, IVY LANE,

PATERNOSTER ROW.

1867.

LONDON:
PRINTED BY VIRTUE AND CO.,
CITY ROAD.

.

PREFACE BY THE EDITOR.

THE great work on Frictional Electricity with which Sir William Snow Harris proposed to close his scientific career he was not permitted to complete. The whole of the First Part up to page 205 of the present volume, had the benefit of his final revision. After his death, his papers were placed in my hands by Lady Harris and Mr. Harris, her son. I found them to be in a confused and unsatisfactory state, but I have done my best to complete the work in as methodical a manner as the materials under command would admit of. All that is given in the Second Part is in the actual language of the author, and in passing it through the press I have had the benefit of Mr. Harris's revision, which has in many cases been of value to the work, while it has given additional confidence to my own labours. Among Sir William's papers I found some notes intended to form the basis of a Preface to this work. I have retained them as they were written. It is probable that had the author revised his Preface he would have exercised his caution in mitigating its severity, or have withheld it altogether; but when a distinguished man is dead, we care more about knowing his opinions than his caution when alive in withholding them.

KING'S COLLEGE, LONDON,
1st *October*, 1867.

BIOGRAPHICAL NOTICE OF THE AUTHOR.

BY THE EDITOR.

THE life of a scientific man seldom calls for much exertion on the part of his biographer. Such a life may be even less eventful than that of a literary man; for not only has the latter a wider and more varied audience, but his work has more individuality and more character; it is more a reflection of the author's mind, more rounded and complete, than a scientific work: the one belongs to the man, the other to nature; the one remains for all time as an exponent of the author's mind, the other is soon absorbed into the general body of science, and its author more or less forgotten. This need not be regretted, since the object of science is the discovery of truth, not individual celebrity; and our love and gratitude are not the less due to the men who have laboured in the cause of truth, even though their work be quickly absorbed, and their books and memoirs soon cease to be read.

The author of the volume now before the reader completed the allotted term of man's life. He made some important discoveries in electrical science, laboured long and successfully to introduce into the Royal Navy his system of lightning conductors, was an excellent sailor, an accomplished musician, a lively intellectual companion, a steady friend, and he died honoured and beloved.

Such was William Snow Harris, and it will scarcely add to our interest to be told that he was born on the 1st of April, 1791, the only son of Thomas Harris, Esq., of Plymouth, whose family had settled in that town as solicitors as early as the year 1600. He was educated first at the Plymouth Grammar School, then for the medical profession, and he completed his studies at Edinburgh. He appears to

have been struck with Dr. Hope's lectures on chemistry, and has
related to the editor several amusing anecdotes of that distinguished teacher. Snow Harris began the practice of his profession
as a militia surgeon, and afterwards as a general practitioner in
Plymouth. But his great love of science, especially of electricity,
interfered with his medical practice; and after his marriage, in
1824, with the eldest daughter of Richard Thorne, Esq., of Pilton,
North Devon, he devoted himself chiefly to the cultivation of his
favourite science. He had already, in 1820, invented a new method
of arranging lightning conductors for ships, with a view to the
defence of the Royal Navy from the destructive effects of lightning. In the electrical and mechanical arrangements of this
system, the metal was permanently fixed in the masts, and
extended throughout the hull, so as to afford the required security
at all times and under all the variable circumstances in which the
ship might be placed.

Harris showed his affection for Edinburgh by laying before its
Royal Society a series of papers, the first of which was entitled,
"Experimental Inquiries concerning the Laws of Magnetic Forces."
This paper appeared in the "Transactions" in 1829, but it is dated
"Plymouth, July 1st, 1827." It contains an account of the Hydro-
static Magnetometer. The second paper, "On a New Electrometer,
and the Heat excited in Metallic Bodies by Voltaic Electricity,"
is dated May 5th, 1831, and was read 19th December of that year.
The third paper, dated April 5th, 1833, "On the Investigation of
Magnetic Intensity by the Oscillations of the Horizontal Needle,"
was read 6th January, 1834. By this time the author adopts the
title, "F.R.SS. Lond. and Ed."

Harris's introduction to the Royal Society arose from his skill
in the improvement and construction of instruments and apparatus,
of which there is abundant evidence in his early papers. While
Sir Humphry Davy was President he had been attracted by the
simplicity and ingenuity of the thermo-electrometer, and invited
Harris to lay an account of it before the Society. This was
accordingly done, and the paper "On the Relative Powers of various
Metallic Substances as Conductors of Electricity, by Mr. William
Snow Harris, of Plymouth, Surgeon," was read before the Society,
14th December, 1826, and appeared in the "Transactions" for
1827.

Harris was justly proud of the favour with which the President and also the Vice-President, Dr. Wollaston, regarded his inventions and discoveries. This was, perhaps, natural in the case of Wollaston, who delighted in manipulative skill; but it was remarkable in the case of Davy, who valued the *end* of scientific research, and rather despised than cherished the *means*.

Harris was elected into the Royal Society in 1831. His papers contributed to the Society in 1834, 1836, and 1839, on the elementary laws of electricity, contain his best work, and display in a striking manner the author's ingenuity and delicate manipulative skill. He was not satisfied with attaining his end by any means, but the means themselves were all-important, and were the subject of long and anxious thought and repeated trials, until at length the best means possible, under the circumstances, had been hit upon. This care in the selection and improvement of apparatus might seem to an ordinary observer to be often superfluous, but it led to success and to the thorough understanding of the conditions of success, so that the failure of an experiment in Harris's hands became next to impossible. The chapter in the present volume on Electrical Manipulation will give some idea of the author's scrupulous care in the management of apparatus. He left nothing to others; even the making of paste for attaching the tinfoil to his jars was a matter of study until the best recipe had been found. But with all this love of apparatus, and of its minute details, Harris had none of the spirit of a mere instrument maker. He knew that the best instrument does the best work under the guidance of the best mind; for, as Liebig has it, *Das Instrument macht ja das Werk nicht, sondern der menschliche Geist.* Nor was his ingenuity confined to his apparatus. There was not a room in his house that did not bear marks of the presence of an original mind. He converted the ceiling of his children's nursery into a planetarium, and the floor into a compass-card. He taught them book-binding and printing; he was ready to repair a roller-blind, or to keep in order his ingenious kitchen-range.

In 1835 the Copley Medal, "the olive crown" of the Royal Society, as Davy loved to call it, was bestowed on Harris, in recognition of the value of his papers on the laws of electricity of high tension. In 1839 his "Inquiries concerning the Elementary Laws of Electricity" formed the subject of the Bakerian Lecture, and

was rewarded by the Society agreeably to the bequest of Henry
Baker, Esq., F.R.S., made in 1774.

But, in the midst of these highly successful researches, Harris
had never ceased to exert himself in behalf of his favourite system
of lightning conductors for ships. So little was the subject known
at the time, that the most erroneous views were held respecting
the phenomena of a thunder-storm, and the action of conductors.
It was considered by many highly-educated persons, especially
among naval men, that conductors did more harm than good—that
they were even positively mischievous by attracting lightning to
the structure they were intended to protect. To set these doubts
and difficulties at rest, a mixed commission of naval and scientific
men was appointed by the Government to investigate and report
on lightning conductors for ships. The Committee met several
times in one of the rooms of the Royal Society at Somerset House,
when Harris gave his evidence, exhibited his experiments, and
answered the inquiries of the Commissioners. Dr. Wollaston took
great interest in this inquiry. On one occasion it was proposed to
send the charge of a large battery through a mixed system of conductors in a curved form. It was asked what the effect would be
if a thin wire was extended along the chord of the curve. Harris
replied that the discharge would be along the chord. Wollaston
then took out of his pocket some of his fine platinum wire, and
joined with it the two ends of the curve. Some difficulty
arose as to how to support the curve. Harris at once said, " I
will hold it." The Committee begged him not to run so much
risk. He replied, " If I were not certain of my principles I
should be ashamed to appear before you. There is no danger."
The discharge went through the platinum wire, which, of course,
disappeared in vapour. It is needless to say that the conclusion
arrived at by the Commissioners was, " that the fixed conductors
of Mr. Harris are superior to all others, and they earnestly recommend their general adoption into the Royal Navy."

Still, however, the official mind was not satisfied. Ten ships
had been ordered to be fitted with the conductors, trials of their
value had been made in various parts of the world, and, although
experience fully justified their use, an order was given to
have the copper conductors stripped off every ship as it came
into dock. About this time some granite chimneys had been

erected in the Victualling Department at Devonport, and the engineer, Sir John Rennie, had fitted one of them with lightning conductors on Harris's plan, and was about to protect the other in a similar manner, when an order was given to remove the conductor from the protected chimney, and to leave the other chimney as it was. On the receipt of this order the engineer waited on Harris in a state of great excitement, and, after talking the matter over, Harris said, " Well, never mind, Nature will avenge us." Never was prophecy more truly uttered, for before the order could be obeyed a thunder-storm passed over Plymouth, the lightning struck the unprotected chimney, and made a great rent in its side, while the protected chimney was of course uninjured. It happened, fortunately for the cause of science and good sense, that the British Association held its meeting in Plymouth soon after this event, and it was a common joke during the week to taunt Harris by pointing to the damaged chimney. Harris, however, smiled, and said nothing, for he knew how fruitful the accident had been to him. His conductors not only got into favour with the Admiralty, but his scientific discoveries were now recognised in the same quarter, and he was recommended to the Government as worthy of an annuity of £300 " in consideration of services in the cultivation of science."

But the conductors were not yet admitted into the Royal Navy. To remove the prejudices which still existed on the subject of protection against lightning, in the minds of naval men and others, Harris published, in 1843, his well-known work on Thunder-Storms. The author requested the editor to see the work through the press, and this led to a friendship which was broken only by death. In the preparation of this work, and of several pamphlets on protection from lightning, &c., subsequently published, the editor was in frequent communication with Harris, either personally or by letter. It was Faraday's theory of induction by means of contiguous particles that first made the phenomena of the lightning-stroke clear, and Harris was the first to take advantage of the theory in his book. In order to be quite sure that he represented the theory fairly, the early sheets of the book were sent to Faraday, who returned them with approval and without a single alteration.

Allusion has already been made to the uncertain opinions held even by scientific men on the subject of thunder-storms. The following extract from a letter from Harris to the editor will show that he at least had clear ideas, and knew how to express them. Referring to his work on Thunder-Storms, then going through the press, he says :—

" I think you will find the last section very interesting, as bearing forcibly upon the results of our experience of the protecting effect of conductors ; and I think I have succeeded in treating satisfactorily all the cases in which buildings armed with conductors are said to have been damaged.

" I read in Arago's *Notices Scientifiques*, at pp. 597, 598, in the *Annuaire* for 1838, an account of the damage done to an angle of the *magasin* at Bayonne having a pointed conductor. This *is a capital case for me*. I disagree with Arago and all of them about it, and think I can show that it proves, beyond a question, the wonderful effect of the conductor in that instance in saving the building. Arago says that the Colonel of Artillery at Bayonne reported on it, and that M. Gay-Lussac subsequently drew up a memoir for a section of the Academy of Sciences. The accident occurred in February, 1829, so that I suppose the account must be found in the volumes of the Academy. As I have the vanity to think I can put the Academy right about this, I am extremely anxious to ascertain at what distance the south-west angle (damaged by the lightning) was from the conductor, and if they have given any diagrams or descriptions sufficiently clear to put the representation on paper. By Arago's account I take it that the arrangement was something of this kind, in which C is the conductor, and A the point damaged. I mean to write to Arago about it. If, however, in the meantime you should happen to be at the Museum, and could just look into the memoirs of the Academy for 1829, you would soon see if such a report is there, and if there are any drawings. I only want two things :—

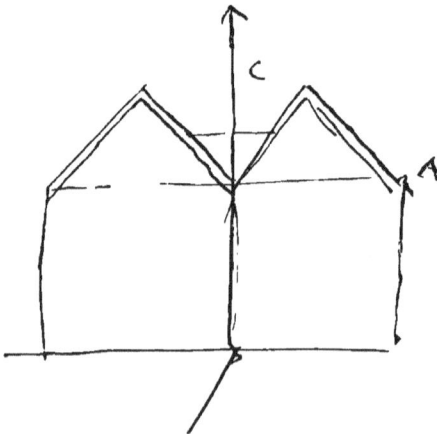

" First. The distance of conductor from the damaged angle.

" Second. A general notion of the position of the conductor on the building, something in the way of the sketch I have just given.

" This building was evidently struck by the bifurcated discharge in two places, viz., on the angle connected with the conductor, and likewise on the conductor itself, and really that the conductor carried off the whole without damage to the building. The point of the angle on which the shock fell would naturally enough be ruffled."

Soon after the date of the above he writes (6th April, 1843) :—

"I have found the notice of Gay-Lussac's Report in the *Annales de Chimie*, tome xl., p. 386—398 ; but it is, to a certain extent, very vague and unsatisfactory, and does not give me the information I want by any means. It is a clumsy affair altogether. I have written to Arago about it. It was of the utmost importance to have given the relative positions of the conductor and the point of the building struck, and the distance.

"I have also found the date of the explosion at Brescia, but I do not know where to find the original accounts.

"I must now make the best of what I have. I think I see the elements of a more extensive and complete work on Lightning than has hitherto appeared. I have obtained an amazing amount of cases from the Logs of the Navy."

The work on Thunder-Storms did not attract much attention, and, indeed, can scarcely be said to have paid its expenses. One of Harris's most intimate friends, Professor Daniell, of King's College,* expressed to the editor his regret that Harris should have treated the subject in a separate volume. " It would," said he, " have made a splendid chapter in Harris's large work on Electricity ;" for even at this period Harris had such a work in contemplation. But his object was not so much the sale of the book as the facility it afforded him of distributing information respecting his system in certain influential quarters, as will appear by the following letter, dated August 10th, 1843, which also shows that he was not indifferent to the circulation of information among the general public ; the first part of the letter referring to a notice of his work which the editor had been requested to write for a popular magazine.

"Nothing can be better than your condensed view of my work intended for the * * * * * *Magazine*. The only fault I find with it is the rather partial reference to myself. However, I am quite content to *put up* with that, especially at your hands. I think you should be cautious, lest it appear that the object of the article was to uphold my system of conductors as applied to the Navy; but I think the rapid and clear condensation of the matter of my book is very admirable.

---

* Professor Daniell included Frictional Electricity in his annual course at King's College, and it was his earnest wish that Harris should accept a Professorship, and deliver these lectures once a year. He felt that it would reflect honour on the College, and that the Council would readily make the appointment; but Harris declined the offer, fearing that it would restrict his movements.

"With respect to the idea I have advanced relative to the suffocating odour attendant on electrical discharges, and that the electrical explosion is accompanied by ignited matter dragged into the track of the discharge, it appears to me worth enlarging on, and made subject for your original thinking. It is stated in the *Nautical Magazine*, p. 439, vol. i., that in the case of the smack *London*, struck by lightning off St. Abb's Head, 'the hold was filled with sulphureous smoke, and a coating of black bituminous matter left on a large extent of surface.' This suffocating smell of sulphureous matter is the most marked on shipboard. You say it is not always compared to that of burning sulphur; but in all the cases in the logs of our ships there is not an instance to the contrary. The sailors always describe the odour as an intolerable stench of sulphur and brimstone. See the last three numbers of the *Nautical Magazine* for my publishing cases of damage by lightning in the Navy.

"And now, having despatched your affair, I proceed to thank you for the plumbago, &c., and to lay myself under fresh obligations to you. First, will you be so good as to inquire of Mr. Parker (to whom I beg kindly to be remembered) whether the book is bound fit to be presented to Prince Albert? If so, to let me know. Second, I wish to send copies of my work to Quetelet, of Brussels, and to Arago; for, although the latter has not treated me very civilly, yet it will be well to send him the book, I think, as I understand that some inquiries on the subject are going on in France. Will you, then, be so good as to get two copies, to be placed to my account? Enclose the note herewith sent, and the book, in a cover, directed to Professor Quetelet, Brussels, and leave it with Mr. Robertson, the Assistant Secretary of the Royal Society. I think, also, it would be well to secure with a little paste to the first blank leaf the strip of paper with the address on it herewith sent.

"I think Robertson could also send the work to Mons. Arago, and I send notes to be served in the same way for this great scientific aristocrat, who has never had the good manners to reply to any one communication I ever sent him. Quetelet is an intimate friend and a most excellent fellow, so I must treat him as such; the other chap I treat on the *noli me tangere* principle."

Another example of the confidence Harris had in his own knowledge of his favourite science is afforded at the meeting of the British Association at Plymouth already referred to. He had to give one of the Evening Lectures illustrated by experiments, before the assembled Association, on his own electrical discoveries, and the day was very wet. Several of the leading members condoled with him on the unfavourable state of the weather, and even advised him to put off the lecture. He accepted their condolence, but declined to act on their advice. He knew his own resources, and relied on them. By means of his peculiar heating-irons he surrounded his apparatus with a warm dry atmosphere, and the result was a brilliant success. He has described to the editor how anxious he was for success; for he had, in the physical section, been opposing some of Coulombe's views, and had tried in vain to shake the sectional President's (Dr. Whewell) belief in the French theory. He has told us how quickly his heart beat as he measured off different quantities of electricity by means of his unit jar, lest the disc, counterbalanced by different weights, should not go down at the exact times required to prove his law, that with a constant surface and at a constant distance, the attraction increases as the square of the quantity of electricity.

Few persons are aware of the long-continued struggle Harris had to undergo to impress upon the public mind the importance of adopting his system of lightning conductors for the ships of the Royal Navy, and few are aware of the varied means used for the purpose. He contributed a number of papers to the *Nautical Magazine* illustrative of damage by lightning; he was always on the watch for the slightest scent of a good case; and he never gave it up until he had tracked it to the ship's log deposited in Somerset House, or obtained an account from the captain or one of the officers of the ship that had been struck. He embodied these cases in letters and pamphlets, which he circulated among Members of Parliament and various persons in authority, including the foreign ambassadors; and it may be mentioned to the honour of the Emperor of Russia, that Harris's system was adopted in the Russian navy before it was fully admitted into our - own. In 1845 the Emperor presented Harris with a valuable ring and a superb vase, in acknowledgment of the merits of his system. Harris also instituted a series of experiments, on a large scale, in Plymouth Sound, showing that he could direct the discharge to any point of the vessel, or to the sea, at pleasure; and he made the points of discharge evident by firing gunpowder. These experiments attracted public notice, and crowds assembled on the Hoe to witness them. This led to a ludicrous circumstance, which Harris related to the editor with great glee. One evening an old woman was passing along the Hoe, when a man called her attention to summer lightning that was flashing in the horizon. "Don't talk to me about summer lightning," remonstrated the incredulous dame; "it is that Dr. Harris playing some of his tricks. If he doesn't take care, he will play them once too often."

Harris not only interested himself in protecting ships from lightning, but he also endeavoured to get his system applied to public buildings. He drew up a long list of buildings that had been damaged, not nearly so full and complete as his list of ships,* but still a formidable indictment against folly and prejudice. He even addressed a memorial to a Church Building Society, pointing out the necessity of protecting every new church that was built. He invited the editor to accompany him to hear the verbal reply, which was to this effect, that the cost of fitting conductors to a church—viz., from £60 to £100—was a fatal objection; for in many cases this additional charge in the estimates would most likely turn the scale against the church being built at all.

At length all difficulties in the way of his long-cherished object were removed or overcome. All the various objections to his conductors had been met: persons in authority had declared that letting the copper bands into the masts weakened them; but Harris proved experimentally that their powers of resistance to flexure were increased. The flagstaff of a ship, placed on the top of the mast above the point where the conductors began, had been struck by lightning and shivered to pieces. This and similar slight accidents, which really proved the efficiency of the system, at length were so clearly understood, that no further doubt remained. It was felt that some public recognition was due to the man who had made our ships safe from the attacks of a destructive foe which had formerly deprived the country of the full services of its navy, killed or crippled its sailors, and wasted many thousands annually of the public money. In 1847 the honour of knighthood was conferred on Snow Harris at the express command of her Majesty the Queen, in consideration of his " very useful inventions," to use the words of Earl Russell. So little was this honour expected, that when Earl Russell's letter arrived at Plymouth, Harris thought it was a hoax; for he, in common with all men of genius, had a strong sense of humour (without which, indeed, genius seems to be scarcely complete, unless *power* take its place, as in the case of Milton and Dante), and this humour was so often let loose upon his friends in good-natured jokes (often practical ones), that no wonder if he were sometimes repaid in his own coin. He took the letter to a gentleman in

* Ordered to be printed by the House of Commons, 1854.

Plymouth, and said, "Have you not a collection of autographs, including that of Lord John Russell?" The autograph in question was produced. He examined it carefully, and said, "No, it is no hoax; the writing in my note is identical with that in yours." But even then he consulted with his friends as to whether he had not better ask leave to decline the honour; but he was nervously anxious not to appear in the slightest degree to oppose himself to her Majesty's gracious wish. He accordingly came up to town and received the well-merited honour. The editor called on him next day to congratulate him, when he gave an amusing description of his own feelings on finding himself for the first time in a court-dress. He expressed himself extremely gratified at the gracious manner of the Queen, and the smile of recognition which he received from the Prince Consort. Indeed, the confidence of her Majesty and the Prince in the perfect safety of Harris's conductors was shown in a request that he would fit up conductors at Buckingham Palace and at Osborne, and also similarly protect her Majesty's yacht. Harris was also employed some years later to design a complete system of conductors for the palace at Westminster. His written instructions, which are full of interest, were ordered by the House of Commons to be printed, and they will be found under the head of "Estimates, &c., Civil Services, for the year ending March 31, 1856."

There was now no difficulty in the way of admitting Harris's conductors into the Royal Navy, and the Government, in consideration of his great services, proposed a vote of £5,000 to the inventor. Sir James Graham, in moving the vote, said that he never voted away money with more pleasure.

In 1850 Sir William was elected an honorary member of the Naval Club at Plymouth, when a number of eminent officers warmly congratulated him on the great service he had rendered to the Navy. In 1854 he was elected an honorary member of the Royal Yacht Squadron at Cowes, by the consent of all the members, in acknowledgment of his public services. Nothing could be more congenial to Sir William's tastes, for he was always more of a sailor than a landsman. His very walk reminded you of the deck of a ship, and the warmth and simplicity of his character of a sailor. It was quite a treat to be with Harris near or on

the sea. There was not a craft that he was ignorant of, and
he was never tired of pointing out the merits and defects of
the vessels around him. He loved to be on the sea, in whatever craft. The editor once accompanied him to the Eddystone
in the lighthouse tender, and on another occasion to Cornwall in a
limestone barge. He had invented a new form of ship's compass
in which he took great interest, and was proud to see it in use.
Being told that a yacht had arrived in the Sound with one of his
compasses on board, he asked the editor to go with him to inquire
how the yachtsman liked it. As soon as he got on board and sent
in his message, the proprietor came up and said, "Oh! I don't
like your compass at all; but I have one here by a man named
Harris that is a great favourite of mine." "I'm Harris," was the
bursting, eager reply, whereupon apologies and warm congratu-
lations ensued.

Sir William had long had a yacht of his own, in the management and sailing of which he took the greatest delight. He was
proud of his nautical skill, and was pleased when some one told
him that a naval man once observing a sailing-boat tacking about
in the Sound, exclaimed, "Egad, the fellow in that boat well
knows what he's about!" It need hardly be said that the fellow
was Harris himself. He had many other characteristics of a sailor,
but there was one point in which he did not resemble Jack Tar:
that is, in his indifference to dancing, although, as we have
said, he was an accomplished musician. Sir William gratified
his love of music in various ways. He had been his children's
tutor in music as in other matters, and the concerts got up in
his drawing-room were almost professional in their finish.*
He was often in London, where, as may be supposed, he had frequent
opportunities of hearing good music; but, in common with the
best minds, he preferred cheap to costly pleasures. He delighted
to go to the pit of a minor theatre when operas were performed.
He was fond of a good play well acted, and reckoned among his
intimate friends Mr. and Mrs. Charles Kean.

Sir William was fond of conversation, and excelled in it. His
sympathies were wide, and he was ready to talk or to listen on
any subject. Captain Lockyer once said of him, "Harris is like

* See note page xxi.

a barrel organ; you may set him to any tune." He was also a man of tender feeling, and many will cherish his memory as of one more than a friend, for he was their doctor and nurse.

In the midst of all his exertions in the cause of lightning conductors, Sir William did not neglect science. When Mr. Weale, the publisher, in the year 1847 was about to start his Rudimentary Series of Scientific Manuals, he obtained the assistance of Sir William, who wrote several treatises on electricity, magnetism, and galvanism, which met with great success and passed through several editions. Sir William's style was clear, because he thought clearly; but it was modelled on that of a former age, when one scientific man seldom named another without giving him all his titles, with "justly celebrated" in addition. He was very fastidious as to the way in which his thoughts were stated; he would write and re-write a page many times, and when at last it reached him in proof he would so twist it and re-arrange it, that the printer's charge for corrections was sometimes higher than that for composing. Sir William would even make organic changes in wood engravings, so that if adopted it would have been necessary to have new blocks. But in all such cases, whether of writing or engraving, he was willing to listen to remonstrance; and indeed sometimes felt it to be a relief to have some one to decide for him. The editor frequently hesitated at this responsibility, and was careful how he advised, lest his advice should be too readily adopted and afterwards repented of. Harris was so grateful for any little services of this kind, that it was a pleasant though a responsible duty to serve him. And then he was so trusting and confiding, so child-like and affectionate to any one whom he loved, that he concealed nothing, not even his own weaknesses, for he had his weaknesses, and had he not he would not have been so lovable a character. He was sensitive as to what others thought or said of him, and sometimes fancied they did not do him justice, or that their discoveries had borrowed too much from his. And when he resumed his labours in the field of original research, and found that his papers were not so welcome in the Royal Society as they had been thirty years before, he could not understand the reasons for their rejection.

He could not admit the fact, still less receive it with calmness
(few men can), that his working days were nearly over; and
that his later work was but a reflection of the light that once
shone so brightly.   His sympathies were with the Bennetts, the
Cavendishes, the Singers, the Voltas of a past age.   Frictional
electricity was his *forte* and the source of his triumphs.   He was
too good a Tory of the old school of science to recognise the broad
and sweeping advance of the new; and he did not feel that he
was behind his age when, in 1861, he presented to the Royal
Society an elaborate paper on an improved form of Bennett's
discharger, and yet later when, in 1864, he discussed the laws of
electrical distribution, and still relied upon the Leyden jar and
the unit measure.

It cannot be said that Harris, with all his ability, had a creative
mind, in the sense that Wollaston's or Davy's was creative.   He
was highly ingenious and inventive, and in his best work one is
always struck with the advance he had made beyond previous
occupiers of the same ground.   He did his work well, and left
his mark on the science of his day ; and although some of his
labours will be forgotten, and others be absorbed and blended
with the branch of physical science that he cultivated, still there
are many points in Harris's character as a man, and in his habits
as a philosopher, that the student may dwell upon with pleasure
and profit.

The few remaining years of Sir William's life were embittered
by bodily suffering.   In August 1861, after returning from a trip
in his yacht, he was seized with an attack of *iritis* which did not
yield to medical treatment during some months.   He was confined
to the house until the following May.   In the autumn of 1862 the
attack returned, and Sir William was advised to undergo two
painful operations ; but the result, after all, was the loss of one
eye and the partial loss of the other.   His general health, too,
suffered from the medical treatment.

When he had partially recovered, he was anxious to bring out
his complete treatise on Frictional Electricity, materials for
which had been accumulating during many years.   He intended
to add to this work short biographical notices of the leading
electricians.   He had also prepared a minute account of the history

of the Leyden jar. The first few chapters of the work now before the reader were sent to the editor early in 1866. There was considerable delay in getting the woodcuts properly executed, so that the work made but slow progress. The first part, however, up to p. 205 of the present volume, was completed under the author's supervision, when he was seized with his last illness, which ended fatally on the evening of the 22nd January, 1867. " He bore his sad calamity during five years and a half with the greatest patience, calmness, and fortitude, and was never heard to murmur," were the words of one who announced to the writer of this notice the loss of her husband and his friend.

[The following note, referred to at p. xviii., has been contributed by a friend of the family, who took part in the concerts referred to.]

Sir William was a musician of rare ability and attainments; indeed, few amateurs possessed such critical acumen or so comprehensive an acquaintance with music, scientifically and practically considered, as he did; for in addition to his extensive study of theory, he was a performer both on the harp and piano, and played with exquisite taste and feeling. With thes' qualifications, Sir William devoted much time to the musical education of his daughters, who became accomplished pianists.

Provided with two of Collard's grand pianofortes, Sir William was fond of collecting a few "strings," and with them enjoying the luxury of concerted music. These parties, small at first, soon swelled into complete orchestras formed of amateurs, to whose aid in cheerful spirit came professors, civil and military, the latter bringing from their bands the necessary wind accompaniments. With such a force Sir William would take the field, and feared not to encounter the most powerful of composers. The names of Beethoven, Weber, Mendelssohn, Mozart, and others, were familiar as household words in his home at Windsor Villas.

Nor was this all, for our enthusiast added to his instrumental productions all the vocal music of which the singers in his neighbourhood were capable. He invited all amateurs found to possess the requisite voices to assist him, and the result was a succession of splendid concerts, not often surpassed in private circles.

To perfect these reunions no efforts were spared; consecutive portions of operas were performed, and descriptive programmes from the ready pen of Sir William were issued to the company, who were thus made familiar with the subject of the pieces produced; to this was added a brief sort of lecture or synopsis of the opera from which the music was selected, and, finally, small scenes were used to bring the more striking incidents of the opera vividly before the audience.

In the year 1851 a wider field seemed to open for the exercise of Sir William's powers of musical organization, and he accordingly took the lead in a committee of management on the formation of a philharmonic society which included subscribers from Devon and Cornwall, who were invited to

participate in or enjoy the really good music intended to be performed for
their gratification.

In this undertaking Sir William was indefatigable, and for two or three
years excellent concerts were produced, aided by eminent singers from
London, amateurs in the neighbourhood, and an orchestra and choir num-
bering 150 persons.

The frequent change amongst the inhabitants caused the maintenance of
this society to become a work of such labour, that the promoters, sanguine
as they were, could not continue their efforts, and Sir William was obliged
to fall back on his delightful home concerts.

In addition to his high appreciation of music, Sir William had a great
admiration for the drama. He was proud to reckon Mr. and Mrs. Charles
Kean amongst his most intimate friends, and often spoke of his acquaintance
with the late Mr. Matthews, whom he professionally attended towards the
close of his life.

THE AUTHOR'S PREFACE.

PERSONS who have been trained more immediately in what may be called the mathematics of electrical physics look generally with much indifference and suspicion at any new experimental processes or tangible investigations which at all run counter to their long-admitted theory. Reposing with a reverential confidence upon a few early experimental deductions, conjoined with analytical forms of expression, they appear to imagine that the limit of our knowledge of the laws of electrical action has been reached, and that to doubt in any degree the truth of what has been termed the Coulombian theory of electricity, and the received laws of the necessary distribution of electricity upon bodies of variable figure, is little less than philosophical heresy. In this sense many profound writers, distinguished for analytical skill, betray an amount of prejudice not very favourable to the advancement of science : they judge of every theory by the quantum of symbolic expression which enters into it, or by the form under which it appears. Thus M. Biot, in his life of Volta,* seeks continually to disparage the quality of mind of that truly great man, upon the ground that his researches and discoveries had not a rigorous mathematical basis similar to the method of Coulombe, and is hence led, although constrained to admit the vast genius of Volta, to refer nearly all Volta effected to previous principles mathematically deduced by Æpinus, although no one ever heard of the condenser or the electrophorus before Volta. Still, we are informed that the principles were foreseen, and their theory given twenty years before in the "Tentamen Theoriæ Electricitatis," by Æpinus, where most likely they would have remained dormant to this day if some great genius in experimental physics had not invented the instruments. The real fact is, that these beautiful instruments having

* *Biographie Universelle,* tome xliv.

been invented and perfected, the theory of Æpinus is applied in explanation of them. Many other theories of electricity could be equally well applied, and it would be just as easy to say that Volta's instruments had been prophesied in those theories as in the theory of Æpinus. If the principles had been so clearly expounded by Æpinus twenty years before Volta invented the instruments, it seems a sad reflection on those who so strongly contend for mathematical rigour not to have applied them.

The real state of the case is, that the events virtually preceded the prophecy, and having taken place, the prophecy was applied to them. M. Biot, evidently biassed by his veneration for a mathematical form of inquiry, and which he calls *la rigueur mathématique*, treats Volta's physics of electricity very slightingly, and so far betrays a great want of appreciation of what Volta advanced, if not a failure of comprehension of Volta's principles. Thus he says, Volta's discoveries of the condenser and electrophorus were for him merely combinations of experience; that he never referred them to their true theory, but to electrical atmospheres, an idea which the most profound geometers could never dissuade him from; that Volta's dissertation upon electrical conductors betrays a total absence of all abstract rigour of research; that he does not fix any of the rigorous elements of the important question of the influence of the general form of conductors upon the conservation or loss of electricity, or upon the energy of their discharges, whilst the rigorous method of Coulombe fixes, and for ever, the exact laws of electrical distribution and equilibrium upon the surface of bodies of different forms.

The substantiality of all this kind of criticism is very doubtful. It is possible that the genius of Volta did not admit of his employing an analytical process similar to that pursued by Coulombe in his experimental inquiries. Yet it does not follow, on that account, that it did not equally involve physico-mathematical accuracy under other forms; and really we have yet to learn the truth of the assertion that his notions of what he calls electrical atmospheres, taken in the sense in which he uses this term, are not perfectly correct, and demonstrable by experiment, or that his views of the variable intensity of a given quantity of electricity, disposed upon conductors of equal surface, but of different extension, are not quite sound. They may not coincide with a given mathematical theory of electricity, but they need not be "vague" on that account. Neither is it so certain that the course pursued by M. Coulombe has really "fixed, and for ever," the exact laws of such variations of electrical force. The experiments upon which Coulombe rests are really very few, and not altogether unexceptionable. The theory of the Proof Plane, upon which depends all the assumed laws of electrical distribution, is extremely doubtful, and its operation very precarious, and in many instances absolutely fallacious; and although a theory of repulsive force, assumed to exist in the particles of a supposed electrical fluid, may be brought to square with a mathematical and analytical application of the observed effects, however fallacious the interpretation, yet that is not evidence of so conclusive a nature as to preclude the reception of other equally general views of electrical force. It would be, perhaps, both an invidious and useless course to seek to oppose to each other the relative merits of simple and pure experimental researches, and of investigations made through the medium of symbolical and abstract mathematical analysis. It is, after all, a liberal combination of the two methods which gives us the most powerful means of advancing physical knowledge. That rigorous mathematical forms and methods of experiment and deduction are most desirable in physical inquiries, few can doubt. The mathematical mind can doubtless see further into futurity than other minds not so constituted, and can thereby save the great loss of time attendant on mere tentative skill; whilst the great beauty and elegance of its analytical processes, and its wondrous developments, lend an increased charm to experimental labours. Nevertheless, we must never forget that it is the experimental labours upon which all our knowledge of nature mainly reposes. Newton himself would not consent to hazard his grand theory of universal gravitation upon any other basis, however near its probability approach to certainty, and, as is well known, withheld it from the philosophical world until correct measurements of the earth had been obtained. The great and justly-celebrated Galileo, who was much more remarkable for a profound geometry than for any skill in symbolical analysis, never failed to verify every assertion by experiment. We have, in treating questions of this kind, to distinguish (1) that class of mind adapted to mere mathematics, considered as an instrument of research ; (2) those who possess great skill in the handling and management of symbols and of general symbolical processes, or geometrical analysis, taken abstractedly ; and (3) those who, with less analytical skill, pursue inventive experimental research solely ; or (4) those who, without the aid of any mathematical processes, repose entirely upon experimental investigation, still possessing intuitively, as it were, a peculiar quality of mind involving really mathematical rigour under another form.   There may undoubtedly exist great and innate powers of thought and philosophical intellect without any of that quality we are accustomed to designate as mathematical. Bacon, for example, one of the most wonderful and accurate thinkers of any age, was totally ignorant of geometry and mathematics.   All the great discoveries and advances in Physics have. for the most part, been derived from men essentially experimentalists, men who either applied great and original powers of thought in the investigations of nature in their own way, without the aid of mathematical processes, or who allowed mathematics to follow, as it were, in the wake of experiments, to correct and assist them, rather than to develop in advance new and, as yet, undiscovered facts.   Franklin, whose researches and discoveries in ordinary electricity were most marked and important, had very little skill in mathematics.   Neither Volta, the inventor of important electrical apparatus, nor Davy, the great discoverer of the bases of the fixed alkalies, can be said to have been led to such discoveries through any mathematical process.   The same may be said of a host of others.   To judge, therefore, of the value of any physical inquiry solely with reference to its mathematical method, or the amount of analytical symbolic representation it may contain, would be not only to apply a fallacious measure of value, but would be really a philosophical injustice.   Mathematical or symbolical analysis is, after all, but an instrument of research of a peculiar kind, undoubtedly most valuable in its operation, but still an instrument which many minds do not require, and which is often very greatly abused and overworked by persons who, conscious of their knowledge of symbolical arithmetic, are rather content to exemplify processes than to investigate truth.   Very little, if any, really useful knowledge of nature is found in the elaborate and interminable pages of symbolic analysis in which many modern philosophical papers abound. As specimens of mere analytical skill they are no doubt valuable, but for any practical result they are frequently valueless. It is the show of learning without the reality. Whilst, therefore, we admit the vast importance of mathematical knowledge and address, and the great advantage of mathematical method in its sound and healthy application to experimental physics, we still think that in considering the exhibition of analytical skill as the great object of philosophy, we sacrifice the end to the means, and thereby fail to arrive at any new result. We substitute a sort of paper philosophy based upon data which are rather taken for granted than definitively proved for the philosophy of nature, and really close the door upon further inquiry. Things are taken for granted upon authority much in the way of the followers of Aristotle, and the result is, that it is difficult to obtain anything like candid consideration of facts which do not fall in with what is denominated *par excellence* THE THEORY. It is well known that, when Galileo asserted that bodies falling to the earth simultaneously from the same altitude would all fall in the same time, whatever their weights might be, the Aristotelians refused to listen to such a heresy, although, in his famous experiment from the tower of Pisa, the ring of the falling bodies on the ground was still sounding in their ears.

ANALYSIS OF THE CONTENTS.

ELECTRICITY IN THEORY AND PRACTICE.

PART I.

ELEMENTARY ELECTRICAL PHENOMENA.

CHAPTER I.—ATTRACTION AND REPULSION.

ATTRACTION, various kinds of, 1; development of electrical force—early knowledge of electricity, 2; terms used in electricity, 3; modes of showing electrical action, 3; electroscopes—trial reed, 4; simple experiments on electrical excitation, 5; list of idio-electrics, 6: Gray's discovery, excitation and attraction, insulation and conduction, 7; list of electrical conductors, 9; insulation, 9; carriers and transfer-planes, 10; limit of the terms insulation and conduction, 10; list of electrics or insulators, intermediate substances, and conductors or non-electrics, 11; remarks on this table, 11; experiments on attraction and repulsion, 12; law of attraction and repulsion, 15; vitreous and resinous electricities, 15; illustrative experiments, 16; positive and negative, or *plus* and *minus* electricity, 17; origin of the two electricities, 18; further illustrations of attraction and repulsion, 19; effects of varying the surface, 19: determination of the kind of electricity, 21; single gold leaf electroscope, 21; double ditto, 22; Bennett's gold leaf electrometer, 23; Henley's quadrant electrometer, 23; Cavendish's electrometer, 23; Harris's electroscope electrometer, 24; Harris's electroscope electrometer of double repulsion, 25.

OCCASIONAL MEMORANDA AND EXPLANATORY NOTES TO CHAPTER I.

Note A. Suspensatory filaments of silk and spider's web, 27.
  „ B. Construction of electroscope needle, 27.
  „ C. Management of leaf-gold, 27.
  „ D. Preparation of amalgam, 27.
  „ E. Construction of sliding insulator, 28.
  „ F. Preparation of glass and gutta-percha tubes, 28.
  „ G. Preparation of gold-leaf electroscope, 28.
  „ H. Preparation of a portable diverging electroscope, 29.
  „ I. Preparation of Henley's electrometer, 29.—Value of the indications afforded by this instrument, 30.
  „ K. Mode of using Cavendish's electrometer, 30.

of metal, wood, cork, &c., 165, 166; method of piercing holes through corks, 166; metallic conducting rods, 167; varnishing, varnishes, and cements, 167; lacquers for metallic surfaces, 168; varnish for wood and French polish, 169; mode of applying electrical varnishes, 169; varnish cup and brushes, 170; varnishing glass rods, tubes, and glass generally, 170; varnishing electrical jars, 171; varnishing glass plates, metallic surfaces, and wooden surfaces, 173; varnishing paper, 175; amber varnish, 176; application of varnish and oil to silk, 177; electrical cements, 177; preparation of electrical plates and cylinders, brimstone cylinders, cones, &c., 179; brimstone plates, 180; cylinders and plates, of resin, wax, &c., 180; electrophorus plate, 181; junction of plates of glass by cementing, 183; colouring matters for cements, 184; value of compound plates 185; isinglass, glue, and size, 185; Japanese cement and paste, 186; mode of covering wood with tin leaf, 187; electrical suspensions, 187; Chinese mode of suspending compass needle, 188; suspension filaments, 188; management and handling of gold-leaf, 189; selection of glass for electrical tubes and electrical machines, 190; precautions required in constructing electrical machines, 191; cylindrical machine, 191; construction of rubber or cushion, 193; plate machine, 194; preparation of amalgam, 195; various recipes for, 196; method of preparing amalgam, 197; method of applying amalgam, 198; *aurum musivum*, or mosaic gold, 199; excitation of cylindrical electrical machine, 200; excitation of plate machine, 201; preparation of the cushions of rubbers, 203.

NOTE TO CHAPTER V.

Note V. Various kinds of lac, 204.

PART II.

ON THE LAWS OF ELECTRICAL FORCE.

CHAPTER I.—BRIEF ENUMERATION OF FACTS AND PHENOMENA OF ELECTRICAL ACTION AS DEDUCIBLE FROM AN INVESTIGATION INTO THE THEORETICAL AND PRACTICAL NATURE, OPERATION, AND LAWS OF ELECTRICITY.

EARLY notions as to the nature of electricity, 207; information obtained only by means of experiment, 207; first great step in the science by Du Fay, in distinguishing two kinds of electricity, 208; Cavendish's researches, as contained in his MS. papers, 208; all the charge on the surface, 208; Robison's and Coulombe's researches, 208; law of electrical force as deduced by Coulombe and others, 209; attempt to define the law more clearly, 209; speculations thereon, 210; how the force varies, 210; physical conditions under which the forces of attraction and repulsion exist, 211; objections to the use of the tangent plane, 211; effect of induced action, 212; effect of varying the form of a charged conductor, 213; experiment with hollow globe and coatings of mercury, 214; action on carrier ball, 214; influence of change of form on electrical tension, 215; ideas of the early experimentalists as to the law of magnetic force, 216; influence of reflected induction, 217; limit of the law of electrical or magnetic force, 217; influence of opposed spheres, 218; objections to the measurement of the force by means of repulsion, 218; operation of torsion balance, 219.

NOTES TO CHAPTER I.

Note A. On the author's views of the Coulombian theory, 219.
  „ B. Details of experiment on hollow glass globe with coatings of mercury, 221.

C

TWO LECTURES ON ATMOSPHERIC ELECTRICITY AND PROTECTION FROM LIGHTNING.

FIRST LECTURE.

SECOND LECTURE.

˙ELECTRICITY

𝔍𝔫 𝔗𝔥𝔢𝔬𝔯𝔶 𝔞𝔫𝔡 𝔓𝔯𝔞𝔠𝔱𝔦𝔠𝔢.

PART I.

ELEMENTARY ELECTRICAL PHENOMENA.

CHAPTER I.

ATTRACTION AND REPULSION.

1. A multitude of varying natural phenomena, continually present, indicates the operation of certain invisible subtle agencies, or physical forces, by which the material universe is apparently controlled. These physical forces or powers, although differing in their modes and forms of action, are still characterised by one remarkable property common to them all, in a greater or less degree. This property has been designated by the general term *attraction*. That is, the agencies or forces, whatever they be, cause particles of ordinary matter, both at finite and at indefinite distances, to approach each other, or tend to approach each other. We perceive this in the falling of a stone towards the ground; in the planetary motions; in the aggregation, or growing together, as it were, of minute particles into solid and compact masses; in the approach of ferruginous matter towards a certain ore of iron termed the *loadstone* or *magnet*; as also in the approach of light substances towards certain bodies when subjected to friction; and in some other remarkable instances. In these phenomena we recognise the operation of one or more subtle natural agencies; and although careful not to regard such phenomena except as effects,

B

we nevertheless usually designate them by such terms as *gravity*, *magnetic attraction*, *electrical attraction*, and not unfrequently refer to them as so many distinct and operative causes, or independent powers of nature.   In treating, however, of one or all of these apparent forces, we do not pretend to deal with occult causes.   We concern ourselves only with such agencies considered merely as force, and with the laws and mode of their operation;   and until these are fully developed and comprehended, all speculation as to the nature of the forces themselves will be defective.

2. The peculiar invisible power, then, which we term *Electricity*, may be viewed as one of those incomprehensible and mysterious natural agencies known to us only by its effects ;   and, although involved in deep obscurity as a source of physical power, it is, in common with other agencies, ever present, often under a form not cognisable by our senses, and is hence termed *latent*.   The slightest change, however, in the attendant circumstances is sufficient to call up this mysterious power, electricity, from its apparently dormant state, and to make us immediately sensible of its presence.   Many substances, for example, on being simply brought into close contact under pressure, and subsequently separated, not only tend to come together again, but seem actually to exert a species of attractive force, both on each other and on other bodies.   A similar result ensues in a great variety of operations, natural and artificial ;   such, for example, as in the formation of rain, hail, snow, mist, and other meteorological phenomena ;   also in the case of variations in temperature, as in the heating and cooling of certain mineral bodies, such as the tourmaline, the topaz, and some others ;   also in changes of form, as in the liquefaction or consolidation of brimstone ;   also in chemical action, as in the decomposition of water, the solution of metals in acids, &c.;   also in mechanical operations, such as cleavage, abrasion, friction, and such like ;   all these are sources of electrical development, and may superinduce on common matter attractive force.

3. The first written notice of electrical action appears to have originated, 600 years before the Christian era, with a Greek philosopher, Thales, a native of Miletus, founder of the Ionic philosophy.   He observed, as a remarkable property of amber, that when subjected to a slight friction it acquired an attractive power, and would draw light bodies towards it.   Thales was so much struck with this phenomenon that he considered the amber to be endowed with a species of animation.

Theophrastus, also a Greek philosopher, subsequently observed, after a lapse of three centuries, a similar property in a stone termed *Lyncurium*, supposed since to have been identical with the tourmaline. This, he says, on being subjected to friction, will not only attract light straws and sticks, but also thin pieces of metal. The elder Pliny noticed the attractive property of amber, and a similar property appears to have been discovered about the time of Pliny, in agate and other precious stones.

4. The attractive power of amber when rubbed may be considered as the basis of our electrical nomenclature; the Greek word denoting *amber* being ηλεκτρον (electron), in Latin *electrum*. The unknown element which, according to Thales, gave it life, has been termed *Electricity*. As our knowledge of the operation of this active principle began to advance, and other substances were found to display a similar property to that of amber, such substances were said to be *amber-like*, or *electrical*, and were hence termed *electrics*. Again, any substance having attractive force superinduced upon it, after the manner of amber, was said to be *electrified*, or *electrically excited*, whilst its peculiar state at this instant was said to be a state of *electrical excitation*. These terms are still retained, and others have been introduced on the same basis. Thus, the attractive force displayed under this excitation is termed *electrical attraction;* and the particular substance itself is denominated an *excited electric ;* while any particular contrivance for better observing and detecting the presence of electrical force is termed an *electroscope*, and when constructed so as to measure its power is termed an *electrometer.*

5. Electrical excitation in any given substance may be observed easily ; it is only requisite to present the excited body to some light substance, when its attractive power becomes immediately apparent. The substances best adapted to the exhibition of electrical attraction are downy feathers, bog down, small balls of elder pith, or the pith of the *Sola* plant, fragments of gold or silver leaf, a thread of cotton, and such like. Light substances of this kind, especially if delicately suspended so as to be free to move in any given direction, are convenient and efficient electroscopes. A thread of fine sewing-cotton two or three feet in length suspended from the ceiling of a room by means of a small piece of sticking-plaster or gummed paper, is very obedient to electrical attractive force, and constitutes a simple and efficient instrument ; any electrified substance, however weakly excited, will, on being presented to the suspended thread, at the distance of some feet, set it in motion towards itself (Note A).

Gilbert employed a delicately balanced magnetic needle for testing the presence of electrical force, a species of electroscope

frequently resorted to by later electricians, in form, at least, if not
in material.   A delicate electroscope needle of this form is shown
in Fig. 1.   It is sensitive and convenient for many purposes (B).

6. A fine cotton or linen thread suspended from an arm or
support, and having a small spherical mass of cotton wool or
bog-down attached to it, about the size of a hazel-nut, as at m,
Fig. 2, constitutes also a delicate and sensitive electroscope.   The

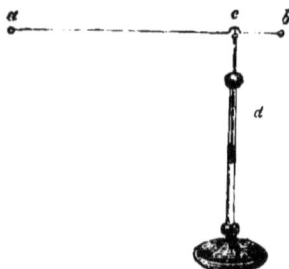

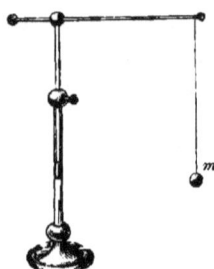

Fig, 1. Electroscope Needle.   Fig. 2. Bob Electroscope.

terminating pendent mass m is sometimes named the *bob*, and serves
to expose a larger amount of attractive surface.   A light sphere
of elder pith, or other light vegetable substance, about one-fourth to
three-eighths of an inch diameter, may be employed for this purpose.
A simple and efficient electroscope is also obtained by means of a
light straw reed, such as is employed in straw bonnet making.
This reed may be attached to a fine fibre of silk or cotton thread

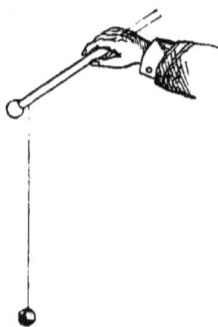

Fig. 3. Trial Reed.   Fig. 4. Gold-leaf Electroscope.

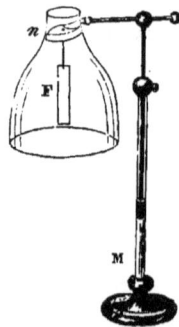

by means of a little sealing-wax or gum, and hung from any
convenient support, as in Fig. 3.   A light pith ball should be
placed on its lower extremity.   This kind of electroscope may be
called a *trial reed*, and if suspended from a light glass or metal
arm, as in Fig. 3, may be held in the hand.   Straw reeds are useful
in detecting electricity.

One of the most delicate tests of electrical excitation is a slip of leaf gold, F, Fig. 4, attached to a short paper holder, and suspended under a bell glass, in any convenient manner, so as to shield it from currents of air. A strip of gold-leaf arranged in this way is obedient to an extremely small force of attraction (C). A great variety of contrivances for detecting electrical force will be duly noted as we proceed.   .

7. The following experiments may be adduced as simple illustrations of electrical excitation :—

Exp. 1. An ordinary stick of sealing-wax, or a roll of common brimstone, is to be gently rubbed with a dry handkerchief of soft silk, or with a piece of flannel. When the excited body is presented to any one of the electroscopes above described, the part free to move will be drawn towards it, even though it be held at the distance of some feet.

Exp. 2. Let a tube of glass, about 15 inches long, and three-fourths of an inch in diameter, be gently warmed, and drawn through the folds of a dry silk handkerchief, held in the hand. It will become so powerfully excited that faint luminous flashes may be frequently observed on its surface in a dark room, and also luminous sparks, termed by the older electricians *electrical fire.*

The excitation of glass is greatly promoted by the application of an amalgam of tin, zinc, and mercury (D). The bisulphide of tin, also termed *aurum musivum*, a friable substance of a yellow colour, may be substituted for the amalgam.

A tube of glass about 2 feet in length and 1 inch in diameter, made very dry, and warm inside and out, excited in this way by warm dry silk, throws out in the dark, vivid sparks and luminous flashes, attended by a crackling noise and a peculiar odour, causing a singular sensation on the face or hand, termed by electricians *spider's web.*

Exp. 3. Two silk ribbons, the one black and the other white, each about a yard in length, are to be applied fairly to each other, and then drawn, with slight pressure, through a fold of silk velvet, or even through the fingers ; they will be found to adhere, although they do not at the moment affect the electroscope ; but directly they are separated, each ribbon will be attractive of the other, and of surrounding matter, and if presented to the electroscope will powerfully affect it.

Such are a few elementary illustrations of electrical excitation as dependent on simple friction and pressure.

8. Every known substance is electrically excitable in a greater or less degree, provided due precautions be observed. But it is

only under certain conditions, and in a certain class of bodies, that we are able, under ordinary circumstances, to develop this property to any extent. We are unable, for example, to excite a metallic tube under the same circumstances and in the same way as we excite a tube of glass or a roll of brimstone (7). Hence has arisen a classification of substances in relation to their electrical excitability. For example : substances which are amber-like—that is, when subjected to friction, display attractive force—have been, from an early period, classed as *electrics*, and considered more especially as electrical bodies, or *idio electrics*, or substances naturally electric in themselves. Later researches, however, do not permit us to recognise electrical excitability as an exclusive property of any particular substance. There is no physical evidence for such a conclusion, and it can only be admitted as being peculiar to certain substances, under certain conditions. With this limitation the classification is not altogether without advantage, and is so far admissible.

The following table contains a list of idio-electrics, or those bodies which have the property of attracting light substances when subjected to friction in the common way :—

TABLE I.

List of Idio-Electrics.

Gum lac, shell lac ; gums of all kinds, including camphor.
Resinous substances of every kind, including common wax.
Gutta-percha.
Bituminous substances, including amber and jet.
Sulphur, and some of its compounds, including vulcanite and other artificial preparations.
The diamond ; nearly all precious stones ; agate.
Mica, and other laminated minerals.
Glass and all vitreous substances.
Porcelain.
All crystalline transparent gems, especially tourmaline.
Silk of every kind and form, especially silk gut.
Dried animal skins, and fur ; skins of living animals in their ordinary state.
Gun-cotton, and wool generally ; hair and feathers.
Baked dry wood, and dry vegetable substances ; very dry warm paper.
Oils and fatty fluids, including turpentine.
All dry gases, including atmospheric air.
Steam of very high temperature.
Ashes of animal and vegetable bodies generally.
Phosphorus ; lime and chalk, deprived of all moisture.
Dry metallic oxides.
Lycopodium.
Dry ice, below the zero of Fahrenheit's thermometer.

9. The most available and efficient electrics for ordinary experimental inquiries are gum lac, gutta-percha, brimstone, amber, animal fur, cat skin, hare skin, glass, silk, vitreous and resinous bodies generally. There are also many artificial compounds of these, which are valuable as electrics, such as the compound of caoutchouc and brimstone, termed *vulcanite*, ordinary sealing-wax, compounded of shell lac and turpentine, to which is added occasionally a little resin. Gun-cotton and collodion, formed from a solution of gun-cotton in ether, are available as electric bodies.

10. Although metallic substances and many other bodies do not display electrical attractive force when subjected to friction in the usual way, they may nevertheless display this property and appear to have attractive force superinduced upon them by placing them in communication with an electrically excited substance (4). This remarkable discovery was made by Mr. Stephen Gray, of the Charter House, London, and others, so long since as the year 1729, and is a fine instance of experimental inductive science.

The history of this discovery is worthy of notice. After many attempts to render metallic substances electrically attractive by friction, the experimenters were led to notice the phenomenon of an attractive power acquired by a cork with which they had closed one of the open ends of a glass tube, T, Fig. 5, subjected to excitation. This led them to insert a light rod of fir wood, c, into the cork b, so as to pass within the tube, while on its exposed extremity was a ball of ivory, d. When the glass tube

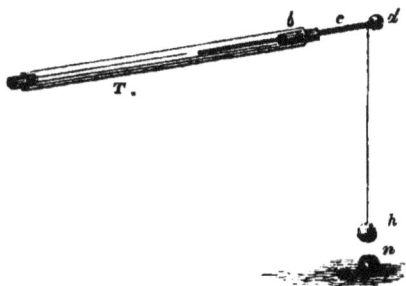

Fig. 5. Example of Excitation and Attraction.

was briskly rubbed they noticed to their great astonishment that both the fir rod and its terminating ball had acquired, in common with the glass, an attractive force, and drew toward it light substances, even more freely than the excited tube itself. The length of the fir rod c, and consequent distance of the ball d, from the source of power, appeared in no way to interfere with the result; so that in suspending a second ball, h, from the rod by means of a line of hemp of greater or less length, they were enabled on exciting the glass tube to attract a light substance, n, from off the pavement, from the summit of a high building.

On extending the inquiry with a view to discover the limit

of distance through which this attractive power might be transmitted, they were led to take a horizontal direction, in which
case it became requisite to suspend the line of communication
with the glass tube and rod, Fig. 5, from a beam, A, Fig. 6,
in a sheltered avenue, in which their experiments were conducted. This they effected by means of a hempen cord of
suspension, m. Under these conditions no evidence of attraction was observable in
the ball h. The cord
of suspension m being
thick, they were led
to imagine that the
electric virtue was ab-
sorbed by it, and car-
ried up to the beam
to which the cord
was appended. This
led them to abandon
the hemp suspension,
and substitute a line of fine silk. The result fully justified
their expectations; the attractive power now appeared in the
ball h, as before. In this way they succeeded by means of silk
suspensions in transmitting the attractive force through a distance
of 765 feet. With a view of still further perfecting their research,
they again changed the silk suspension for a still finer suspension
of metal wire; but this, so far from improving the electrical
transmission, destroyed it altogether, as in the case of the hempen
cord. It was hence evident that the silk cord had satisfied
some important condition in this interesting experiment, and
that the result they arrived at rather depended on the *nature* of
the substance of the cord of suspension m than on its *size*. In this
way a most important discovery was made, namely, the *conducting
power* of certain bodies. Both the hempen cord and the small wire
had evidently caught up and transmitted the excited force to the
beam A, from whence it had passed away to other bodies; whilst
the silk having no such power, had caused the electricity to be
retained on the line of electric communication between the excited
tube T and the ball h.

11. This fine elementary discovery speedily led to a new and distinct class of phenomena dependent on what has been since termed
electrical conduction, and to another classification of substances
considered as *conductors* of electricity, in contradistinction to the
class enumerated in Table I., termed *electrics* or *non-conductors*.

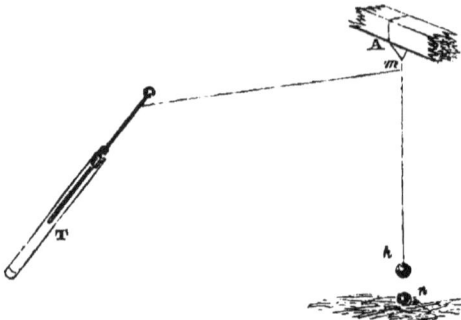

Fig. 6. Illustration of Insulation and Conduction.

The substances which fall more especially under the class of conductors are placed in the following Table :—

TABLE II.

ENUMERATION OF ELECTRICAL CONDUCTORS.

All the metals and metallic bodies.
Plumbago.
Well-burned charcoal.
Concentrated acids and saline fluids generally.
Water, including aqueous vapour.
Ice and snow above 10° of Fahrenheit's thermometer.
Animal solids and fluids.
Vegetable fluids and most vegetable solids.
Flame and smoke.
Alcoholic and other spirituous vapour.
Vapour of ether.
Highly rarefied air.
Nearly all earths and stones.

12. Many substances owe their conducting power to the moisture contained in them. Thus, baked wood is an electric, and does not easily transmit electrical force ; but wood in a natural state, containing moisture, is a conductor. By exposure to very great heat, the volatile parts of wood are driven off, and, if air be excluded, its indestructible base, *charcoal*, remains. This is a conducting substance. But charcoal further exposed to heat with an access of air undergoes combustion, and is converted first into a volatile gas known as *carbonic acid*, or carbonic *anhydride*, while the solid mineral part of the wood remains in the form of ash. The carbonic acid gas and the charcoal are both electrics. The influence of heat in this process is not very intelligible ; it being difficult to determine the more immediate sources of electric and conducting power. We must, however, conclude as a general fact that all electric substances (Table I.) are non-conductors, and reciprocally all conductors (Table II.) are non-electrics.

13. We infer from the preceding facts (10), that in order to retain electricity on any given conducting body, as on P, Fig. 7, it is requisite to support the body on one of the non-conducting substances enumerated in Table I., as for example on a glass rod, N. By this means we imprison, as it were, the electrical agency, and confine it to the given body, where otherwise it would not remain (10). A conductor of electricity thus circumstanced is said to be *insulated*, and is termed an *insulated conductor;* when electrified it is said to be *charged* (E).

14. By means of a small conducting sphere, or thin disc, *u*, Fig. 7, insulated on a long slender rod of varnished glass, or some

other electric substance, we are enabled to transfer electricity
from a charged insulated conductor, P, to another insulated conductor, M, not so charged. This may be effected by first touching
the charged body P with the small insulated neutral sphere
or disc, *u*, which body now participates in the charge, as if it
were a portion of the electrified body P. We then proceed to

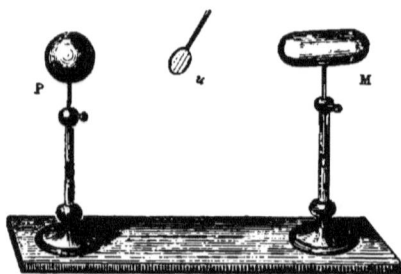

transfer the electricity thus
taken up by the transfer
sphere or disc *u* to the given
neutral body M, by bring-
ing it into contact with
M. The insulated body M,
in virtue of the contact, shares
in its turn the electricity of
the transfer ball or disc *u*, and
thus by repeated contacts or

Fig. 7. Insulated Conductors.

transfers we are enabled to bear off a great portion of the electricity
of the charged body P, and deposit it on a second and distant insulated conductor, M.

The insulated transfer sphere or disc *u* has been termed a *carrier*,
and is largely employed in electrical research. A small thin
circular metallic disc, termed by the French philosophers a *tangent
plane*, or *plate*, is frequently substituted for the carrier ball, as
being better adapted to certain peculiar methods of experiment.

As insulation (13) is the great desideratum in this experiment
particular care is necessary in perfecting it. The insulator should
be small, long, and nicely varnished.

15. Although the march of experiment up to an advanced
period of electrical inquiry has been such as to completely
separate the phenomena of *insulation* and *conduction* as dependent
on two distinct and opposite principles, yet according to Faraday there is really no substance which absolutely arrests, or
which perfectly transmits electrical action; conductors and
insulators being so far identical in physical constitution, that
the electric or insulating power and the non-electric or conducting power differ in degree only. Admitting this as highly
probable, if not positively demonstrable, we have still to observe
that the great practical value of the division of bodies into
electrics and conductors (as first proposed by Desaguliers so
long since as the year 1740) is unquestionable. The extremes
or limits of the differences are so wide apart as really to admit of
substances being considered generally either as conductors or
insulators, allowing an intermediate class of substances partaking

of either property, and hence considered as imperfect either as conductors or as insulators.

If we tabulate bodies in respect of their conducting or insulating power, we find metals, charcoal, and concentrated acids to be the best conductors at one extremity of the series; shell lac, brimstone, silk, vitreous and resinous bodies generally, the best electrics at the other extremity of the series. In the intermediate or transition class we have bodies such as marble, chalk, earths, and stones.

In the following Table the substances are systematically arranged in the order of conducting and insulating power. If taken in relation to insulating power, the series commences in the first column with resinous bodies and terminates with metallic substances in the last. If taken in the order of conducting power, the series must begin with the last substances of the third column—that is, with metallic bodies—and must be read in the reverse order, terminating with the resinous bodies in the first column.

TABLE III.

Electrics or Insulators.	Intermediate Substances.	Conductors or Non-Electrics.
Lac, and all resinous bodies.	Spermaceti and fatty substances.	Flame.
Brimstone, gutta-percha.	Dry metallic oxides.	Saline substances.
Amber and bituminous substances.	Dry vegetable bodies.	Aqueous fluids.
Camphor and gums.	Wool and hair.	Water.
Gun-cotton and collodion.	Feathers.	Saline fluids.
Mica.	Phosphorus.	Aqueous vapour.
Glass, and all vitrifications.	Wood.	Alcoholic vapour.
Tourmaline and jet.	Paper, parchment.	Snow and ice at 32°F.
Diamond and other gems.	Leather.	Concentrated acids.
Silk, and silk-worm gut.	Marble.	Plumbago.
Animal furs and skins.	Stones and earth.	Charcoal.
Oils, especially turpentine.	Ice above 10° F.	All metals.
Dry air and gases.	Rarefied air.	
High-pressure steam.	Steam.	
Fish-glue or isinglass.	Smoke.	
Dried animal gut.		

16. The order of the series in the above Table can only be considered as a useful approximation, since we really cannot, in the present state of electrical knowledge, assign in the scale of conducting power the precise position of every kind of substance, natural and artificial. Besides which, many circumstances arise that tend

greatly to embarrass the judgment. Thus, in determining the
position of charcoal as a conductor, we must look at the process of
its manufacture, and ascertain from what kind of wood it was
prepared, for there is an amazing difference in the amount of ash
in different woods, and this has an influence on the conducting
power of the charcoal. ' On the other hand, when we examine the
insulating or electric power of glass, we have to consider in like
manner its manufacture and the kind of glass, whether made with
potash or with soda, whether the silica in its composition is free
from iron, or contains that or any other metal, and so on. We
therefore must not regard these bodies with a very rigid eye; for
many substances owe the greater portion of their conducting power,
as already observed, to the water contained in them, whilst many
insulators are dependent in great measure on the state of the atmosphere. In any order of sequence therefore that may be attempted
we necessarily repose on a large generalisation. Nevertheless the
position of most bodies, as electrics or as conductors, is pretty well
determined. Thus the metals, although differing amongst themselves in degree of conducting power, are certainly upon the whole
the best electrical conductors, and stand first in the conducting
series; on the other hand, resinous and vitreous substances are
the best insulators, and come first in the order of electrics, but
last in the sequence of conductors. Such bodies as marble, parchment, wool, &c., are certainly to be classed as intermediate or
transition substances. The best conductors are silver and copper.
The best insulators are amber, lac, brimstone, and resinous and
vitreous bodies generally. The liability, however, of glass to condense aqueous vapour on its surface, by which its insulating power
is diminished, renders it necessary to cover the surface with a
thin solution of shell lac in naphtha, dried off by heat. Resinous
bodies not being so liable to aqueous deposition, this precaution
is not so immediately called for.

17. Attractive force, although the first, and commonly the most
evident result of electrical excitation, is not the only exhibition of
power consequent on the electrical state of common matter. On
a further examination of the attendant phenomena, new facts
present themselves. If the electrical excitation be vigorous,
and the electroscope, or other attracted substance, be insulated, the
attractive force appears to vanish at the instant the two bodies
come into contact, and an opposite or apparently repulsive force
succeeds, so that the insulated electroscope now recedes from the
excited or charged body with considerable violence, and cannot be
again attracted until it has been touched by some conducting non-

electrified body, of such magnitude as can deprive it of the electricity it has acquired.

18. The following experiments are strikingly illustrative of this singular and important fact:—

Exp. 4. Suspend the electroscope, Fig. 2 (6), by means of a filament of unspun silk, by which its ball will be insulated. This may be termed the *insulated electroscope*, in contradistinction to the first electroscope suspension, Fig. 1; or if the needle electroscope, Fig. 2, be employed, it is to be sustained on a varnished rod of glass, in which case it will be an insulated needle. Excite the glass tube T, Fig. 8, furnished with a conducting rod and ball *d*. Present its electrified ball to the insulated electroscope *m*, the electroscope will be attracted and drawn toward the ball *d*, and if allowed to come into contact with it the electroscope ball *m* will instantly rebound from it as indicated by the dotted line *m'*, and will not again come near the charged ball *d* of

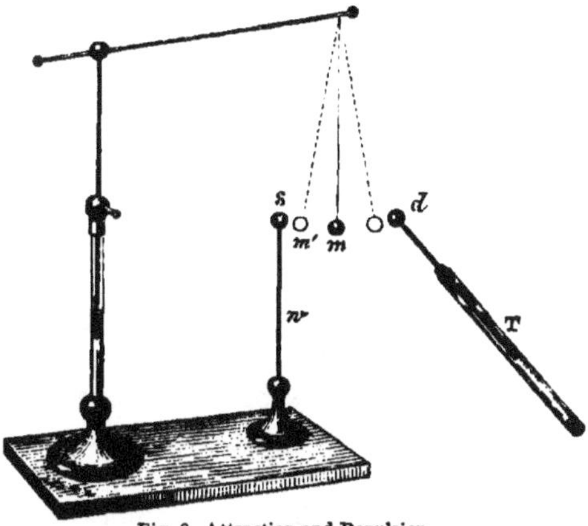

Fig. 8. Attraction and Repulsion.

the excited tube T until touched by the finger or some freely conducting substance.

Exp. 5. Let the metallic sphere *s*, Fig. 8, about 2 inches in diameter, sustained on a conducting rod, *w*, be placed on one side of the pendent ball *m* of the insulated electroscope, and present to the ball *m* on the opposite side the excited tube and ball *d*. The pendent ball *m* of the electroscope will be first attracted and then repelled by the charged ball *d*, and will fly to touch the opposite uninsulated ball *s*, to which it will communicate its acquired charge (14); it will then be in a condition to be again attracted by the charged ball *d*, and in this way will vibrate for a certain time between the two balls, depending on the force or amount of electricity in operation.

This experiment may be repeated under a very simple form. It is only requisite to bring the charged ball of the excited tube near any light substance or fragment of conducting matter resting on a deal table, and a series of attractions and repulsions of a similar kind, exerted between the table, the light body, and the charged ball, will be immediately apparent. A light feather, a fragment of paper or of leaf gold, may be employed in this experiment.

19. The repulsive force thus evinced is better displayed by a charged insulated conductor than by a simply excited electric. If the attractive force be weak, the attracted substance frequently remains, as if adhering to the excited surface, and does not immediately leave it. It may probably remain attached to it, but it will be always thrown off after repellent power can operate between the bodies at any very small distance.

20. On a further investigation of the electrical repellent power we find a most important relation of electrical force to the different kinds of ordinary matter, a relation which, in the early progress of our knowledge of this wonderful agency, could scarcely have been anticipated. We find, for example, that when certain excited electrics repel the electroscope ball, other excited electrics actually attract it, so that when repelled by some excited electrics it is attracted by others, a result evident in the following experimental illustrations.

Exp. 6. The glass tube T, Fig. 8 (18), being powerfully excited, let it attract and repel the insulated electroscope ball *m*, the metallic sphere *s* being removed. Whilst in this repulsed state withdraw the repellent ball and tube T, and present to the electroscope ball *m* an excited roll of brimstone, gutta-percha, or gum lac. These substances do not continue to repel the electroscope ball; on the contrary, the ball flies toward them with remarkable facility.

Exp. 7. Let a gutta-percha tube of an inch or more in diameter, and about 18 inches in length, be prepared and fitted with a conducting rod and ball as in the case of the glass tube T (10). Excite these tubes, the glass one with dry silk, the gutta-percha one with dry woollen cloth. Bring the ball *d* of the excited glass tube on one side of the insulated electroscope ball *m*, and the ball of the excited gutta-percha tube on the opposite side, as at *s* and *d*, Fig. 8. The electroscope *m* will for a moment or two vibrate between the two balls in a similar way to that shown in Exp. 5 (18).

21. A simple and useful kind of instrument, in which both kinds of excitation may be produced, is obtained by coating a glass tube with a solution of shell lac through one half its length, leaving the other half free, and then inserting a conducting wire and ball into each end of the tube. This instrument, Fig. 9, supplies the place of the two separate tubes just noticed, according to the

portion we excite, whether the glass or the lac. It is desirable to guard the centre c, when held in the hand, by rolling a few turns of silk ribbon round it, and covering the ribbon with a short piece of varnished glass tube ; we thus obtain a convenient insulating handle ; and this enables the operator to excite either side with more freedom. The exciting tube may be about 2 feet in length ; either half may be excited with dry soft white silk made moderately warm (F).

22. The foregoing phenomena (18) led the early electricians to the following important deduction, viz., excited electrics develop two kinds of electricity, attractive of each other, but repulsive of themselves. We are indebted to M. Du Fay, Intendant of the gardens of the King of France in 1730, and Member of the Royal Academy of Sciences, for this remarkable fact. His valuable memoirs, printed in *L'Histoire de l'Académie*, from 1733 to 1737, were of great importance to the progress of electrical research at that period of its history.

Conceiving that the opposite electricities developed by vitreous and resinous bodies (20) were peculiar to these substances, M. Du Fay inferred the existence of two distinct electric principles. "Excited glass," he observes, "always repels excited glass; whilst, on the contrary, it attracts all bodies the electricity of which is of the nature of amber. Reciprocally, amber, and other substances of the same electricity as amber, repel each other, but attract the electricity of excited glass." M. Du Fay considers that his fourth memoir established two great facts : first, that electrically excited bodies begin by attracting all other bodies, that when they have rendered bodies electrical by communication (18), then they repel them, but not before ; second, that there exist two kinds of electricity differing essentially from each other, one produced by excited glass and other vitreous bodies, the other by excited amber and other resinous bodies ; these he styles the one *vitreous* the other *resinous* electricity, as depending on the sources from which he supposes them to be derived, whether from vitreous or resinous bodies.

23. This notion of vitreous and resinous electricity did not, however, long survive the discovery on which the doctrine of M. Du Fay was based. Indeed, M. Du Fay himself began at last to countenance an hypothesis more or less prevalent at the time, which assumed an identity of the two forces, the observed difference being merely a difference in degree of power by which the stronger overcame the weaker.

Fig. 9. Compound Glass and Resin Tube.

That the two supposed electricities are not either of them peculiar properties of vitreous and resinous bodies, was soon made apparent by the fact, that both may be obtained in the excitation of either a vitreous or a resinous substance. Moreover both are developed in every case of electrical excitation, and are co-existent.

Exp. 8. Let the resinous half of the electrical tube (21) be excited by dry soft silk, or woollen cloth : it attracts the electroscope when charged with vitreous electricity, that is to say, when repelled by the glass half, excited by silk; but will repel the electroscope if charged with resinous electricity, that is to say, when repelled by the glass half, excited with cat skin ; evidently showing that both vitreous and resinous electricity have been developed in the excitation of the glass by silk and by cat skin. (29) Table IV. This is further shown by exciting the glass tube T, Fig. 8 (18), with silk, and when it freely repels the electroscope charged with vitreous electricity, exciting the tube with the fur of cat skin, made warm and dry ; the tube instead of repelling, now attracts the electroscope, when charged with vitreous electricity.

Exp. 9. Excite the gutta-percha tube Exp. 7 (20) with dry, warm silk, it will repel the electroscope charged with resinous electricity—that is to say, when attracted by the excited glass tube T, Fig. 8. Excite now the same gutta-percha tube with the rough side of dry oiled silk, smeared over with the metallic amalgam already referred to (7) (D) ; the same gutta-percha tube now attracts the electroscope when attracted by the excited glass ; or in other words, the resinous surface will have developed both the electricities. (29) Table IV.

Exp. 10. A wide silk ribbon is. to be rolled round a long cylindrical cork, c, Fig. 10, and a varnished glass rod is to be

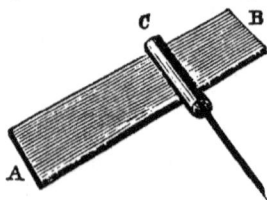

Fig. 10. Silk Rubber and Glass Plate.

passed into the cork to serve as an insulating handle. Let now this insulated silk rubber be passed with slight friction a few times in one direction over a dry, warm slip of window glass, A B, retaining the silk rubber by its insulating handle in one hand, and the glass by one of its extremities in the other. Both the silk and the glass will attract the uninsulated electroscope (6), and each may be caused to attract or repel a charged insulated electroscope (18), but they will not both attract or both repel at the same time. The phenomena will be precisely the same as in Exp. 9. The silk will attract when the glass repels and conversely, and the effects of Du Fay's vitreous and resinous electricities be obtained.

24. It is quite evident by these experiments that both electricities are produced by the same substance; that they are both developed in the process of excitation, and consequently co-existent, although not always both apparent. The reason for this is that in ordinary cases the rubber is not insulated, but is in communication with the ground, through the medium of the hand, or of some other conducting body. Hence the electricity of the rubber is continually passing away, or being otherwise neutralised. For a similar reason we fail under ordinary circumstances in endeavouring to excite conducting bodies such as the metals. In order to obtain the excited electricity of such bodies, it is requisite to insulate them (13).

We have further to observe that in this process of excitation by friction, the electrical development is by no means perpetual, except under certain conditions. If both the rubber and the electric be insulated, the development after a certain time ceases. It would seem as if the rubber could only develop a certain quantity of the opposite electricity. It is not, therefore, until we have restored these bodies to their normal state by contact with free conductors, that we are enabled to continue the excitation. In the case of the excited tubes (20) the hand, being a conductor, continually preserves the rubber in its normal state, whilst the insulated rod and ball, projecting from within the tubes, operates continually in abstracting the excited electricity of the tubes, and in giving it off to other bodies. It is in this way we are enabled to obtain a continuous succession of sparks from the conducting ball.

25. The terms vitreous and resinous, however, as applied to opposite electricities, displayed in the foregoing experiments, carry with them an hypothetical deduction not warranted by the facts of the case. It has been hence considered more philosophical to characterise the opposite electrical forces by the terms *positive* and *negative*, and to represent them by the algebraic signs + and −, or *plus* and *minus*. By this means we disembarrass the question of an illusory form of expression, and are enabled to reason from the effects of the unknown causes of these phenomena, as if the causes were known; much in the same way as the astronomer calculates the movements of the bodies in planetary space, without at all understanding the nature of the occult cause of their movements.

26. A striking and important result of electrical excitation by friction is, that so long as the electric and rubber remain in contact, no electrical development appears to result. It is only when the

c

two bodies are separated that electrical force becomes apparent in either.

Exp. 11. P w, Fig. 11, is a circular plate of glass about 10 in. in diameter, sustained in a slit ball, *n*, by means of a compressing screw. R is a circular rubber in contact with the glass plate.

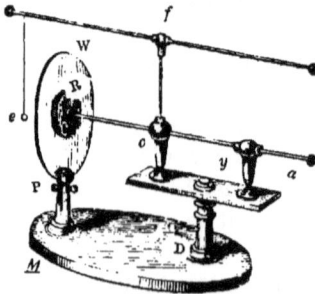

Fig. 11. Glass Plate and Rubber.

R is fixed at the extremity of a long glass rod, *a* R, movable in the supporting balls *c y*, so as to be easily turned round against the glass. The glass rod and rubber are secured in position upon a convenient base and pillar, fixed on an elliptical stand, M D; an electroscope ball, *c*, suspended by a silk thread from a light insulating rod, *f*, is hung opposite the glass plate P w, so as to indicate any electrical force developed in the plate by the movement of the rubber.

The glass being warm and dry, and the rubber R lightly covered with electrical amalgam (7), turn the rubber round against the glass so as to produce a continuous friction. Not the slightest effect will be observable on the electroscope ball *c*. Break the contact between the rubber and the glass by withdrawing the glass rod *a* R through its supporting balls *c y*, an immediate electrical development on the glass plate ensues; the electroscope ball is forcibly attracted towards the glass plate.

It has been inferred from these phenomena that the development of electricity by friction is the result of the separation and renewal of contact between dissimilar bodies, and has been further elucidated by a simple hypothetical exposition. The electrical agency, in whatever it consists, may be supposed to be distributed through bodies in quantities proportionate to their affinities for it. Let now the affinity of glass for electricity be represented by the number 6, and the affinity of the rubber by the number 4, making together 10. When the two bodies are brought into contact these relative affinities are supposed to be changed: the affinity of the glass being increased from 6 to 8, the affinity of the rubber decreased from 4 to 2, the total 10 being still unchanged. When we separate the rubber from the glass or break the contact, the original affinities 6 and 4 are restored, whilst the new distribution remains. The quantity of electricity in the glass, therefore, has been increased by 2. Hence the glass remains positively electrified, or plus 2, and the rubber negatively electrified, or minus 2.

27. The important results so completely established in the course of the preceding experimental investigations, lead directly to the conclusion that substances in opposite electrical states, one plus and the other minus, attract each other; substances in similar electrical states, both plus or both minus, recede from each other. This great general and characteristic law of electrical force, is concisely expressed in the following formula :—Opposite electricities attract, similar electricities repel. This prominent feature of electrical action is ever to be kept in view, since it is found more or less involved in every species of electrical research.

28. An instructive and elegant practical illustration of this elementary law is seen in the following experiment:—

Exp. 12. Excite, as before, the glass tube P, Fig. 12, and having selected a sheet of leaf gold, such as is commonly used by gilders and bookbinders, blow it gently into the air from off any convenient plane surface. The leaf will for the moment float as it were in free space, as indicated in the figure. Whilst thus floating, present to the leaf the charged ball P of the excited tube. The leaf will be immediately drawn toward it ; but directly the bodies touch, the leaf recedes with astonishing rapidity from the ball, and may now be pushed through the air, by a distant impulse of the positive electricity of the tube, in any direction. Whilst thus driven away from the excited glass, present to the leaf the ball of the excited gutta-percha tube (20) ; the gold leaf will now fly rapidly toward the ball of the gutta-percha tube, illustrating in a very elegant and striking manner the opposite character of the two electrical forces ; clearly showing that they are attractive of each other but repulsive of themselves. In this experiment the forces may be observed to operate at great distances, which gives it a very definite character and renders the experiment a fine example of electrical action.

Fig. 12. Excited Tube and Gold Leaf.

29. The two opposite electrical forces thus called into play, in every case of electrical excitation are mainly dependent on the relative condition or state of the surfaces or bodies in contact, or of the surface of the excited electric, in respect of what is usually termed the rubber. We find, for example, as already observed (24), that opposite electricities may be obtained from the same substance. A polished glass surface, for example, when excited by dry silk, develops positive electricity; but when rubbed with cat skin, negative. If we change the condition of the polished glass surface by abrasion with emery powder, then the glass excited by means

c 2

of woollen stuff develops negative electricity. So again, if we give a shell-lac surface to a glass tube, by coating it with a solution of shell-lac in naphtha (as in the positive and negative exciting tube), (21), laid on as in varnishing, we obtain, on exciting this surface with dry silk stuff, negative electricity ; but if we smear the silk with metallic amalgam (7), we then obtain from the same lac surface positive electricity. The method of rubbing produces also, according to Faraday, in some cases very considerable difference, although, as he remarks, it is not easy to say why. If we strike a feather lightly against dry canvas, it becomes electrified negatively ; if drawn with pressure between folds of the same canvas, it is electrified positively. Friction applied across the grain of a substance may develop an opposite electricity to that of friction in the direction of the grain. The colour, or probably the colouring matter of a substance, will often determine the kind of electricity developed. Thus black and white silk frequently give opposite electricities, although excited by the same substance and in the same manner.

30. The following are some general experimental deductions relative to the development of electricity by the process of friction :—

1. Cat skin is positive in all cases of friction with other substances.

2. Smooth or polished glass is positive with all other substances except cat skin.

3. Rough glass is positive with dry silk, but negative with woollen stuff.

4. Resinous bodies are negative with non-metallic substances, but positive in most cases with metals.

Table IV., originally drawn up by Cavallo, and improved by modern electricians, exhibits a concise and comprehensive view of the results of experiments of this kind. But although this table may be considered as pretty correct, we must remember that many sources of error are inseparable from the experiments by which it has been determined : anomalies are liable to arise in examining the particular electrical state of substances. For example, it is stated by Faraday that one part of a cat skin may, after excitation, exhibit positive electricity, and another portion negative electricity. The practical manipulation with delicate electrometers, and the manner of treating them experimentally, may also give rise occasionally to apparent discrepancies. The general deduction, however, that vitreous matter develops positive electricity when excited by most other substances, and resinous bodies negative electricity, is doubtless fully established.

TABLE IV.

SHOWING THE KIND OF ELECTRICITY EVOLVED IN THE FRICTION OF
VARIOUS SUBSTANCES, ONE WITH THE OTHER, UNDER DIFFERENT
CIRCUMSTANCES.

Substances.	Positive, by friction with	Negative, by friction with
Cat skin.	Every other kind of matter.	. . . .
Smooth glass.	Every substance except cat skin.	Cat skin.
Rough glass.	Dry oiled silk, sulphur, metals.	Woollen-cloth, paper, wood, sealing-wax, the hand.
Tourmaline.	Amber.	Diamond, the hand.
Baked wood.	Silk, paper, metals.	Flannel, hare skin, polished glass.
Amber.	. . . .	Silk, and most other substances.
Resinous bodies.	Many metals.	Substances generally.
Black silk.	Sealing-wax, glass.	White silk, hare skin, ferret skin.
White silk.	Black silk.	Cat skin.

31. The practical determination of the kind of electricity in operation being of great importance, it may be as well to observe that this can always be ascertained by presenting to an insulated electroscope (18), charged with either of the electricities, the particular substance the electrical state of which is to be determined. If an excited body be of the same kind as that with which the electroscope is charged, repulsion will ensue; if of an opposite kind, attraction (27). Electrical attraction and repulsion have hence been extensively employed as fundamental principles in the construction of electroscopes and electrometers; and since a correct determination of the force and kind of electricity which an electrified body evinces is of paramount importance in every kind of electrical investigation, it may be as well to advert to the construction and application of some of the most elementary and useful of these instruments.

32. Gold leaf, as already observed (6), is an extremely sensitive and efficient test of electrical action. Fig. 13 represents a gold leaf electroscope of a simple and convenient kind. A single gold leaf, R, is suspended within a glass receiver, mounted on a firm foot, E. It has a lateral opening, q, to admit of bringing any substance to operate on the leaf, which will be found extremely obedient to electrical force. The object of this species of electroscope is to ascertain first whether a body evinces any electrical sign whatever, or is electrified

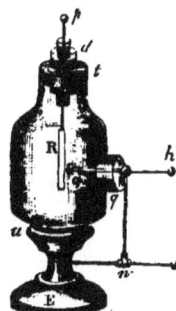

Fig. 13. Gold Leaf Electroscope.

in any minute degree. Secondly, to determine the kind or quality
of the electricity. In order to ascertain the first, we set a small
metallic disc, g, fixed at the extremity of a sliding rod, $g\,h$,
within a short distance of the suspended leaf R. We then bring
the given subject of experiment in contact with a communicating
plate or ball, p. If the substance to be tested be in the smallest
degree electrically excited, the leaf R is attracted toward the disc g,
which should be in this case uninsulated; or, conversely, we may
make the given substance under examination touch the ball h of
the insulated rod $g\,h$, so as to communicate with the disc g. So
sensitive is this instrument, that the mere touch of a silk handkerchief on the plate or ball p causes the leaf to strike the disc.
To determine the kind or quality of the electricity we set the disc g
at a convenient distance from the leaf; and having communicated
to the leaf through the ball or plate p a weak charge of either
positive or negative electricity, so small as not to be sensible to the
disc g at a given distance, we then make the body under examination touch the ball h of the insulated rod $g\,h$. If the electricity
be of the same kind as that communicated to the leaf through the
plate p, the leaf will be repelled and conversely (27). (G.)

A glass receiver, with one lateral opening, q, as seen in Fig. 13,
suffices for ordinary operations. A second lateral opening, on the
opposite side, however, as at q', Fig. 14, gives greater experimental
facility, since it enables us to oppose the substance itself under
examination directly to the suspended leaf through either of the
lateral openings $q\,q'$, or enables us to place the leaf between two

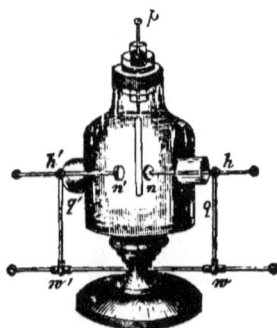

Fig. 14. Single Gold Leaf
Electroscope.

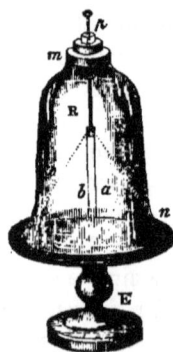

Fig. 15. Double Gold Leaf
Electroscope.

discs by means of a second slide and insulator, w', similar to the first
already described; the tubular rod w being passed clear through
the foot, for the support of the additional slider w'. Fig. 14 represents the single leaf electroscope as thus constructed with two
lateral openings.

33. An extremely sensitive electroscope with two gold leaves is represented in Fig. 15. Two strips of gold leaf, *a b*, are suspended within a glass receiver, R. Electricity is communicated to the leaves through a light conducting rod and ball, *p*. When electricity is communicated to the ball *p*, the leaves *a b* immediately diverge, as indicated by the dotted lines *x y*. If the electricity of the substance under examination be different from that with which the ball *p* has been charged, then on making it touch the ball the leaves will immediately close. If, on the contrary, the electricity be the same as that with which the ball *p* is charged, the leaves will have an increased divergence (27), and thus the kind, and in some degree the force, of the electricity in operation is easily observed.

The application of leaf gold to the purpose of an electroscope was first suggested by the Rev. Abraham Bennett, F.R.S., Rector of Wirksworth, in Derbyshire, who gives a description of what he terms his *Gold-leaf Electrometer*, in a letter to Dr. Priestley, read before the Royal Society, 7th December, 1786, and printed in the seventy-seventh volume of the Philosophical Transactions. Bennett's celebrated gold-leaf electrometer, as originally constructed, consisted, as in the present case, of two delicate slips of leaf gold suspended from a conducting rod within a glass receiver (H).

34. *Henley's Quadrant Electrometer.*—This species of electroscope was invented by Mr. Henley, F.R.S., so long since as the year 1772. It is represented in Fig. 16. A light reed, *c d*, terminating in a pith ball, *d*, is mounted in a delicate axis, *c*, attached to a vertical conducting rod or stem, A B. This axis is in the centre of a graduated quadrant or semicircle, also affixed to the same stem. The stem has a ball below at B, against which the terminating pith ball *d* of the reed reposes, when in a vertical position. But when electrified by being connected with a charged body, the reed rises, and denotes on the graduated arc the angle of divergence by which the comparative amount of charge acting on the reeds may, in some cases, be estimated. The instrument as commonly employed is not very available as an accurate quantitative measure. It is, however, practically useful, and may by a little care and management be occasionally employed as an electrometer with advantage (I).

Fig. 16.
Henley's
Quadrant
Electrometer.

35. *Cavendish's Electrometer.*—The Hon. Henry Cavendish also employed the principle of diverging reeds as a measure of electrical force. The kind of instrument used by him is represented in Fig. 17, and the following description of it is, for the most part, taken from

the manuscripts of this distinguished philosopher, in the possession of the author. Two wheaten straw reeds, a A and b B, each about a foot in length, are set on fine steel pins as axes, movable in notches in an insulated metallic plate, P, or on centres in a ring of support, so as to turn with the axis. The straws are left open below, and carry two light cork balls, or balls of pith, A B, about one-third of an inch in diameter. In order to increase the force with which the straws tend to close when diverging, the lower open ends of the reeds are occasionally loaded with short pieces of wire. Now it is not difficult to find, on ordinary mecha-nical principles, the force with which the reeds, considered as levers, will tend to the perpen-dicular from any given angle of inclination, when light and when loaded. We may thus estimate the relative repulsive power required to maintain the same angle of divergence in these two cases. For example, let the force tending to the perpendicular from a given angle of divergence be four times greater when loaded than when light; we might infer that if divergent by electrical repulsion, four times the repulsive force must be exerted to sustain the loaded reeds at the same angle as the light reeds. If we suppose the force of gravity in this kind of instrument to be concentrated in the balls themselves, "the force required to separate them will be as their weights directly," so that by a careful manipulation we may, in many cases, with this kind of instrument, measure the repulsive force pretty accu-rately (K).

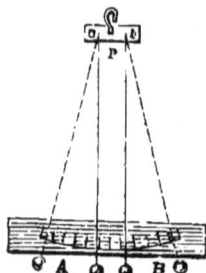
Fig. 17. Cavendish's Diverging Electrometer.

36. *The Electroscope Electrometer.*— Fig. 18 represents a simple and available instrument invented by the author, and termed by him an Electroscope Electro-meter, as combining, to a certain extent, the object of both instruments.

An electroscope reed, $m\ n$, about a foot in length, terminating in small pith balls, m, n, is mounted on a delicate axis, set on fine points within an elliptical metallic ring, T. This electroscopic reed consists of two parts, m T, n T, united through the medium of a central axial-pin at right angles to the axis. The ring, with the electroscope $m\ n$, is joined by means of a short projecting

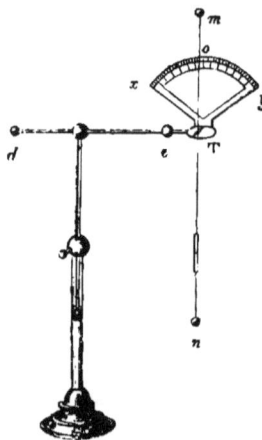
Fig. 18. Harris's Electroscope Electrometer.

stud and ball to the extremity of an insulating glass arm, *d e*. The ring and axis allow the electroscope *m n* free motion in a vertical plane in either direction, which is determined to the vertical by a reed slider on its lower arm; this slider enables us to adjust the balancing power in such a manner as to measure, with some degree of accuracy, the force in operation. The attractive or repulsive force is developed in opposing a charged body at a given distance, to either ball, *m n*, of the electroscope, either in a neutral state or charged with one of the electricities. The electroscope ball will incline toward the charged ball or recede from it, according to the attractive or repulsive force developed. In either case it will be found more convenient to operate upon the lower ball *n* rather than on the upper one, *m*, although either may be employed, according to the circumstances of the experiment. In the case of attractive force the electroscope is neutral. In the case of repulsive force the electroscope is charged either positively or negatively, which may be easily done by communicating positive or negative electricity to the ball *e*. The amount of attractive or repulsive force is shown by the deviation of the electroscope reed from the perpendicular. In order to estimate the angular deviation of the arm *m n*, a small quadrantal arc, *x y*, about 4 inches radius, is fixed to the elliptical ring т, immediately behind the central axis. This arc, with its radial arms, is of light mahogany, the arc itself being faced with a similar arc of thin cardboard, divided into 60 degrees on each side of its centre, marked °. The Zero line coincides with the index arm of the electroscope, when at rest in a vertical position. The amount of attractive or repulsive force may hence be measured in degrees of the arc included between the Zero line and the line of deviation of the index arm. We thus not only determine the kind of electricity in operation, whether positive or negative, but we have also, in a certain degree, a measure of its amount (35). All we require to know is the distance of the attracting or repelling bodies, the amount of charge in operation, and the relation of the force to the degree of angular deviation, all of which may be calculated experimentally (I). It will be sufficient in any case to take the distance between the charged body and the ball of the electroscope. This simple instrument is efficient and convenient in ordinary electrical elementary inquiries.

37. *Electroscope Electrometer of Double Repulsion.*—This instrument, Fig. 19, is constructed on similar principles to the Quadrant Electrometer of Henley (33). A small elliptical metallic ring, *c*, is set obliquely on a short brass rod, *c m*, sustained on a sliding in-

sulating support, s; two light metallic wires, ca, cb, proceed in opposite perpendicular directions from the extremities of the long diameter of the ring, and these wires terminate in gilt balls of pith or cork, $a\ b$. A delicate axis is set on fine points in the direction of the short diameter, which, by a fine central transverse axis, carries two light reeds, cd, cf, forming together, as in the preceding instrument, Fig. 18, one long index, df. This index also terminates in gilt pith balls, df. When unelectrified, the balls df of the index repose against the balls ab of the vertical wires ca, cb; when electrified, either alone or by connecting the rod cm with a charged conductor, the index is repelled in opposite directions by a double repulsion, above at d and below at f. The amount of divergence is estimated by a graduated quadrant fixed in the centre of the axis of the index behind the elliptical ring c. The tendency of the index to a vertical position is regulated by short reed sliders on the opposite arms of the index, and which move on them with friction so as to admit of being placed at various distances from the centre of motion. The electroscope, therefore, admits of a similar measurement of force, as in the case of the electroscope of Cavendish (35), when loaded and when light; thus the index diverges with an extremely small force. This kind of electroscope, although convenient and useful in many instances, is, nevertheless, of very limited application in others. We

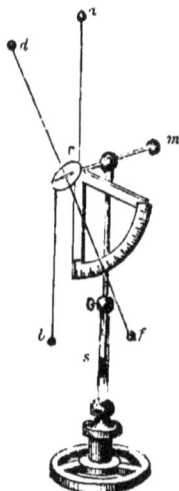

Fig. 19. Harris's Electroscope Electrometer of Double Repulsion.

are certainly enabled to say that, *cœteris paribus*, when the angular deviation is the same, the same quantity of electricity is in operation; with a greater divergence there is a greater quantity, with a less divergence, less, but how much greater or less is not easy to determine, since we have to take into the account the diminishing force of repulsion as the distance increases, and the simultaneous increasing force of gravity at different angles, together with all the different and varying distances from the centre to the extremities of the repelling arms, and also the variable oblique action, all tending to complicate any mechanical calculation to which we can subject it, and render it more or less dependent upon empirical experiment.

OCCASIONAL MEMORANDA AND EXPLANATORY NOTES.

(A) A fine filament of unspun silk from the cocoon of the silk-worm, with a light pith ball at its extremity, is very sensitive of electrical force. A still more delicate test of electrical attraction is a similar suspension by the spider's thread, which may be twisted round some thousand times without exhibiting the least torsion or reactive force. It requires, however, great manipulative skill in the construction of an electroscope of this kind. If we should be fortunate enough to catch a spider in its descent from the ceiling, leaving its thread behind it, there is the electroscope required; the spider itself being the attached bob, or substitute for the pith ball. By a little care we may cut the spider adrift from its thread, and attach by a little gum a light pith ball to the thread itself. The method of securing light spider's threads for delicate electrical suspensions is very simple. A wood spreader should be prepared after the manner of a pair of compasses, so as to admit of being extended to any required distance, the points of the spreader being touched with a little weak gum, the observed spider's thread is carefully caught between them, and the thread subsequently attached to the electroscopic pendulus body, and to the given point of suspension.

(B) This needle is constructed of fine silver wire, about one-twentieth of an inch in diameter, and 10 inches in length. The centre of motion is near one extremity of this wire, dividing the needle into two arms, $c\,a$, $c\,b$, of unequal length. At the centre of motion c the wire is bent into a small curve, and a fine point soldered to the middle of the curve, projecting downwards, so as to allow the system to turn freely on it, preserving at the same time the point of suspension above the centre of gravity of the mass. The extremity of the long arm of the needle carries a light ball of the pith of elder, about three-tenths of an inch in diameter, fixed on its point; the extremity of the short arm carries a balance-weight, b, which may consist of sealing-wax neatly run into a large globule of sufficient size to preserve the needle when suspended in a horizontal position: the whole is sustained on a hard metal centre, united to a small vertical support of brass or of glass, d. Any electrified substance presented to the pith ball a, at the extremity of the long arm of the electroscope needle, affects it powerfully from a considerable distance.

(C) The strip of leaf gold may be about 3 inches long and from a quarter to half an inch wide. It should be prepared, and cut fair upon a gilder's cushion. The most simple and extemporaneous way of suspending such a strip of gold is to attach it to a short slip of moistened gummed paper; and insert the paper-holder in a cleft stick of what is called *skewer-wood*, which, being neatly tapered, may be passed into a bung of fine cork; the bung being pressed upward in the open neck n of a bell glass, the gold leaf downwards. The whole may now be sustained from a short arm of support by means of a slight clip of wood or metal, as shown in the figure, being sustained on an insulated rod and foot, M.

(D) A good metallic amalgam for promoting the excitation of glass is obtained by melting together in a clean iron ladle 1 part tin, 2 parts zinc, and 4 parts mercury, the proportions being 1, 2, and 4, or double of each other. The zinc should be just heated, and when nearly at the point of fusion the tin should be added; the two metals will run freely together. The temperature of the mercury being raised to somewhat short of its boiling point, should be poured on the fused metals, stirring the mass at the same time with an iron rod. The resulting amalgam, after being allowed to cool somewhat, should be now poured into a covered wooden box, and kept in a state of agitation until cold, when it will be commonly found in the state of a black powder, or otherwise in a soft metallic mass, which after a time will harden, and may be pulverised in an iron mortar. The triturated substance should now be passed through a muslin sieve. When wanted for use a little of it should be rubbed up with some purified lard, and applied to the surface of the rubber with which the glass is to be excited. For a more especial account of the preparation and application of metallic amalgams and other substances in promoting electrical excitation we must refer to the chapter on Electrical Manipulation (85).

(E) Insulated supports of various kinds and altitudes are frequently called for in Experimental Electricity. Fig. 20 represents an insulating support, the altitude of which may be varied to any required extent, without interfering with the insulation. It may be termed a *sliding insulator*. It is a very efficient and convenient auxiliary electrical instrument. Its construction is as follows:— *n o* is a solid cylindrical rod of glass or vulcanite, of any convenient length and thickness. It is movable within a varnished glass tube, *t*, capped by a small ball of wood, *v*, through which the rod passes. This insulator is movable within the glass tube *t*, after the manner of a piston-rod, having a compressible cylinder of fine cork at its lower extremity, *o*, covered by a layer or two of silk ribbon, which enables it to slide freely in the tube. There is a small stop-screw, *v*, in the wooden ball through which the insulating-rod passes, by which it may be secured at any given height. The body to be insulated is placed on the upper extremity, *n*, of the insulated rod, by means of any convenient holder attached to it. For the support of spherical bodies a concave holder—such as an inverted watch-glass—will be found convenient. Plane surfaces may also be sustained upon inverted spherical segments; and thus the altitude may be regulated to any convenient height. The glass tube *t* is fitted in a neat foot and holder, *a* (or other basis), furnished, if necessary, with a ring of lead to give it stability.

Fig. 20. Sliding Insulator.

For general purposes the rods of these sliding insulators may be about a foot in length and from one-eighth to one-quarter of an inch in diameter. The glass tubes within which they move may be from 5 to 8 inches in length and from ·3 to ·8 of an inch in diameter.

(F) Two hollow cylindrical tubes, one of glass, the other of gutta-percha (20), are very useful in Practical Electricity, since they furnish a ready means of producing the opposite forces. These tubes should be well selected and carefully prepared; they may be from 20 inches to 2 feet in length, and a full inch in diameter. They can be powerfully excited, and pretty strong sparks may be obtained from them. The old electricians confined their experimental researches within the limits of the powers of excited glass tubes. These were, in fact, their *electrical machines*, and they took especial care of them, and enclosed them, when not in use, in paper or woollen envelopes. Glass tubes for electrical excitation should be rather thin than otherwise. When in good condition a gentle excitation with a cloth smeared with metallic amalgam (7) will be sufficient.

(G) The glass R, Fig. 13, within which the leaf is suspended, may be of the form given in the figure. It has two contracted necks, with open mouths, *t u*, about 2¼ inches in diameter, and a lateral mouth opening, *q*, about the same diameter. The dimensions of this circular glass screen may be about 6 inches in diameter, and 8 inches in height. It should be mounted on a firm foot, E, so as to admit of a small tubular rod of brass, w, being passed horizontally and partially through it beneath the glass. This rod carries a short slider, w, and insulating support, upon which is sustained a light movable conducting rod, $g\ h$, which enters the receiver through the lateral opening q. The rod $g\ h$ has fixed to its inner extremity the gilt disc g, about half an inch in diameter and one-tenth of an inch thick, and terminates externally in a conducting ball, h; by means of the slide w the disc g may be set at any convenient distance from the surface of the suspended leaf. The leaf is appended from a wide metallic pincer, at the extremity of a short metallic rod, passing through an insulating glass tube, d, sustained within the receiver through a varnished cork in its upper neck, t. The suspension rod terminates above in a brass ball, p, about three quarters of an inch diameter, or in a small brass plate, about 2 inches diameter, and through which electricity may be communicated to the leaf R.

(H) *The Portable Diverging Electroscope.*—A simple and useful modification of the Double Gold-Leaf Electroscope of a portable kind is obtained by substituting two light diverging reeds or pith balls for the gold leaves. A small brass rod, $m\ n$, about 4 inches in length and about one-tenth of an inch in diameter, is supported through a central ball, d, insulated at the extremity of a short glass rod, $c\ d$; the extremities of the brass rod are capped by small metallic balls, $m\ n$, about one-eighth of an inch in diameter. Two light threads, or very light straw reeds, $m\ o$, $m\ p$, terminating in small pith balls, $o\ p$, are suspended from the ball m; these are very delicately hung to the ball, either by very fine wire ring joints or small silk filaments; so that

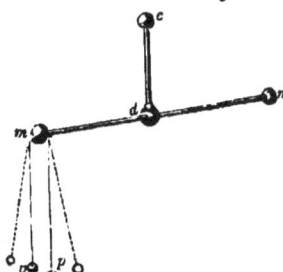

Fig. 21. Portable Diverging Electroscope.

when the ball n at the opposite extremity of the brass rod $m\ n$ touches an electrified body (the instrument being held by its glass insulator, $c\ d$), the reeds or threads $m\ o$, $m\ p$, diverge, or, if divergent, tend to come together; thus showing not only the excited state of the body touched, but the kind of electricity with which the given substance is affected.

(I) Henley's Electrometer is often roughly and ill-constructed, and is hence generally a somewhat rude instrument. The most refined method of construction is as follows:—The stem A B, Fig. 16, should be a small silvered brass tube about 6 inches long and one-eighth of an inch in diameter, terminating at its upper extremity in a small silvered or wood ball, A, about half an inch in diameter. A light straw reed, $c\ d$, about 5 inches long, is affixed to a delicate axis, in the centre of an elliptical ring, c, attached to the stem A B, about an inch below the terminating ball A. The axis is set on fine watchmaker's points, so that the reed has great freedom of motion; and has a short pin passing centrally through it. The under portion of this pin carries the electroscope reed $c\ d$; the upper portion is furnished with a small counterpoise ball about one-tenth of an inch in diameter. A light quadrantal arc, q, divided into 90 degrees—whose centre is the centre of the axis carrying the reed—is fixed behind the elliptical axial ring to the stem A B. This arc should be an open segment about $2\frac{1}{2}$ inches radius and a quarter of an inch wide on its circumference. It may be of ivory or varnished cardboard glued upon a light wood segment. A pith ball, d, of about a quarter of an inch in diameter, is fixed on the extremity of the reed, and reposes, when the reed is in a vertical position, against a small metallic ball on the lower part of the stem. The whole when in use is sustained on a stout metal point, proceeding from the electrified or charged body. As the charge accumulates the reed rises and marks upon the graduated arc the progress of the charge, as already observed.

Much elaborate investigation has been expended upon the value of the indications of this instrument, but not with any great advantage. We may, however, put some confidence in the following general principles:—It is a well-known mechanical fact that the momentum or moment of the force with which a pendulous body delicately suspended tends to the vertical, is measured by the weight of the body multiplied into the distance of its centre of gravity from point of suspension, multiplied into the sine of the angle of inclination. Having in this way ascertained the value of the mechanical force in operation for any given angular divergence of the reed, we are enabled so far to employ the instrument as an electrometer of measure. The centre of gravity of the reed with its pith ball and axis may be determined, as practised by Cavendish, by balancing the reed across a fine knife edge. If the reed, therefore, be furnished with a small slider set at different distances from the centre of motion, the distance of the centre of gravity from the point of suspension will vary, and thus by a simple graduation of the reed we may estimate the mechanical force at a given angle. This is very simple, but whether it is all we require it is not easy to determine. The inclination, for example, of the repellent reed to the vertical pillar undergoing continual change, the electrical force not only acts at variable distances from the vertical, but also acts more or less obliquely. Hence it does not at all follow that the quantity of Electricity producing a double divergence would be also double. In order, therefore, to arrive at an exact quantative measurement we are in a great degree dependent upon empirical experiments. Cavendish states in his manuscripts that when this electrometer is considerably elevated on a long stem above a charged conductor on which it is usually placed, the indications will be different from those obtained when close upon it. In the first case the electrometer is more sensible at the beginning of the motion; in the latter case it is less sensible. The difference by experiment with similar quantities of electricity was found to be considerable; thus, when close upon the conductor the divergence was only 5 degrees, when elevated the divergence amounted to 21 degrees. On the other hand, towards the close of the motion the sensibility of increase was found to be less in the elevated than in the low position. The safest method, therefore, perhaps is to determine experimentally for each particular instrument the angular quantity corresponding to a given charge, and to estimate accordingly.

(K) In applying this electrometer, Fig. 17, we suspend it at a distance of about 6 inches before a strongly-marked pasteboard scale, the eye being situated about 30 inches before the scale, so that by means of an eye-piece the angle of divergence is easily observed. The straws a A, b B, reach nearly to the bottom of the cork or pith balls A B, but not quite; the lower ends of the small wires, therefore, with which the reeds are occasionally loaded may be just even with the surface of the ball, being retained in their places by a little soft wax. If, instead of loading the reeds with weights, we furnish them with small graduated sliders, it would be easy to accommodate the centre of gravity of the reeds to any given angular divergence. The electrometer as employed by Cavendish, when loaded with the wires, required, according to his calculation, three times the force to maintain the same angle of divergence than when not so loaded; in which case Cavendish inferred, on the principles already noticed (I), that with the same angle of divergence the repulsive forces were in these cases at 3.1.

CHAPTER II.

INDUCTION.

38. The preceding elementary facts being fully understood, we are prepared to enter on a more advanced class of phenomena, remarkable for their singular interest and importance, namely, electrical action at a distance, or *Induction* as it is called, or by the French, *Electricité par influence*, also *Electricité dissimulée*. Hitherto our attention has been limited to the development of certain electrical forces of attraction and repulsion, either by simple excitation (7), or by immediate communication (10). We now proceed to consider a new species of electrical force, consisting, apparently, in the development of similar powers by the influence of excited bodies on other bodies beyond the limits of contact; in short, a species of apparent electrical emanation or sympathetic action, operating between bodies at sensible, and even at considerable distances.

39. This new species of electrical action is well exemplified in the following experiments :—

Exp. 13. Let an insulated neutral conductor, Q, Fig. 22 (which may be a light hollow cylinder of gilded wood, about 5 inches in length, and from 3 to 4 inches in diameter), be directly opposed to a similar insulated conductor, P, charged either positively or negatively (30),—suppose positively. Under such circumstances an action, apparently of a sympathetic kind, arises at a distance between these conductors, P and Q. The result is to change the actually existing electrical state of the two bodies P and Q. The previously neutral body Q, without any direct communication with P, evinces a state of excitation so that a delicate neutral electroscope, *e* (36), placed near the remote surface *p'* of the body Q, becomes attracted towards Q, as indicated in the figure. On the other hand, a similar electroscope, *f*, repelled from the remote face *q* of the body P, falls back towards P in a greater or less degree, thereby showing a decrease in the repellent force of P at its distant extremity, *q*.

The state thus indicated in Q is a peculiar state ; its near face, *n*, immediately opposed to the charged body P, assumes an opposite electricity to that of P, whilst its remote face *p'* indicates the same electricity. Thus, supposing P to be charged with positive or vitreous electricity, the face *n* of the body Q will be negative, and the face *p'* positive. This change or disturbance of the neutral condition of Q, however, is not all ; it is further attended by a reciprocal or similar influence reflected back from the

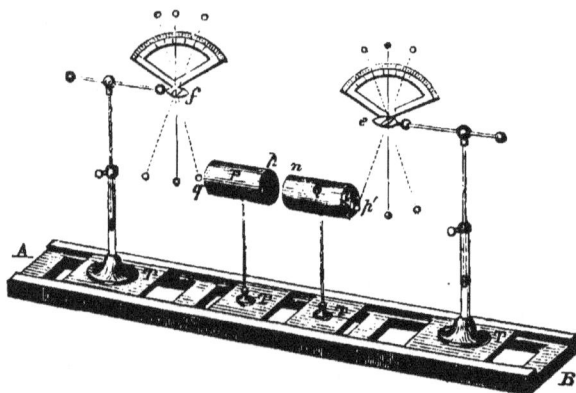

Fig. 22. Apparatus for showing some effects of Induction.

body Q, upon the charged body P, shown, as just observed, in the diminished divergence of the electroscope *f*. This reflected influence, therefore, is such as to increase the force in the charged body P, at its face, *p*, directly opposed to the face *n* of the body Q, and diminish the force of its face, *q*, distant from the face *n* of the opposed body, so that the immediate result of the induction is a succession of comparative alternate positive and negative states. These states may be tested by means of a carrier (L).

40. This succession of positive and negative force by inductive influence, may be still further, and perhaps more directly, demonstrated in the following manner :—

Exp. 14. Let two cylindrical conductors, P, Q, Fig. 23, similar to the former, be fitted with movable faces, *q*, *p*, *n*, *p'*, sustained in position on slender rods of varnished glass, *a*, *b*, *c*, *d*, inserted in small movable feet upon sliding bases. Let the two bodies be opposed to each other at some given distance, as indicated in the figure, the body P being positively charged. When so charged withdraw in succession the faces, *p'*, *n*, *p*, *q*, examine their electrical states through the medium of one of the gold leaf electroscopes already adverted to (32), weakly charged with positive or negative electricity. These faces will evince, as just observed, a succession of positive and negative states; that is to say, the

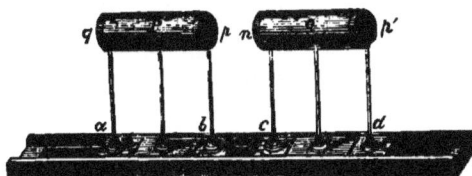

Fig. 23. Induction Apparatus.

face p' will be positive, the face n negative, the face p again positive, the face q less positive (M).

41. We see by these experiments (39, 40) that when an electrified body is directly opposed to a neutral body, two inductive actions ensue. First, we have a direct induction of the charged body P upon the neutral body Q, changing its neutral state into a state of excitation; secondly, the neutral body thus influenced reacts in its turn upon the charged body, and changes the actual condition of the charged state of that body. The first of these actions may be considered as *direct* induction; the second, or reactive state, may be termed *reflected* induction. Both these inductions will be more apparent in some inverse ratio of the distance between the bodies, in some direct ratio of the quantity of electricity in operation, and the superficial extent of the two bodies. Thus, if the body Q opposed to the body P (39), instead of being a small cylinder of equal dimensions, have very great extension, its reflected induction upon the body P will be considerably increased. If Q be put in communication with the ground, so as to give it unlimited extension, or inductive capacity, the influence upon P is the greatest possible. Thus the diminished repulsive force upon the electrometer f, by the influence of the body Q upon P, will be the greatest possible when Q has unlimited extension; that is to say, when its distant face p' is connected with the earth; so that the influence of a neutral on a charged body would in great measure depend upon what may be termed its *inductive power*. Reciprocally, the electrical change which the charged body P undergoes by the influence of the body Q, excited by direct induction, will mainly depend upon the extent of its charged surface.

42. The ordinary phenomena of electrical attraction are inva-

D

riably attended by these different inductive actions. When, for example, a charged body (39) exerts an attractive force upon a neutral body, it first changes the neutral state of that body into an excited state. It calls up in the near surface of the neutral body an opposite electricity to that of itself, while the neutral body, thus excited by induction, reacts upon the near surface of the charged body. The two approximated surfaces being hence in opposite electrical states, the bodies—if all impediment to motion be removed—immediately approach each other (27).

43. The phenomenon of electrical repulsion is attended by similar inductive changes in the repellent bodies.

For example, let two bodies (39) be charged with the same electricity—suppose positive—and be directly opposed to each other at a given distance. Inductive actions immediately ensue between the two bodies similar to those just alluded to. One of the bodies endeavours to reverse the electricity of the other, and to excite in its near surface an opposite electricity to that with which it is already charged. The two bodies resist these changes, both being charged alike; so that, if the existing forces are sufficiently powerful to withstand the new inductive actions brought to bear on them, the two bodies recede from each other, and apparent repulsion ensues, there being no impediment to motion. Thus it is that similar electricities are said to repel each other (27). We see, therefore, that in the phenomena of electrical attraction and repulsion, bodies are first rendered attractable or repellent of each other by induction, and then attraction or repulsion follows as a subsequent result. Hence we may infer, as an especial characteristic of electrical force, that the attracted or repelled bodies are first rendered attractable or repellent before attraction or repulsion ensues.

44. Electrical attraction and repulsion, together with their attendant inductive phenomena, admit of the following elegant and striking experimental illustrations.

Exp. 15. N, Fig. 24, represents a light disc of gilt wood about 3 or 4 inches in diameter, and one-twentieth of an inch in thickness, insulated on a slender rod of glass or vulcanite, at the extremity of a balanced cylindrical rod, *m* N, of light wood. This rod is mounted on a delicate vertical axis, *x*, set in an elliptical metallic ring, so as to be susceptible of circular movement in a horizontal plane. A radial arm, *b d*, projecting from a movable centre, revolves about a central pivot-foot fixed in a strong base of support, R s. The foot carries a central vertical insulating rod of support, *a b*. A second equal similar disc, P, insulated on a slender glass rod, at the extremity of the projecting radial arm *b d*, is opposed to the disc N. Each disc has a strip of gold leaf, *n'*, *p'*, attached to its distant face, after the manner of an electroscope leaf, so as to hang freely and parallel along the faces of the discs, and diverge readily when the discs are electrically excited. The disc P being turned away at a distance from N, and charged either positively or negatively—say positively—its gold leaf, *p'*, will immediately be repelled from the disc, as indicated in the figure. Whilst thus divergent, turn the arm *b d*, and cause the charged disc P to approach the neutral disc N. As the discs approach each other, the

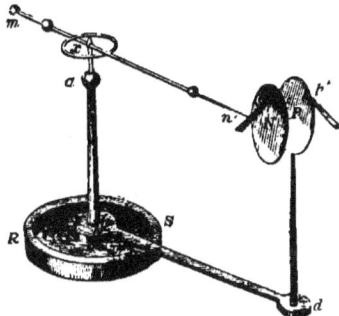

Fig. 24. Induction and Attraction Apparatus.

leaf *n'* of the disc N will evince a state of excitation by induction. At this instant attraction will ensue, and N will move toward P; and thus the simultaneous operation of the inductive and attractive forces are elegantly and experimentally demonstrated. The reflected induction (41) of N upon P will be also more or less apparent as the two discs approach each other, so that the gold leaf *p'* will tend to collapse as the leaf *n'* diverges. This last, or reflected inductive action, however, is not so apparent as the direct induction, owing to the limited extension of the attracted disc N. In order to render both the inductive actions very sensible, the attracted body N should have great extension (41). This it is difficult to effect without inconveniently loading the balanced rod *m* N. We may, however, remedy this inconvenience by substituting for the disc N a very light gilt paper cylinder, about 4 inches long and 3 inches diameter, in which case its reflected induction upon P will be more sensible.

The following experiment, demonstrative of the reflected induction, will be perhaps equally, if not more satisfactory than the preceding, as being more sensible in its operation.

Exp. 16. Turn the discs P, N, away from each other, as at first. Let the disc P be connected with the ground. Charge the disc N with one of the electricities, say positive. Let the disc P, thus neutral and free, be caused to approach the charged disc N. The gold leaf, *n'*, will immediately tend to collapse, in consequence of the reflected induction of P upon N, at which instant attraction

D 2

ensues between the two discs. If, instead of placing the disc P in a free state by connecting it with the ground, we give it considerable extension (41) by connecting its charged face with an insulated cylinder about 4 inches in diameter and 5 or 6 inches long, both the direct and reflected inductions may be experimentally observed oy the electroscope gold leaves attached to the respective discs, which will be simultaneously affected.

Exp. 17. Place the discs P, N, far apart. Charge the one, P, positively, the other, N, negatively. The leaves will then be repelled from the discs, one with positive, the other with negative electricity. Let the disc P approach the disc N, as before; reflected inductions from the discs upon each other will arise, and the leaves tend to collapse by the reciprocal action of the positive and negative electricities, but at a greater distance than in the former experiment, and attraction will again ensue.

Exp. 18. The two discs being neutral, and at a distance from each other, charge them with the same electricity, either positive or negative—suppose positive. The electroscopic leaves will diverge equally. Let the discs P, N, thus charged, approach each other. The electroscope leaves will exhibit an increased divergence; and at the same instant the disc N will recede from the disc P, the two discs being apparently repulsive of each other. It is, however, a question with some electricians whether this be an instance of pure repulsive force, or whether the two discs do not recede from each other in consequence of their respective attractions and inductions upon surrounding neutral matter more immediately influencing the distant surfaces of the oppositely charged discs.

If we substitute for the conducting discs P, N, light discs of mica or other non-conductor little susceptible of electrical induction, the attractive or repulsive force is not so apparent. Hence it follows that electrical attraction is most sensible between bodies having large inductive capacities (41).

45. Such are a few of the more elementary phenomena of electrical action at a distance, as exhibited in opposing to each other a neutral to a charged body, or otherwise two charged bodies, either charged with the same or with opposite electricities. These phenomena, although of extensive application, constitute a portion only of the whole inductive process. In order to understand more fully this kind of electrical action, and the importance of its application in the propagation of electrical force, so as to cause it to assume the appearance of action at a distance, it becomes necessary to take note of what is going on in the insulating medium, such as the air, by which conducting bodies, said to be charged with elec-

tricity (13), are necessarily surrounded and separated from each other.

Exp. 19. Let s, Fig. 25, be a spherical or other conductor, insulated in free atmospheric space. Charge this insulated body with either positive or negative electricity—suppose positive. We have then a stratum of atmospheric particles, $+ + + +$, &c. &c., immediately surrounding its surface, which, if the electricity communicated to the sphere s be positive, will also be positive. The stratum of electrified particles, $+ + + +$, &c., now operates by-induction upon the next stratum of adjacent particles, causing in them a development of the opposite or negative electricity, $— —$. The negative electricity thus called forth excites, in its turn, positive electricity in the next adjacent particles, until the electrical wave, or alternating series of positive and negative particles, reaches distant neutral matter, A B C D, causing electrical force to appear there of an opposite kind to that of the sphere s (40); so that, in fact, we have a series of disturbances between the charged conductor and distant neutral matter, in the form of an alternating series of positive and negative forces, and which may be represented as in Fig. 26, in which P N may represent portions of opposed conducting bodies, with an intervening insulating medium, such as the air. Let *a b c d*, &c., stand for and represent adjacent or contiguous particles of the insulating air.

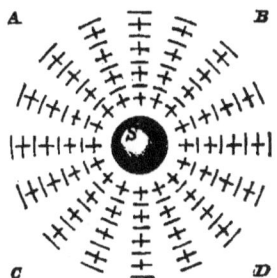

Fig. 25. Diagram to illustrate Induction.

Fig. 26. Diagram to illustrate Induction.

If, in this case, P be charged positively, then we have some such change as that just alluded to; that is to say, an alternate series of positive and negative forces, as indicated by the black and white portions of the particles *a b c d*, &c.; these reaching eventually the distant surface N, cause a force, N *n*, to appear there, opposite to that of P *p*, from which the inductive action is supposed to proceed.

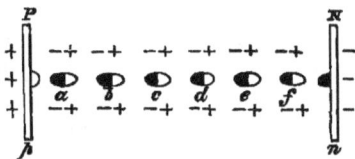

46. In the charged sphere s, Fig. 25, this kind of action, supposing the sphere to be in free space, extends equally all around.

Exp. 20. Place around and at equal distances from the insulated metallic sphere s, Fig. 27, about 5 inches in diameter, four or more insulated conducting discs, *a b c d*, each about 3 inches in diameter, with a light electroscope reed and pith ball appended to each of them, as shown in the figure.   Charge the sphere either positively

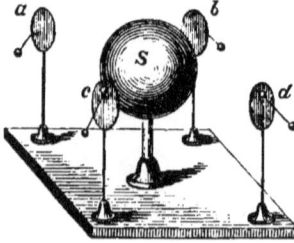

or negatively.   Each of the discs will be acted on inductively, and the electroscope reeds will all equally diverge, evidently showing a sort of radiation of power from the central sphere s in all directions.

Fig. 27. Sphere and Electroscopes.

47. With a view of more clearly identifying these several actions, Faraday, with his usual precision, has applied to them a comprehensive and appropriate species of nomenclature.   First. It being important, when two insulated conductors are in inductive relation to each other, to distinguish the charged body, or surface originating and sustaining the induction, from the opposite or neutral body, in which the action terminates, or to which it extends (45), it is agreed of the two bodies, to call the charged body the *inductric* or *inductive* body ; and the opposed body in which the action terminates, the *inducteous* body.   Second. The medium separating the bodies is denominated the *dielectric medium ;* the term dielectric being employed to designate any kind of insulating substance through, or across, which the electric forces operate.   Third. The peculiar induced state (45) of this medium is termed a *polar* state ; and the medium itself is said to be *polarised.* Electrical *polarity*, therefore, is such a disposition of force as gives to the same particle opposite electrical powers. Induction, then, consists in a polarised state of the particles of a dielectric medium, caused by the inductric body sustaining the action ; and in which medium positive and negative forces become symmetrically arranged.   By *contiguous* particles, we are to understand particles next to each other, whether we imagine them in actual contact, or whether there be something or nothing between them.

48. Under ordinary conditions, induction may be supposed to consist of a species of electrical force resulting from the action of matter charged with electricity through insulating matter, tending to produce in the surrounding dielectric particles, a series of opposite electrical states, and this it does by first polarising the particles next it. These, in their turn, polarise the next particles, and so on, until the action propagates itself from the charged surface to any distant mass, where the contrary force is caused to appear. Hence it is only through, or across, insulators that induction is

sustained. We have, according to this view, no such thing really as electrical action at a distance, or any kind of sympathetic force between particles or masses separated from each other by empty space.

Electrical influence is an actual propagation of force through the medium of intermediate matter. The distinguishing difference in the polarisation of an electric substance, and the polarisation of a conducting substance, consists in this : the first is a polarisation of forces, which do not communicate their forces to each other ; the second is a wholesale polarisation, as it were, of the entire mass (40), so that opposite portions of the body exhibit opposite electricities ; and this arises from the circumstance that conducting particles will not admit of the same constrained state as non-conducting particles. It is to be further observed that induction upon conducting bodies does not require sensible thickness for the development of the positive and negative states above alluded to ; the thinnest leaf gold may by induction become positive on one surface and negative on the other, without any interference whatever of the opposite electricities with each other ; that is, so long as the source of the induction remains. Conducting bodies are, in fact, in mass, as regards induction, what electric or non-conducting bodies are in particles.

49. Reverting for a moment to a further consideration of the inductive phenomena with opposed conductors, we find them all resolvable into the propagation of electrical force across an insulating or dielectric medium (48). If we take a positively charged conductor standing alone in free atmospheric space, then, as in the case of the charged sphere s (45), we have a polarisation of the surrounding atmospheric medium in all directions, which would finally extend, like a wave, to other conducting matter, however distant, in which an opposite electricity to that of the charged body would eventually appear. Cavendish traced this from a small electrified globe placed in the centre of a room sixteen feet square to its surrounding walls. Faraday traced it from a ball, suspended in the middle of a large apartment, to the walls distant twenty-six feet. Such would be the case supposing the charged body to be alone in free atmospheric space. Directly, however, we bring into the field a second body, and place it near the charged body (39), then the induction becomes determined, more especially upon the second or near body ; and the action is so confined to the adjacent surfaces, and the insulating medium immediately between them, that the induction in other directions frequently becomes so small as to admit of being neglected.

The decreased divergence of the electroscope f (39) is strikingly illustrative of this exclusive reciprocal action. This diversion, or compensation of force, by the reflected influence of near neutral matter upon a charged body, is characteristic of electrical induction, or electrical action at a distance.

Exp. 21. Let a neutral body, o, Fig. 28, in a free state, be caused to approach a charged mass, M, whilst attracting a distant body, *n*. The influence of the near body o is such as to weaken or neutralise the attractive force between the body M and the distant mass *n*, so as frequently to supersede the action between M and *n* altogether.

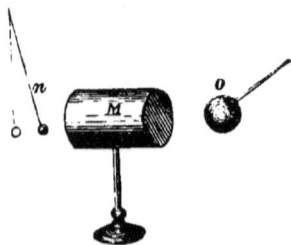

Fig. 28. Induction Apparatus.

50. The simultaneous presence of two opposite electricities in every instance of electrical excitation (24), is a striking and very important feature in the operation of electrical force, however produced. These two apparently constituent elements of electrical agency (in whatever that agency consists) seem to be inseparable, one having no existence independently of the other. Thus, in the communication of electricity to simple insulated conductors, or in the development of electricity by friction, or in electrical accumulation by any other means, it is impossible so completely to sever the two electricities as to get rid of one of them, and leave the other free to act alone. We may, it is true, electrify a body either positively or negatively, and in this sense may be said to have charged the body with one of the forces only; but this is not an independent condition. If, for example, we excite a glass tube, we develop positive electricity on the glass. This positive electricity, which commences at the points of friction, becomes at once related to an equal amount of negative electricity developed in the rubber itself (23), and also to distant negative electricity called up by induction in surrounding matter.

51. This elementary and wonderful fact has been well shown by Faraday, and is apparent in the following experiments :—

Exp. 22. Let *a c d*, Fig. 29, be a cylindrical vase of thin metal, about 7 inches in diameter and 1 foot deep, suspended from any convenient support by means of fine insulating lines of varnished silk gut, or other insulators, *a m, d n*. Let P be a light metallic ball about 3 inches in diameter, suspended by a fine insulating line, P *h*. Let E be a double leaf electroscope communicating with the outer surface of the vase $a\,c\,d$. Raise the ball P by its insulating line P h out of the vase, and communicate to it a charge of either electricity, suppose positive (21). Then return the ball to its place within the vase. As the ball descends into the vase, the gold leaves of the electroscope E begin to diverge, and continue to diverge, until the ball is about 3 inches below the edge of the vessel, when the divergence remains constant. Now, the exact position of the ball P within the vase is of no moment whatever, whether it be near or distant, from either the sides or the bottom, provided it be not so near as to discharge its electricity against the vase. Under this condition, place it where you will, the divergence of the leaves remains unchanged. The sum of the forces of induction, therefore, is the same constant quantity.

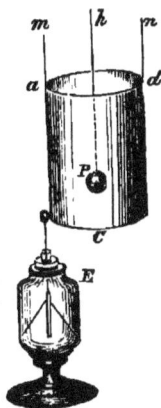

Fig. 29. Vase and Electrometer.

Again: raise the charged ball P by its insulating line, P h, clear of the vase. The leaves of the electroscope E will now close, but will again diverge on restoring the charged ball to its former position.

If we now examine the kind of electricity communicated to the electroscope E from the outside of the vase, it will be found to be the same as that of the ball P; that is to say, positive. The inductive influence of the ball P, therefore, has been such as to develop on the outside of the vase the same electricity as that of the ball itself; namely, positive. If we now examine the interior surface of the vase by means of a small insulated carrier (14), raising the electrified ball P without the vase whilst the carrier is in contact with its interior surface; then, on removing the carrier, we find the interior surface of the vase negative. We see, therefore, that the two sides of the vase are placed by the inductive influence of the ball P in opposite electrical states; the surface next the ball being negative, the distant or outer surface positive; phenomena strictly in accordance with the teaching of (40). That this is a pure result of inductive influence from the charged ball P is evident from the fact that everything becomes restored to its original state on withdrawing the ball P, or discharging its electricity.

Exp. 23. The apparatus being the same as in the last experiment, but in a neutral state, charge the ball P whilst raised clear of the vase, positively, as before; and when so charged, lower it

carefully into the interior of the vessel, so as to remain suspended
about its centre. The leaves of the electrometer will again diverge.
Whilst thus divergent, cause the charged ball P to touch the interior surface, so as to communicate to the cylindrical vase its own
proper positive charge. Not the least change will take place in
the divergence of the gold leaves. They will still remain divergent
to the same extent. If, therefore, the ball P be merely suspended
in the interior of the vase *a c d*, it acts upon the vase by induction, and evolves electricity of its own kind on its distant or outer
surface. The electricity, therefore, evolved upon the outer surface
a c d, by the inductive action of the ball P, must be exactly equal
to that with which the ball P is actually charged, without which
the divergence of the leaves of the electroscope E must necessarily
change. This is further evident from the fact that, on withdrawing the ball P from the interior of the vase, after touching its
inner surface, the ball P comes away uncharged and perfectly
neutral, which could not be the case if it had not parted with all
its charge to the vase itself. The effect, therefore, of the contact
of the charged ball P with the interior surface of the vase, has
been to unite the two opposite electrical forces, existing upon the
ball and upon the interior surface of the vase; the charged ball P
being positive, and the interior surface of the vase opposed to it,
negative; and these two electricities must be co-existent, and
exactly equal in amount, since the ball P, after contact with the
interior surface, comes away perfectly neutral.

As this electrical change, however, produces no effect upon the
leaves of the electroscope, we may infer that the electrical charge
induced by the ball P, and the electrical charge existing in P, are
exactly equal in amount.

52. These phenomena do not appear to be affected by any number
of cylindrical surfaces interposed in series between the charged
ball and the surrounding walls of the vase.

Exp. 24. Let several concentric vases, *r*, *s*, *t*, *u*, Fig. 30, be
suspended, by fine insulating lines, one within the other, so as not
to touch in any point. The charged ball P, suspended within this
system of concentric cylindrical vases, produces the same effect as in
the case of the single vase, so that the intervention of many conducting surfaces causes no difference in the amount of the inductive
effect. If we connect the two interior cylinders, *t* and *u*, by means
of a wire suspended from an insulated thread, and insert it between
them, the leaves still remain in a state of divergence, as before;
and continue to do so, if the cylinders *s* and *t* be connected in a

similar manner, and so on. The result, therefore, in this case is precisely the same as if the interval between the outside of u, and the inner surface of the exterior cylinder r, were a solid mass. Supposing, however, the interior cylinders, r, s, t, u, to remain insulated from each other, and the ball P to be in the centre of the system, then we have a succession of positive and negative surfaces (40); that is to say, the inside of u, next P, is negative, and the outside positive; and so on for the successive cylinders, until the induction reaches the outside of r, where we find a charge by induction of the same kind as that of the ball P, and of exactly the same amount, as already shown (51).

If the charged ball P be now made to touch the interior surface of the cylinder u, then, as before (51), the opposite equal, positive, and negative states neutralise each other; that is to say, the positive electricity of the ball unites with the negative electricity induced upon the inner surface

Fig. 30. Concentric Vases and Electrometer.

of u, so that the charge originally on the ball P, and all the inductions dependent on it throughout the system, vanish; and the ball, on being withdrawn from without the system, comes away perfectly neutral, leaving the amount of its charge upon the exterior surface of the cylinder r, as in the preceding experiment.

It appears, therefore, by these experiments, that the divergence of the electroscope leaves E remains the same, whether there be only one cylinder, or whether there be a series of concentric vessels, r, s, t, u, insulated one within the other; or whether the interval between the ball and the external cylinder be a solid mass, so that a certain amount of electricity upon a ball, acting within the centre of a vessel, exerts precisely the same power externally, whether it operates by induction through the space between the ball and the surrounding wall of the vessel, or whether it be transferred by communication through solid matter to that wall, so as absolutely to annihilate or neutralise the induction within by the union of the opposite electrical states, which we have shown to be exactly of the same amount, and of opposite kinds. We may conclude, therefore, from these experimental inquiries, that, whether

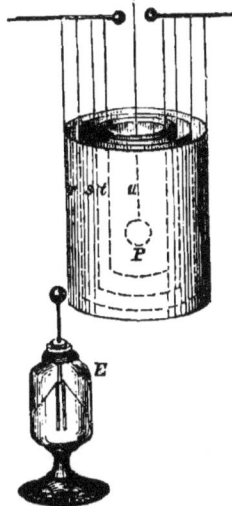

a conducting substance be a solid mass, or merely a hollow shell, the amount of electricity which can be accumulated upon it is solely dependent upon its external surface.

53. It appears by the foregoing results that neither induction, nor attraction, nor repulsion, has any relation to solidity or quantity of matter, or even to the kind of matter of which substances consist. Whether a body be metallic or not, hollow or solid, the quantity of electricity it can finally receive and support, together with the subsequent induction, attraction, or repulsion it exerts, is in each case the same; the only element which appears to enter into the development of the force is time. Inferior conductors, such as wood, require a somewhat longer time for accumulating or abstracting electricity, or for operating inductively, than more perfect conductors, such as the metals, the action of which is apparently more instantaneous. The French philosopher Coulombe found that an equal division of electricity took place between substances of similar form and equal surfaces, whatever difference existed in the kind of matter of which they were composed; or whether hollow as a mere shell, or solid.

In fact, in communicating electricity to an insulated conductor, as, for example, an insulated metallic sphere (45), we cannot be said to electrify the metal itself bodily. The metallic surface merely enables the electricity to expand upon the dielectric medium surrounding it, and in such way as to cause the surface of the metal to become enveloped by dielectric particles adhering so closely to the metal as to admit of being considered an electrical extension of the body itself. It is this envelope of dielectric particles, supported by the metallic surface, which sustains the induction, and through which the action reaches distant matter. If it were practical to remove the conducting sphere without disturbing the surrounding medium immediately in contact with it, there would remain a hollow globular vase of dielectric particles.

It may be hence further inferred that accumulated electricity exists on the surfaces of bodies, and has little or no relation to their mass, so that as much electricity can be accumulated on a hollow metallic sphere as on a solid sphere. In no case does an electrical charge ever penetrate the substance of the charged body; or, so far as yet ascertained, does it even affect the particles beneath the surface in any very sensible degree. What we require, therefore, in accumulating electricity on insulated conductors, is a large extent of inducteous surface, together with free inductive action upon other matter; a deduction abundantly verified in the following experiments:—

54. Saussure, in the year 1766, observed, as an interesting experimental fact, that an electrical charge tended to the surface of the charged body.

Exp. 25. Charge an insulated conducting mass, R, Fig. 31, in which there are deep pits, *s, t, u.* Introduce a small carrier ball, *p, p',* into these pits, and withdraw it without touching their edges. Then, as found both by Saussure and Coulombe, not a particle of charge will have been taken up by the carrier. Bring the carrier now in contact with the surface of the body; it will immediately receive an electrical charge.

55. The Hon. Henry Cavendish directed his attention to this question, and, as we find by his manuscripts, contrived, so long since as the year 1775, the following elegant and conclusive experiment, by which he anticipated not only all the great facts which were discovered in 1785 and the following years, as recorded in

Fig. 31. Saussure's Apparatus.   Fig 32. Electrified Spheres.

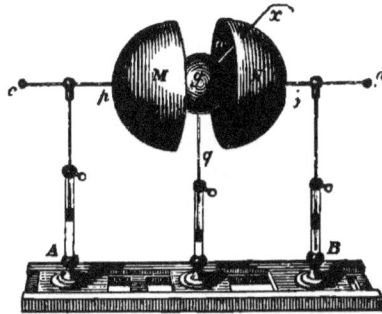

the Memoirs of the Royal Academy of Sciences in Paris, but also in other Memoirs of learned societies of a more recent date.

Exp. 26. The intention of this experiment (says Mr. Cavendish) is to find whether, when a hollow globe is electrified, a small globe contained freely within it, and communicating with the outer globe by a wire or other conducting substance, becomes at all over-charged or under-charged (as he terms it); that is to say, positive or negative. To this effect, a light, metallic, insulated globe, *g,* Fig. 32, was enclosed between two insulated, hollow, metallic hemispheres, M, N, leaving a space between the interior and exterior shells, and constituting, when joined, an external globe of about 13 inches in diameter.

The two hemispheres, M, N, were insulated in rectangular frames, which admitted of being turned back so as to expose the interior

globe g, and leave it quite free of the two hemispherical envelopes, an operation indicated in Fig. 32 by means of the insulating supports, A c p and B d r, and the sliding bases A B.
Thus circumstanced, a temporary conducting communication was
established between the inner and the outer globes by a short
brass wire, w, attached to an insulating rod, x, and by which
the wire could be easily removed. Mr. Cavendish then communicated an electrical charge to the enveloping globe M N; in
which case, as is evident, if any part of the charge tended to per-
vade the system as a mass, it could freely do so by means of
the communicating wire w. "I drew out the wire w," says
Cavendish, "connecting the inner with the outer globe, which, as it
was drawn away by an insulator, could not discharge the electricity
either of the interior globe or its exterior envelope. I then instantly withdrew the hemispheres, and applied a pair of small pith
balls, suspended by fine linen thread, to the inner globe, to see
whether it was at all over or under-charged;" that is, either positive
or negative. This electroscope he further explains as being fixed
at the extremity of an insulating glass rod, covered with a little
tinfoil in that part intended to touch the globe. "The result was
that, though the experiment was repeated several times, I could
never," says Cavendish, "perceive the pith balls to separate, or
show any signs of electricity." Every precaution seems to have
been taken in this experiment. On withdrawing the hemispherical
shells, their electricity became speedily discharged, so that no subsequent electrical action could possibly arise.

Cavendish further endeavoured to discover how small a quantity
of electricity, not sensible to his electroscope in its ordinary state,
might be made apparent; for which purpose he communicated to
the electroscope balls a weak positive or negative charge. In this
way he found he could render sensible on the inner globe a
quantity of redundant electricity less than the one-sixtieth part of
that on the outer; and hence concludes that in this experiment
the redundant electricity, if any, existing on the interior globe,
must "certainly have been less than the one-sixtieth part of that
on the outer globe;" but thinks "there is no reason to believe that
the inner globe was at all charged."

The double gold leaf electroscope (33) would be a still more
delicate test than the pith balls.

Although this simple and perfectly conclusive experiment fully
proves the tendency of electricity to the surfaces of bodies, yet the
following, which may be taken as the converse of Cavendish's
process, is by no means unimportant.

Exp. 27. Charge the interior globe *g*, Fig. 32, either positively or negatively, the hemispherical envelopes M N being withdrawn, as in the figure, and the communicating wire *w* removed; when so charged, replace the enveloping hemispheres, and insert the communicating wire by means of its insulating handle, so as to make a conducting communication between the inner and the outer globes; again withdraw the wire, and separate the hemispheres, all the electricity will have left the inner and previously charged globe, and be found on the exterior surface of the hemispherical shells, charged, in fact, with the electricity of the interior globe.

These experiments may be performed with a small sphere of 2 inches diameter, and light hemispherical shells of sheet copper.

56. An experiment somewhat similar to that of Mr. Cavendish has been described by M. Biot in his *Traité de Physique*, which, if the student should fail to verify, he must not be disappointed. In this experiment a conducting ellipsoid, M, Fig. 33, has closely fitted to its surface two similar superficial ellipsoidal

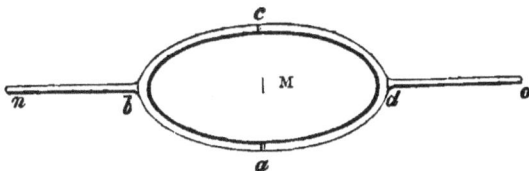

Fig. 33. Ellipsoidal Covers.

envelopes, *a b c*, *a d c*, each furnished with a light insulating holder, *b n*, *d o*, and without leaving any sensible space between. These shells being removed, the interior body M is charged with electricity, after which the envelopes are replaced. On being again withdrawn, all the charge, it is said, will have passed into the superficial shells, and the interior body M will be found perfectly neutral. Now it is quite clear that this experiment cannot possibly succeed unless the envelopes be rapidly and simultaneously withdrawn, so rapidly, indeed, as to exceed the rapidity of the electrical expansion over any small space or opening upon the surface of the interior body M, which will necessarily occur between the envelopes at the instant of their separation, and so simultaneously, that one shall not be in any sensible degree after the other, for if it be, the charge will certainly expand from the remaining envelope over the exposed part of the interior ellipsoid M. This experiment, therefore, is much more precarious and inconclusive, and certainly far inferior to that of Cavendish just described (55).

57. Volta likewise, in 1779, instituted the following conclusive experiment, illustrative of the tendency of electricity to the surface of a charged body.

Exp. 28. Let s, Fig. 34, be a thin hollow metallic sphere, about 5 inches in diameter, and having a circular opening at *d*, about 1½ inch across. Place this sphere on a long insulating support, *v*. Let *a* be a carrier ball about three quarters of an inch in diameter, and insulated on a slender glass rod. Charge the carrier, and introduce it within the interior of the shell s, but without touching the edge of the opening at *d*. Having brought this electrified ball in contact with the interior of the shell, again carefully withdraw it clear of the opening *d*. Every particle of the charge will have left the carrier *a*, to appear on the outer surface of the sphere s, which will have now become attractive and repellent of a delicate electroscope. · By a few repetitions of the experiment, the exterior charge upon s may be rendered very powerful; for notwithstanding the previous communication of electricity to the inner surface of s, the small ball *a* will be continually and completely robbed of its charge.

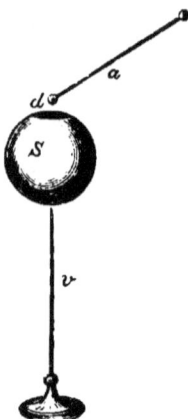

Fig. 34.   Hollow Metallic Sphere.

If the hollow shell s be originally charged, and the insulated ball *a* in a neutral state introduced within it, then, as in the experiment by Saussure (54), it will not, on being removed, exhibit any sign of electricity.

Exp. 29. Let v, Fig. 35, be a metallic can, sustained on an insulator, ɪ, and having attached to it an electroscope of divergence, *e*. Let *a c* be a fine metallic chain, reposing at first in the bottom of the can, having a silk cord, *p*, attached to it, extending outside the can, so that the chain *a c* may be raised out of the can without interfering with any electrical charge that may be given it.

The chain reposing at the bottom of the can, charge the can positively, until the electroscope *e* becomes divergent. Now raise the chain by the insulating silk line *p* until it is freely suspended in the air. The electroscope *e* will now decline, showing that a portion of the charge had quitted the can to expand upon the elongated chain, thus confirming, in a remarkable way, all the preceding experiments. This is known as Franklin's " Can and chain" experiment, and was contrived by him so long since as the year 1754.

Exp. 30. A conical muslin bag, *a c b*, Fig. 36, rendered sufficiently stiff to preserve its figure, but without materially affect-

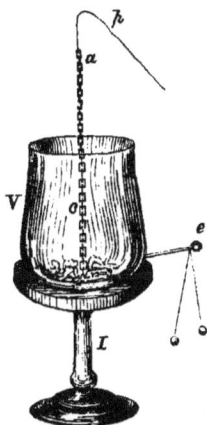

Fig. 35. Can and Chain Experiment.

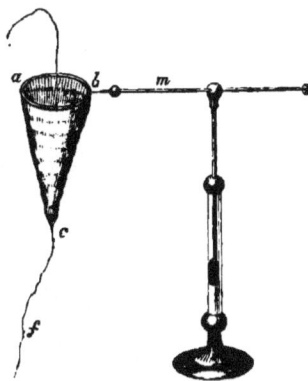

Fig. 36. Insulated Muslin Bag.

ing its pliancy, is fixed to a wire hoop, *a b*, and insulated on a glass rod, *m*; a silk line, *f*, is attached to the vertex *c* of the conical bag, so as to pull it inside out, or again reverse it. If this have an electrical charge communicated to it, and the interior be then tested by a carrier in the usual way, no electrical indication is apparent, whilst the external surface *a c b* attracts and repulses the electroscope freely. Reverse the surfaces of the bag by means of the silk line attached to the vertex of the bag, so that the interior may become the exterior surface; then we have precisely the same result; the electricity is all found upon the surface externally. If we again reverse the surfaces, we restore the bag to its first condition, in which case the original electrical development is again restored. This elegant experiment is due to Faraday.

Exp. 31. The most general, and perhaps the finest experiment on electrical induction, was carried out in the lecture-room of the Royal Institution by the same philosopher, in November, 1837. Having constructed a light cubical chamber, c, Fig. 37, of 12 feet cube, the walls of which consisted of slight wire, covered with metallic paper, and insulated by adequate suspending insulators, *m, n, o, p,* he placed himself within it, used lighted candles, very delicate electroscopes, and other tests of electricity, but although the cube was placed in communication with a powerful electrical apparatus, and charged so highly through an insulating conducting rod, A,

E

as to throw off powerful sparks and brushes of light from the
exterior surface, yet not the least effect was produced upon the
electroscopes or other bodies within.　The cube was now put into
communication with a perfect discharging train, and the air

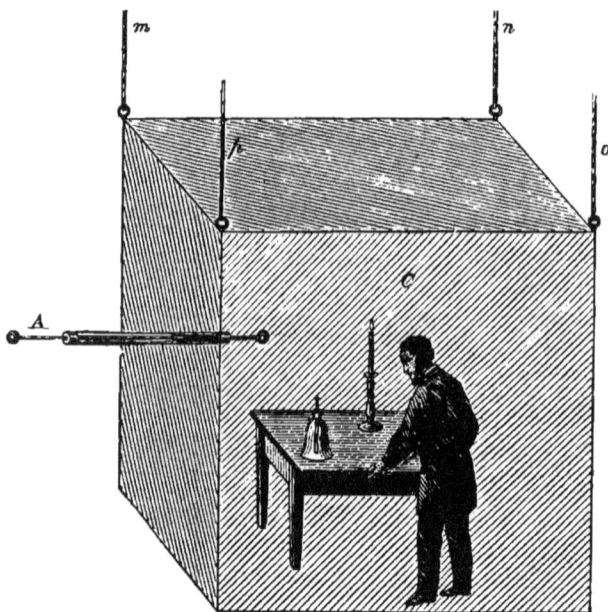

Fig. 37.　Cubical Electric Chamber.

within strongly electrified by means of a conducting wire passing
into it from the electrical apparatus; but every attempt to charge
the air bodily and independently with either electricity failed.
The conclusion arrived at was, that non-conductors, as well as
conductors, have never yet had communicated to them an absolute and independent charge of either electricity, and that to all
appearance such a state or condition of matter is impossible.

58. Induction appears to be the essential function both in the
first development of electrical force and all the subsequent phenomena attendant on it.

Insulation and common conduction seem to consist in an action
of contiguous particles (47), and to depend upon the two opposite
forces developed in them.　These forces are the particles in a state
of polarity (47), in which state they have a greater or less degree
of power in communicating their forces one to the other.　Bodies
are better or worse conductors, or better or worse insulators, according to the greater or less degree in which the developed forces
can re-unite (47).　Induction and conduction appear, therefore,

so far, to be in principle the same, though generally assumed to be totally different (15). If the particles maintain their polarised state (47), insulation is the result in a greater or less degree. If they communicate their forces more or less freely to each other or to other bodies, conduction ensues, and the phenomenon termed *electrical discharge* necessarily follows. Retardation of electrical discharge, therefore, is virtually insulation. If the induction, therefore, across a dielectric remains undiminished, then insulation is the consequence, and the higher the degree of polarisation, that is to say, the greater the amount of disunion of the opposite electrical forces, the greater the polar state, which state or disunion has been usually characterised by the term *tension*, or recombining power.

If the contiguous particles do not maintain the polarised state, but communicate their forces in a greater or less degree one to the other, this state of tension falls, and, as just observed, is followed by *conduction* and *discharge*.

The reduction, therefore, of two effects, insulation and conduction, so long held distinct, to an action of successive particles obedient to one common law, is no unimportant result.

When the polarisation of the intermediate particles is raised to a degree which they cannot support, the forces recombine with a sort of convulsive effort, attended by an evolution of light and heat, the result being the ordinary *electrical spark*. The distance through which discharge can take place between two conductors (45) is called their *striking distance*. When we continue to diminish the extent of surface originating an electrical spark, and finally arrive at a small terminating point, very curious effects are observable.

59. The influence of a pointed body in promoting the depolarisation and discharge of a polarised dielectric (47), was first noticed by Franklin in 1750. He showed that when a pointed conductor was presented to an electrically charged body, the point rapidly caught up the electricity of the charged body, even at considerable distances. He charged an iron ball, about 4 inches in diameter, and observed that on presenting to it an uninsulated pointed needle, the attractive force of the ball on a small thread immediately ceased, as if its electricity had vanished. He further observed that this influence of a pointed body was also exerted when the pointed body was projecting from the charged body itself, the charge being rapidly dissipated by the projecting point. At the same time a current of aerial particles set off apparently from the electrified point, so that if the latter were free to move, it would recede

in a direction opposite to that of the aerial current, or *electrical aura*,
as it is called.   On this principle the little toy known as the electrical fly, acts.   It consists of a wire with the ends pointed, and bent
in opposite directions, and a cap in the middle for supporting it
on a vertical wire attached to the electrified body.   The aura,
or wind, which apparently escapes from the points of the fly, will
cause it to rotate in an opposite direction.   So also the electrical
aura may be made to drive light models, representing water-
wheels, &c., for which purpose they are made to move on centres,
and are furnished with vanes, against which the stream from the
point plays.

60. It is not difficult to explain, upon the principles of inductive action just adverted to, the more immediate effects of a
pointed body in dissipating the electricity of a charged body.
According to the theory of induction (41), it is assumed that the
particles of the dielectric are in a certain state of tension (58),
which rises higher and higher in each particle as the induction is
raised higher and higher, either by the closer approximation of
the inducing surfaces, increase of the charge, or variation of
form, &c. (41).

The spark-striking distance (58) will therefore be dependent on
the discharge of a few particles of the dielectric, occupying a very
limited space, in consequence of which the induced polarised
state of the whole series is lowered, and the molecules return to
their previous or normal condition, in the inverse order in which
they left it, whilst their powers to propagate or continue the
discharging operation from the point where the subversion first
occurred, become united.   A good mechanical illustration of this
may be derived by setting up on end and near each other a series
of thin rectangular pieces of wood ; if we overturn one at the
end, the next, the next, and so on, must fall, because each in succession becomes pressed upon by the united forces of the preceding,
which now complete the downfall of the whole series in a given
direction.   The few particles originating the discharge are generally next one of the terminating conductors ; in this point of
subversion, however, they are not merely pushed aside, but they
assume for the time an extreme tension, and the powers discharge throughout the series with violence and explosive force :
the ultimate effect is the same as if we had put a discharging
wire in place of the dielectric particles, and operated by conduction immediately between the limiting conducting surfaces.
The tension of the particles of the dielectric next the points in
the limiting conductors being greater than those in the middle

of the series, it is hence in these points that the discharge commences; so that when these conductors terminate in mere points
or small surfaces, the tension upon the particles of the dielectric
in contact with them is excessively increased; in fact, all the lines
of inductive force may be supposed to concentrate upon the point.
If a spherical uninsulated conductor be placed some distance from
a point projecting from a charged conductor, then the lines of
inductive force will concentrate, as it were, upon the point, and
the latter will originate an active mechanical force, and preserves its predominance over the other portions of the conductor, behind it, by a continued discharge of the accumulated
electricity. Hence currents of wind arise by the recession of the
charged particles of air, and which are, in every way, favoured
by the shape and position of the conducting surface in the rear.
If the point be more or less central to the walls of a room, without any more immediately opposed conductor, or be exposed to the
induction of any other substance in its vicinity, still the same
result ensues, since there is no distance so great as to limit the
operation of this inductive action (45).

The theory applies, by the converse of this, to an uninsulated
point, opposed to a charged body.

The conditions, therefore, requisite for the production of an electric spark are, two opposed conductors (45), charged, one positively
and the other negatively, with an intervening dielectric medium.
If the opposite forces be continually increased in strength by
exalting the electric state of the two conductors, or by bringing
the conductors nearer together, or by diminishing the density of
the intervening dielectric, a spark at length appears, discharge
ensues, and the two forces combine, and are reduced to a state of
neutrality. The opposed conductors may be considered as the
termini of the inductive action, all the effects prior to discharge
are inductive, and the degree of tension required before the spark
passes, is the limit of the influence of the dielectric in resisting
discharge,—that is to say, of its conservative power,—and, therefore, a measure of the forces in operation.

61. The phenomena of electrical induction, which we have been
hitherto considering, are calculated to throw much further light
on the theory and nature of electrical force. According to the
hitherto received doctrines, electricity has been referred to an
invisible, imponderable, highly elastic fluid, intimately associated
with common matter, and pervading the solar system. The discovery of Du Fay (22) led to the idea that this assumed fluid
consisted of two primary elements, possessing distinct and opposite

properties; these elements, as already observed, have been termed
vitreous and resinous, or positive and negative electricities; they
are further assumed to be highly repulsive of themselves, but
attractive of each other. When combined they are neutralised,
and electrical repose is the result. When disunited, each becomes
active, and is the source of electrical phenomena. This has been
termed the *double fluid* or *French theory.*

According to Franklin, Watson, Cavendish, and other English
philosophers, electrical phenomena depend on the presence of a
single agency in a greater or less degree, repulsive of its own
particles, but attractive of the particles of common matter. When
distributed in bodies in quantities proportionate to their capacity
for it, such bodies are said to be in their natural state; the result
is electrical quiescence. When we increase or diminish the quantity
in any substance, we disturb this equilibrium, and the bodies,
according to Cavendish (55), are said to be under or over-
charged, and to be either positively or negatively electrified, in
which case a powerful action ensues, arising out of the tendency
of the bodies to regain their natural state. This it does by taking
away electricity from other bodies if its own quantity be diminished, or by throwing off electricity upon other bodies if the
quantity be increased.

Faraday's researches, however, throw a further light on the laws
and nature of electrical action. Without entering upon any assumption in explanation of the occult nature of electricity, Faraday
regards it simply as force; much in the same way as Newton considered gravity, without pretending to explain its nature. According to this view, then, all electrical phenomena are resolvable into
induction (58). In every case of electrical action two opposite forces
are invariably present; one cannot exist without the other; we cannot charge a body with one of the forces only, independently of the
other, such a condition of matter being impossible (50). It is
assumed that the particles of every kind of matter are as wholes
conductors; that not being polarised (47) in their original state,
they may become so by the influence of neighbouring charged
particles; that the particles, when polarised (47), are in a tense or
forced state, and are constantly endeavouring to return to their
normal condition; that the *ready* communication of the opposite
forces between near particles constitutes *conduction*, and the *difficult* communication, *insulation;* that induction can only take place
through or across insulators. The leading feature of the electric
power is, that it is limited and exclusive, and that the two forces
always present are exactly equal in amount (51), the forces being

related to each other in one of two ways, either as in the ordinary state of an insulated uncharged conductor, or in the charged state of an insulated conductor, which last state is a state of *induction*.

62. There are two terms, namely, *intensity* and *tension*, frequently employed in treating of electrical force, and more or less involved in the phenomena of electrical induction, that may require a little further explanation, especially as they are frequently employed in a loose and indefinite manner. In order fully to comprehend their real signification and immediate application, it should first be observed that the term *quantity*—a term sufficiently intelligible—designates the actual amount of the unknown agency, whatever it be, constituting electrical force, as referable to some arbitrary standard of measure, in terms of which it may be expressed. The term *intensity*, in its usual acceptation, has been referred to hypothetical views of the occult nature of electricity itself, often vague and unsatisfactory. Intensity has also been used to signify the peculiar state of an hypothetical elastic fluid, the electrical agency being supposed capable of changing its condition, just as we may imagine a spring to have greater or less elastic power. It will be found, however, upon a critical investigation of the facts upon which this assumption rests, that the term intensity applies more especially to electroscope or electrometer indications, and is but another term for the quantity of electricity operating at a given point in a given direction ; as, for example, in the direction of the electro-scope.

For example, let A, C, S, Fig. 38, be three insulated conducting

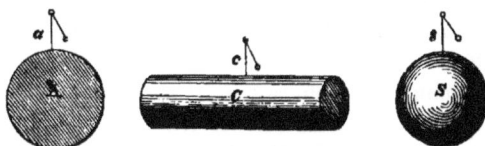

Fig. 38. Mutual action of Insulated Conductors.

bodies of different forms and magnitude. Let A be a circular plate of small thickness, c a cylinder, and s a sphere, and suppose each furnished with a delicate electroscope of divergence, *a, c, s* (the three electroscopes being precisely alike). Suppose also these three conductors to be charged with electricity to such an extent as will bring each of the electroscopes to the same angle ; in this case, as is well known, the quantity upon each of these bodies will be very different, notwithstanding the sameness of the electroscope indication. Here the electrical intensity of each of the bodies is said to be the same, and the electrical charge to have the same density in each. Again, take two insulated conducting spheres, M, N, Fig. 39, of unequal diameter, and suppose the surface of the one, M, to be three times that of the other, N, having delicate electroscopes, *m n*, attached to them as in the former case; then, if we charge each of the globes with the same quantity of electricity, the electroscope on the small sphere N will have a much greater angular divergence than the electroscope upon the large sphere M, from which it has been inferred that the electricity upon the small sphere has greater elastic power than the electricity on the large sphere; or, in other words, that we have varied the intensity without changing the quantity. A further investigation of these phenomena, however, leads to a more definite and simple explanation. When we place different quantities of electricity upon the three bodies A, C, S, Fig. 38, without affecting the angular divergence of the electroscopes, it is the total quantity which varies, and not the quantity in any given point of the surface; that is to say, the absolute quantity affecting the electroscope; the quantity affecting the electroscope is really the same in each case, although the total quantity upon each of the three bodies may greatly differ. It is, in fact, quite evident that since the three bodies vary in extent of surface, the greater that extent the greater the number of points the surface will contain; the quantity, therefore, in any one point of either of the bodies may still be the same, notwithstanding the quantity in each of the bodies is different.

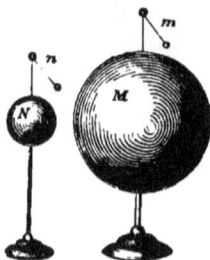

Fig. 39. Insulated Con-
ducting Spheres.

The total quantity or charge must be taken to act upon an electroscope projecting from its surface, as it would do supposing the electricity were equally distributed; this is evident from the fact that, in whatever point of the charged surface the electroscope be placed, its angular divergence remains the same. Intensity, therefore, in this case, when rigidly interpreted, is nothing more or less than the quantity of electricity at a given point acting on the electroscope, and cannot be taken to express any difference in the actual state or condition of the electrical agency itself.

Let us now take the case of a different angular divergence, the quantity of electricity being the same. Take for example the case of the two unequal spheres M N, Fig. 39, each furnished with a delicate electroscope, and let the same quantity be accumulated on each. The electroscope of the larger sphere will, in this case, as already observed, have a much less angular divergence than the electroscope of the smaller sphere. Here we perceive that the quantity in any one point of the two spheres is not the same, since the total quantity being alike in each sphere, and the surface over which it is expanded in the larger sphere three times as great as the surface over which it is expanded in the smaller sphere, there would be a greater number of points in the larger sphere for the reception of the charge, taking the surfaces, as supposed, 3 to 1. Hence where there is one particle of force on a point of the larger sphere, there will be three particles of force on a point of the smaller sphere. Here again, intensity, when correctly interpreted, is nothing more than the quantity of force operating at a given point; but it does not follow that the agency in operation has a higher amount of elastic power in the one case than in the other, or is necessarily more or less dense. If it exhibit a greater degree of energy in any given point of the small sphere than it exhibits in any given point of the large sphere, it may be because there is a greater amount of mere force in operation in that point. Until we have a clearer conception of the nature of electrical force, we cannot say whether that force is susceptible of change in quality or constitution or not. Any inference therefore of a change in electrical density must be mere assumption. The idea of a difference of density in the two cases of the small sphere and the large sphere, supposing the electrical agency to be material, and subject to the laws of ordinary matter, is certainly in accordance with the deduction that where there is one particle of electricity acting in a given point of the large sphere, there are three particles in a given point of the small sphere. So far this corresponds with the hypothesis of density, or as better expressed, perhaps, by the French philosophers, with "thickness of stratum," but as already observed, since we have no knowledge whatever of the occult nature of the electrical agency, we can scarcely venture to rely upon any hypothesis of this kind, but must be content to consider electricity as mere force, without assigning to it any specific elementary condition, much in the same way as we accept gravity as a mere force, without troubling ourselves as to its occult nature.

It follows from this that there is no such element as intensity independent of quantity; and that what we are to understand by intensity, is only the greater or less amount of force in a given point operating on the electroscope.

63. We have already remarked (49), that supposing a charged body to be alone in free space, there is a propagation of force in all directions. Directly we place another body near the charged body, then the inductive action becomes shared, more especially with the near body. For example, let the cylinder c, Fig. 38, be charged with a given quantity, so as to give to its electroscope c a given angular divergence. Let now an insulated neutral body, s, furnished also with an attached electroscope, s, be brought near the charged cylinder c; the angular divergence of its electroscope c becomes sensibly less, whilst the electroscope s of the insulated neutral body s begins to diverge, evidently showing a propagation of force in that direction. The electrical intensity of the cylinder is in this case said to be diminished by the proximity of the neutral body, a phenomenon already noticed (49). It is easy to perceive in this case that the angular divergence of the electroscope of the cylinder is not diminished in consequence of any change in the assumed elasticity of the electricity with which it is charged ; but in consequence of the force which at first operated exclusively on the electroscope of the cylinder being shared with another body, and therefore operating in two directions instead of one ; that is to say, in the direction of the neutral body, as well as in the direction of the electroscope. It was on this principle that Volta succeeded in rendering sensible minute quantities of electricity not appreciable by the electrometer. The term tension is more especially applicable to the polarised state of electric particles, or to the state of the particles of any dielectric medium intermediate between a positive and negative surface, as already explained (58).

Tension may therefore be taken to represent the recombining power of the disunited electricities in the polarised particles of a dielectric medium, interposed between a positive and a negative surface, whilst *intensity* is more especially referable to electrometer indication.

We may further observe that whilst intensity, as measured by direct attractive force between a charged and a neutral body, is as the square of the quantity accumulated, tension, as measured by the striking distance (58) of the electrical spark, is as the quantity itself,—phenomena perfectly consistent with each other, and the laws of electrical force.

64. *Specific Induction.*—Cavendish (as appears by his valuable MSS. of the year 1771, confided by the Earl of Burlington to the care of the author and the late learned Dean of Ely) observed a great difference in inductive action when transmitted through

different kinds of glass and some other insulators. Thus, coated crown glass, for example, differed in its charging power from coated window-glass. Coated resin or bees-wax differed from coated brimstone or shell-lac. Faraday, in the course of his valuable researches in Experimental Electricity, printed in the Philosophical Transactions for 1837, was led to treat this subject in a novel and instructive way, under the title of *Specific Induction.*

The question whether different dielectrics have any degree of influence in promoting inductive action through them, is a question of singular interest and importance. The question may be thus stated. Suppose M, Fig. 40, to be an electrified plate of metal, insulated in free air, and x y two exactly similar but uninsulated plates at equal distances, parallel to and on each side of it. The charged plate M will, as is evident, operate equally by induction on x and y (46). If some other dialectric than air, such as shell-lac, be introduced between the plates, the question is, will the induction remain the same? Will it depend solely on the distance between the plates, without relation to any intervening matter (as was once imagined by Coulombe and other philosophers to be the case), or will it vary? Now, it is to be here observed that what is termed electrical charge, is, in fact, a polarisation of dielectric particles between opposed conductors. If, therefore, induction be related to the particles of the surrounding dielectric, then it is related to all the particles under the influence of the inducing conductors, and not merely to the few particles next the charged body. The nature and properties, therefore, of the dielectric intervening between opposed conductors would be very likely to be attended by what we have termed *Specific Induction;* that is to say, the induction, for example, across air might possibly differ from induction across shell-lac; and such is really found to be the case. Faraday compared together the inductive capacities of different dielectric media, and found shell-lac to evince greater facility in propagating inductive action through its substance than air. Assuming the specific induction of air to be 1, shell-lac would be 1·5. The conclusion that shell-lac exhibits a higher specific inductive capacity appeared irresistible. Glass was found to have little or no specific induction, although different kinds of glass, according to Cavendish, exhibited different charging powers. Sulphur, as compared with air, is found to be as 1 : 2·24,

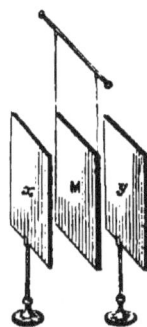

Fig. 40. Specific Induction Apparatus.

being greater than any substance yet tried. An extensive experimental inquiry, with a great variety of substances—including gases,
various fluids, air, rare and dense—may be adduced to show that
most bodies have a specific inductive capacity of greater or less
extent; but in the case of gases and vapours, the specific inductive
capacity seems to be pretty much alike in all. A complete experimental investigation, however, of this interesting question is
difficult, and occasionally precarious. Glass and many other insulators are found, on examination, totally unfit for the purpose of
accurate experimental research. Thus, glass, for example, principally in consequence of the alkali contained in it, though made
warm and dry, always has a certain degree of surface conducting
power, mainly depending on the state of the atmosphere. This
renders it unfit for a delicate test experiment. It is evident that a
very feeble degree of surface conduction may tend to produce effects
indicating a greater inductive capability than that of a more perfect insulator. Shell-lac and sulphur, of all other substances, appear
to be the most free from this objection.

Some experimental inquiries by the author, relating to Specific
Induction, will be found in the Philosophical Transactions of the
Royal Society for 1842, Part ii.

The given substance to be examined being disposed under the
form of a circular plate of 1 foot in diameter, and about four-tenths
of an inch in thickness, tinfoil coatings, 6 inches in diameter, were
applied to its opposite surfaces, after the manner of coated glass.
One of the coatings being connected with the attractive disc of an
electrometer, it was easy to determine, by a simple and direct experiment, the three following elements requisite for the elucidation
of the question of Specific Induction :—First. By insulating the
system and depositing a given measured quantity of electricity on
the coating operating on the electrometer, we determine the intensity
of the given quantity (63). Secondly. By connecting the opposite
coating with the electrometer, we determine the direct induction
between the coatings through the intervening dielectric. Thirdly.
By connecting the one coating with the earth and charging the
opposed coating, we determine, by means of the electrometer, the
sensible portion of the charge. In this way we may examine any
dielectric medium, whether solid, fluid, or gaseous, contained
between the metallic coatings, and compare their respective
influences over the degree of induction which takes place through
them.

The following are some results of the author's inquiries. Calling the specific inductive capacity of air 1, we have for *resin*, 1·77 ;

for *pitch*, 1·8; for *bees-wax*, 1·86; for *glass*, 1·9; for *brimstone*, 1·93; for *lac*, 1·95.

The results in the case of *lac* and *air* very nearly coincide with those of Faraday, who found (Experimental Researches, 1270) the relation of *lac* to *air* as 2 : 1, or very nearly. He also found a very high inductive capacity for sulphur, although the specimen employed in the author's experiments did not give a higher capacity than lac, as appears to be the case in the experiments of Faraday. The author has given in this paper many interesting practical observations relative to experiments on specific induction.

65. The general question of specific inductive capacity may be reduced to an experimental form through the medium of an instrument (Fig. 41), which may be termed a *Specific Induction Electrometer;* or by Faraday, a *Differential Inductometer. m, n* are two electroscope gold-leaves suspended from two small rounded gilt conductors within a glass receiver, κ, as in the gold-leaf electroscope. The leaves are from 3 to 4 inches in length, and about four-tenths of an inch in width. The gilt conductors from which the leaves are suspended

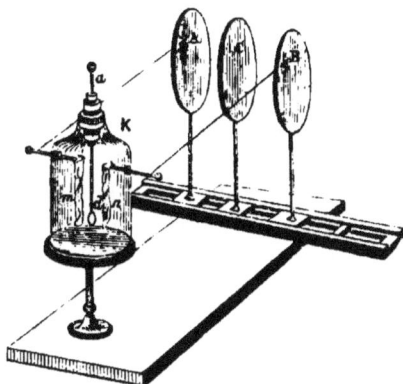

Fig. 41. Specific Induction Electrometer.

are fixed at the extremities of small stout brass wires, passing through fine corks inserted in holes drilled through the sides of the glass receiver κ, by which the distance of the projection of the leaves within the glass may be varied. *a d* is a small vertical metallic rod passing through an insulating cork in the mouth of the receiver, and terminating in a thin gilt disc, *d*, about half an inch in diameter, immediately between the leaves; so that on slightly charging the disc through its vertical metallic rod, the leaves may be acted upon by the disc on either side.

Exp. 32. Connect one of the plates A with one of the gold-leaves *m*, and the opposite plate B with the other leaf *n*, and adjust the distance of the plates A and B at 1¼ inch from the middle plate C. Let the gold-leaves *m* and *n* be 2 inches apart. Slightly charge the centre plate C positively, the plates A and B, with their gold-leaves, being at the same time connected with the ground, that is to say, uninsulated; which being done, immediately re-insulate them. In this state of things c remains positively charged, and A and B negatively, by induction (39), the same dielectric air remaining in the interval between the plates C A and C B, the gold-leaves m and n hanging parallel to each other in a relatively unelectrified state. Thus circumstanced, let a plate of shell-lac, three-quarters of an inch thick and 4 inches square, perfectly insulated, be introduced between the plates c and A: the electric relation of the three plates becomes immediately changed; the gold-leaves m, n, now attract. Remove the shell-lac-plate from between the plates c and A: the attraction ceases. Introduce now the shell-lac between c and B: attraction again takes place, but again ceases on removing it; evidently showing that induction through *air* is different from induction through *lac*. Similar results are obtained with a plate of sulphur. By means of this instrument (Fig. 41) Faraday was enabled to show the difference of the specific inductive capacity between a space partly filled with air and partly occupied by thin shell-lac-plate.

Thus circumstanced, the gold-leaves m, n, become as delicate a test of specific induction as they would be of ordinary electrical charge in Bennett's Electrometer.

OCCASIONAL MEMORANDA AND EXPLANATORY NOTES.

(L) The opposite electrical states of the terminating surfaces of the bodies P, Q, Fig. 22, are best examined by the gold leaf electroscopes adverted to (32, 33). The body P being charged positively, apply a well-insulated carrier plate (14), about three-quarters of an inch in diameter and one-tenth of an inch thick, to each of the faces, $p'\, n$, $p\, q$; and after each contact, transfer the carrier to the double gold leaf electroscope (33), or other of the electroscopes already described (33), weakly charged with positive or negative electricity,—suppose positive; the divergence of the leaves or balls will tend to increase if the carrier plate has been applied to the faces p or p' of the bodies P, Q, and diminish if it has been applied to the face n of the body Q. If applied to the face q of the body P, we have a comparative difference in the actual states of p, q,—q will be less positive than p.

(M) The mechanical arrangement referred to, (39) is well adapted to experimental purposes in electricity.

The bodies P Q, $e\,f$, are insulated and supported on small wooden platforms, T T T T, about 5 inches square, movable in an open rectangular frame of wood, A B, about 4 feet in length by 8 inches in width. This rectangular frame is constructed of mahogany battens 2 inches square, stiffened by cross pieces, in order to preserve the sides straight and parallel. The platforms T T T T are accurately fitted within the sides of the frame, so as to slide freely between guide pieces throughout its whole length. The bodies P Q, $e\,f$, may hence be conveniently placed at measured distances from each other. P Q are light hollow cylinders of gilt wood, each of them having an adjacent electroscope electrometer, $e\,f$, to indicate any electrical change which they undergo.

(N) The rod m N, Fig. 22, carrying the disc N, is about 14 inches long, mounted on a light axis, and counterpoised by a small balance ball, u. The axis is set on fine centres, counterpoised by a small balance ball, u. This disc is set within an elliptical brass ring, x, 6 inches long and 1 inch wide, so as to admit of a very free motion of the rod in a horizontal plane, the rod being counterpoised by a small ball. The whole is supported by means of a short projecting pin, on a small wood ball, a, insulated on a glass rod, $h\,b$. The radial arm $b\,d$, sustaining the insulated disc P, is attached to a circular base, b, movable about the pivot foot b of the insulating rod $a\,b$, supporting the elliptical ring x and movable arm m N. We can thus make the two discs, P N, approach or recede from each other. The whole is supported on a convenient base, R S. The electroscope leaves p' n' are applied to the discs P N by means of a small piece of cork, which causes the leaves to hang freely, and at short distances from the surfaces of the discs.

66. *The Electrical Machine.*—It having been found essential to the progress of electrical inquiry to produce and accumulate the two opposite electricities (22) in great quantity and in a more easy and expeditious way than could be effected by means of the excited tubes already referred to (20), or other simple methods, electricians were led to a mechanical arrangement, more especially termed an *Electrical Machine.* The elements of any arrangement of this kind are these—1. An electric to be excited; 2. A rubber for excitation; 3. An insulated conductor, or conductors, for collecting either the electricity of the rubber, or the electricity of the excited electric (23). The glass and gutta-percha tubes already referred to (note F) may be considered as electrical machines of the most simple form. We have, in the glass or gutta-percha tubes, the electric to be excited; the silk or other substance held in the hand, the cushion or rubber; the projecting rod and ball, the insulated conductor for collecting the electricity of the tube. We are not enabled to collect the electricity of the rubber, because the rubber, being held in the hand, is in this case not insulated, and, hence, its electricity becomes dissipated.

67. For the purpose of subjecting electrics to powerful friction, it occurred to the old electricians to mount them upon an axis in a wooden frame, and turn them round, by means of a winch, against a fixed cushion, or even against the hand. Otto Guericke, so long since as the commencement of the seventeenth century, was the first to resort to this method of excitation. He mounted a globe of brimstone upon an axis, and turning it round against his hand, evolved negative electricity in abundance (30). Since his time, other electrics have been employed, including vitreous and resinous substances generally. Glass, however, has been commonly used in the form of hollow globes, cylinders, or plates.

Electrical machines, as at present constructed, are principally of two kinds, termed the *cylindrical* and the *plate* machines; the electric to be excited being either a hollow cylinder of glass, or a glass plate.

68. *Cylindrical Electrical Machines.*—A machine of this kind of improved construction is shown in Fig. 42, in which A *m* B is a hollow glass cylinder, the diameter being about two-thirds the length of its side. This cylinder is blown with a short open projecting neck, A B, at each extremity of the axis; these open necks are closed air-tight with fine cork, the air within the cylinder being previously made perfectly dry, and the glass within freed from all deposit of moisture. The necks are covered with neat brass caps, or caps of varnished mahogany cemented over them. The caps have short

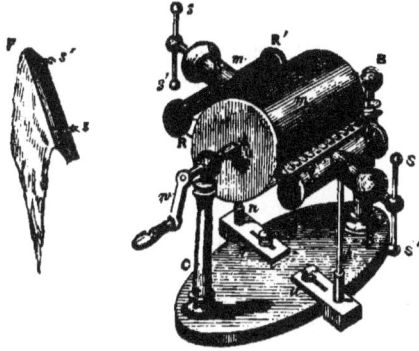

Fig. 42. Harris's Cylindrical Machine.

axial projections, A B, of hard wood, on which the whole is mounted on two mahogany or glass pillars, C D, so as to admit of the cylinder being turned round by a winch, *w*, or a wheel and band, against a fixed rubber, R R'. The rubbing cushion is freely sustained with pressure against the glass cylinder by an insulated cylindrical conductor, N, termed the *negative* conductor, insulated on the glass pillar N *n*. This insulating pillar is based on a sliding piece, *t*, having a stop-screw to fix the slide piece at any required point. The rubber has a flap of oiled silk, *m*, attached to it, extending from its upper edge over the cylinder to about one-third of its circumference. The silk is oiled on one side only, the rough side being next the glass. The cushion R R', with its flap *m*, is attached to a wooden back, and is held in place by means of holes drilled in the conductor and two projecting pins, as shown at F, which represents the detached cushion and its flap. Two small spiral springs, *s s'*, project from the back of the rubber. These springs, under compression, set nearly fair with the surface of the conductor N, and hold the cushion with variable gentle pressure against the glass cylinder. The application of a flap of silk to the electrical machine is very important. We owe this improvement to Dr. Nooth,* who in this way sought to cut off the advancing portion of the excited glass from all electrical communication with the cushion it had just left, and thus prevent any retrograde movement of the developed electricity.

* Phil. Trans. 1774. Vol. lxiii., pt. 2.

69. The rubber R' consists of several layers of thick woollen stuff, or some other elastic substance, supported, as just observed, by a wooden back, from which the holding pins project, and is faced up by fine morocco leather. The silk flap consists of very thin silk, termed Persian, oiled on one side, the opposite or rough side being next the glass (68). The construction of the cushion and flap, although of a very simple form, is by no means a matter of indifference. The best material for the stuffing of the cushion is a modern manufacture termed *spongeopiline*, a combination of sponge and wool ; it is thick, pliable, even, and elastic.*

70. The positive electricity generated by the movement of the cylinder against the rubber and under the silk flap is received upon a row of metallic points, q q' (59), projecting from the insulated or positive conductor P, termed the *prime conductor*, immediately opposite the termination of the silk flap m. This conductor is called the *positive* conductor. It is insulated on the glass pillar P p, which is also set upon a sliding-piece, u, similar to that of the insulator of the negative conductor N, and is steadied in position by a stop-screw. The two conductors P N are cylindrical, and of a T shape. Each conductor consists of two elongated portions, P and q q', well rounded at their extremities ; one portion, q q', of the T being parallel to the side of the cylinder, its length being three times its diameter ; the other, a shorter portion, P, is set at right angles from its centre, and has a rounded extremity, from which projects a brass ball of an inch in diameter set in a sliding brass tube. This ball carries a sliding rod, s s', terminating in light metallic balls. The sliding rods s s', with their terminating balls, can be turned aside to any convenient angle. As it may be occasionally desirable to effect a large electrical accumulation, a cylindrical conductor with full rounded extremities of about twice the length of the glass cylinder and half its diameter is employed. This, when required, is united to the projecting central portion of the positive or negative conductors, according to the kind of electricity to be accumulated. In this way we obtain a powerful and dense electric spark ; the brass tubes and balls projecting from the centres of the conductors being in this case removed to admit of the junction of the new conductor. The electricians of the last century were in the habit of employing prime conductors of very large dimensions.

When the cylinder is made to revolve freely, under friction of

---

* For the most approved method of preparing the cushion and flap, see the Chapter on Electrical Manipulation, Nos. 83 and 84.

the cushion and silk flap, the conductor N being connected with the ground, a surprising stream of positive electricity is obtained from the insulated conductor P, in the form of a current of bright sparks or scintillations. If the conductor P be connected with the ground and the cushion insulated, we have a copious evolution of negative electricity from the conductor N ; thus we see that both electricities are produceable by the same machine (24). If we desire to obtain positive electricity we connect the negative conductor N with the ground, and take sparks from the positive conductor P. If we wish to obtain negative electricity we connect the positive conductor P with the ground, and take sparks from the negative conductor N. [Note O.]

71. The most powerful and complete cylindrical machine of past

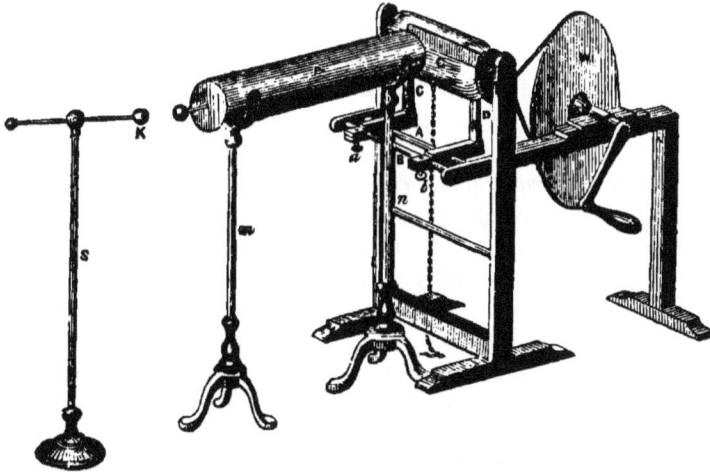

Fig. 43. Nairne's Electrical Machine.

days appears to have been constructed about the year 1773 by Mr. Edward Nairne, F.R.S., of London, mathematical instrument maker. This instrument having been celebrated in the pages of electrical history it may be worth while to describe it here. It is represented in Fig. 43. The glass cylinder G is 1 foot in diameter and 19 inches long between the shoulders. The rubber is 14 inches long and 5 inches wide, is insulated on two horizontal glass rods, A B, and is acted on by two wood springs, C D, so as to press the rubber against the glass cylinder. The prime conductor P is of large size, it being 5 feet in length and 12 inches in diameter. It terminates in a short brass rod and ball, and is insulated on two solid glass pillars, m n. A receiving ball, K, is opposed to the ball at the end of the conductor, and is fixed on the extremity of a brass tube, movable in a hole in the insulating stand s, which supports it, and by which it may be caused to communicate with the ground. The cylinder is set in motion by a multiplying wheel and band, w. With this machine electrical sparks of great power, from 12 to 14 inches in length, were obtained. Nairne employed two prime conductors ; a large one, P, for dense and a smaller one for lesser accumulations.

72. It is important in the construction of these instruments that the interior of the glass cylinder be perfectly free from moisture, otherwise little or no result is obtained. Hence it is that many well-formed cylinders fail in exciting power until made warm by means of heated irons. In constructing the instrument the cylinder should not be closed air-tight until the atmosphere be in a perfectly dry state and the barometer at 30 inches. The older electricians were in the habit of coating the interior of their cylinders with melted sealing-wax, or some other resinous sub-stance, which was found to be efficacious in promoting excitation, doubtless in consequence of the resinous coating protecting the interior of the cylinder against a deposition of moisture.

73. *Plate Electrical Machines.*—Dr. Ingenhousz, about the year 1764, proposed to substitute a plate of glass for the hollow globes and cylinders in use up to that time. This kind of machine, which soon became general throughout Europe, consisted of a circular glass plate of a foot or more in diameter, mounted on a horizontal axis, between two strong vertical supports. The plate was made to revolve between two pairs of cushions. Mr. Cuthbertson, a celebrated mathematical instrument maker in Holland, and afterwards of London, greatly improved this appa-ratus, about the year 1770. He constructed plate electrical machines of 2 feet or more in diameter, and of such power that they exceeded any kind of electrical machine previously designed. Since that period plate machines have been constructed from 2 to 8 feet in diameter.

74. An improved plate machine by Cuthbertson is represented Fig. 44, in which c d is the basis upon which the instrument rests, A c the framework within which the plate revolves, x y z the horizontal axis carrying the plate, z the winch handle, q v the cushions set in light spring frames, so as to apply with pressure to the glass plate by means of compressing nuts and screws, s p silk flaps (68) projecting from the cushions, A w b the prime conductor, projecting from the framework in front of the machine, and terminating in two curvilinear branches, armed with points, for collecting the excited electricity of the glass (59) as it

passes out from under the silk flaps. The conductor A w b is insulated on a stout glass rod, x, attached by a sliding dovetail piece, u, to the front upright of the framework of the machine. The conductor consists of a light hollow centre piece, w, and two tabular curvilinear branches, A w, w b, extending across a horizontal diameter of the plate. When the plate is caused to revolve between the cushions, the excited electricity accumulates on the conductor A w b, from whence it is evolved in bright sparks and scintillations. To remedy the inconvenience arising from the absence of the negative conductor of the cylindrical machine (68), the whole apparatus is insulated upon glass legs, m, n, o, p, fixed in a firm base, F H. A conducting rod and ball, s, projects from the base c d of the framework, as a negative conductor. By these means the entire framework of the apparatus is rendered negative

Fig. 44. Cuthbertson's Plate Machine.

when the machine is in motion (24), the framework being immediately in connection with the rubbers. In order to perserve to the framework a sufficiently perfect insulation, the plate is turned by an insulating varnished glass handle, z. Cuthbertson subsequently constructed his instrument with two equal glass plates supported on the same axis, by which its power was greatly increased. The two plates were rubbed by four pair of cushions, and the excited electricity taken up by collecting points passing between the plates.

75. The defect of these machines appears to have been their liability to fracture, and the absence of an especial negative conductor immediately connected with the cushion, as in the cylindrical machine. The method of converting the entire framework into a negative conductor, although allowing negative sparks to be obtained from the operation of the rubbers, did not succeed very well. Professor Copland, of Aberdeen, found that, although insulated, the winch handle emitted a flash of negative electricity to the pillar of the machine at every revolution of the plate, perfectly visible in the dark, and the effect of which he felt nearly to his shoulders.*

76. Van Marum, in 1791, remedied this inconvenience by a machine of new construction (Fig. 45). The plate was 2 feet

* Nicholson's Journal, vol. xxvi., p. 10.

7 inches in diameter. This machine, as well as Cuthbertson's double plate, was of an ingenious, although of a rather heavy and somewhat complicated character. The cushions A B are each separately insulated on stout glass pillars, C D, and are applied nearly in the direction of a horizontal diameter of the glass plate. The prime conductor P is a large ball, to which is attached a light semicircular rod, M P N, terminating in two metallic cylinders, M N, each 6 inches long and 2½ inches in diameter. The semicircular branch, M P N, may be turned from a vertical to a horizontal position, so as to bring the metallic cylinders M N into contact with the rubbers, in which case the prime conductor exhibits negative electricity. In order that the rubbers may communicate with the ground, when positive electricity is required, there is a second semicircular branch, *v*, on the opposite side of the plate, nearly at right angles to the first, and connected with the ground. When negative electricity is required, the semicircular branch, M P N, is brought into contact with the cushions, and the opposite semicircular branch *v* brought into a nearly vertical position, so as to carry off the electricity of the glass. The axis W of the plate is supported on a single column, G, for which purpose a bearing piece, Z, is provided, furnished with two brass collar guides. With a view to counteract friction in the collar guides, the axis is balanced by a counterpoise of lead, E. The conductor P is provided with a long curvilinear copper tube and ball, II, movable like a radius about a stem projecting from the conductor, so as to communicate electricity in any required direction. The middle of the conducting part of the axis is a cylinder, *x*, of baked wood capped with brass. The glass plate is secured to the front face of this central cylinder by means of a hole drilled through the glass, and a projecting screw and nut, being guarded from undue pressure by means of intervening rings of felt. The whole plate thus turns freely with the axis and winch handle.

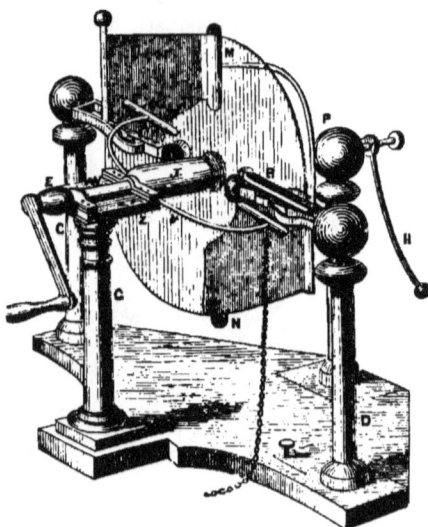

Fig 45. Van Marum's Plate Machine.

77. In the plate machine about to be described, we have all the advantages of the cylindrical machine and the plate machine of

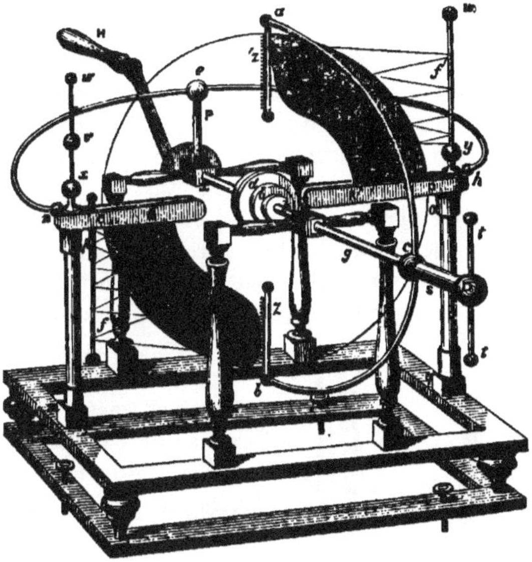

Fig. 46. Harris's Improved Plate Machine.

Van Marum, under a light form and simple construction. P, Fig. 46, is a circular glass plate 3 feet in diameter, mounted on a horizontal brass axle, $x l$, and sustained in a strong mahogany

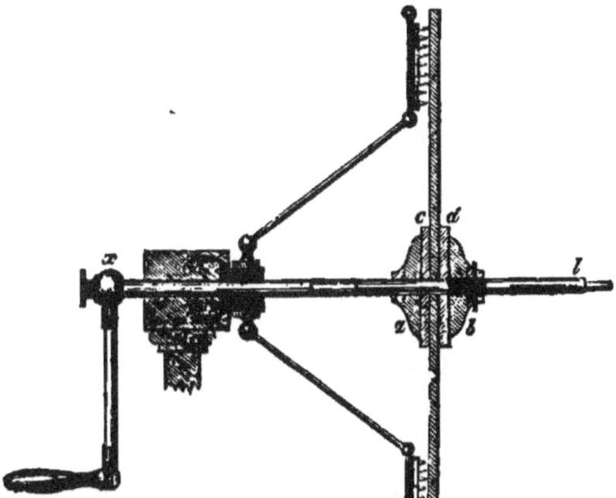

Fig. 47. Method of Mounting Plate, &c.

frame, $m q$. The glass plate is held fast upon its axle $x l$, between two hollow brass flanges, $a b$, Fig. 47, guarded by intervening collars of leather or felt; one of the flanges, *a*, is fixed, the other, *b*, mov able, as a sort of screw nut, so that the plate revolves with the axle and handle. As in turning the plate the screw flange *b* is liable to press forward upon the glass, and thereby cause it to crack through the centre, a serious objection to the ordinary plate machine, a small stop screw is passed at *b* through the projecting edge of the flange, into the substance of the axle, which effectually prevents the pressure from increasing beyond a given point requisite for the movement of the plate. For more effectual security against pres- sure two circular glass plates, *c d*, 6 inches in diameter and one- quarter of an inch thick (drilled through their centres for the passage of the axle) are attached by thin varnish to the centre of each face of the plate. These circular plates serve to thicken and strengthen the large plate at its centre and effectually preserve it against fracture. The plate is excited by two pairs of rubbers, R R', Fig. 46; each rubber is 12 inches long by 2½ inches wide and about half an inch in thickness, constructed in the manner of the cushion of the cylindrical machine (69). The rubbers R R', Fig. 48, are loosely enclosed between light spring pieces of mahogany, *z z*. These pieces are fitted by dove- tails *k k'*, in a wooden block, B, and are compressible against the brass plate by means of short brass

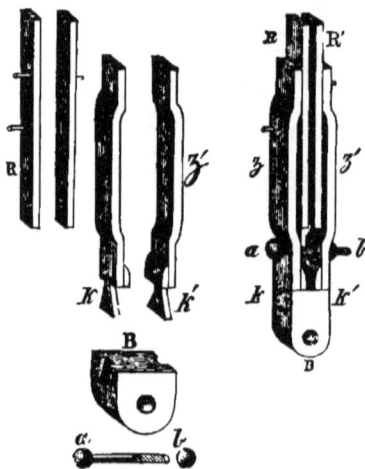

Fig. 48. Details of Cushion.

rods, *a b*, passing through the spring pieces immediately before the block. The rods have screw terminations and small globular nuts, *a b*. These press the rubbers gently toward the plate from opposite sides. They are insulated upon stout glass pillars, *n m*, *o q*, as in Fig. 42, one at each extremity of a diameter of the plate; the blocks and spring holders, in which the rubbers are fixed, are secured to the insulated pillars by means of brass caps, with projecting screws and globular nuts. Each rubber, R R', has an oiled silk flap, R *s*, R *u*, Fig. 42, attached to it, shaped to the plate, the silk being prepared as directed in the Chapter on Electrical Manipulations, and applied with the rough side next the glass. Each pair of flaps is united over the edge of the plate by narrow silk binding. They are prevented from dragging upon the plate by silk cords, *f f*, carried round a descending rod of glass, *f*, on one side, inserted in a sliding brass tube passing through the block of the rubber, and on the opposite side, round a light perpendicular tube of brass, f' y, inserted in the ball confining the block to the insulator, as shown in Fig. 46.

78. The prime conductor a c b, Fig. 46, is placed in a vertical position in front of the plate, and consists of two curvilinear tubular branches a c, c b, proceeding from a large central hollow ball, c. A cylindrical brass socket, s, Fig. 49, 3 inches long, proceeds from this ball, and receives within it a similar cylindrical socket, T, terminating in a second hollow ball, D, the two sockets fitting so closely as to admit of an easy circular and sliding motion. This last ter‑minating ball, D, carries a central

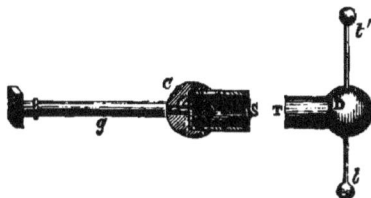

Fig. 49. Telescopic Joint Conductor.

sliding-rod, t t', having a metallic ball at each extremity. The rod t t' may be turned to any angle by means of the movable telescopic socket T. The experimentalist may hence lead off connecting wires from the conductor in any convenient direction. The curvilinear branches a c, c b, of the conductor a c b, extend to within a short distance of the glass plate, and have affixed to them light vertical rods, z z', Fig. 46, about 8 inches long, and three-eighths of an inch in diameter, terminating in balls of varnished wood. These vertical rods are furnished with thin knife-edged pieces of brass plate, about half an inch wide, and 6 inches long, carrying small collecting points (59). The points project about one-quarter of an inch beyond the knife-edge, immediately opposite the terminations of the silk flaps R s, R u, attached to the cushions. The vertical conductor in front of the plate is insulated on, and supported by, a glass rod, g, of 1 inch in diameter, and 1 foot in length, furnished with two brass caps, one of which connects it with the brass socket within the hollow of the ball c, so that the ball with the conductor can be turned about upon the brass cap of the insulating rod g, if required; the other cap is connected with a dovetail plate fitted into a groove attached to the centre of the front cross-bar supporting the axle, as in the figure.

79. The negative conductor n e h, Fig. 46, passes behind the plate, and consists of two metallic tubular curvilinear branches, e n, e h, three-quarters of an inch in diameter. These connect together the insulated blocks R R' that carry the rubbers. The curvilinear branches e n, e h, are united over a piece of flexible metal tube, through a brass ball, e, insulated, on a glass rod, P, 6 inches long and half an inch in diameter. This insulating rod is fixed by a brass

cap upon the centre of the back cross-bar of the framework of the machine, immediately over the axle of the plate. To render this negative conductor perfectly efficient, and further cut off all communication with it, the plate is turned by a strong insulating winch-handle, H, of varnished glass, about 1 foot in length and a full inch in diameter. The negative conductor $n\ e\ h$ is supported upon the blocks carrying the rubbers by means of spherical sockets, attached to two circular pieces of brass about half an inch thick, which resist the pressure of the balls, securing the blocks of the rubbers to the insulating pillars. The brass ball x, Fig. 46, which fixes the rubber on the side x to the insulator, carries a vertical tube of brass, $w\ x$, terminating in a small ebony ball, w. The vertical brass tube proceeding from the ball x, carries a sliding ball, r, upon it, for the convenience of attaching connecting rods to the negative conductor, when required. The brass ball y, uniting the rubber on the side y to the insulator, carries also a light vertical brass tube, $f'\ y$, terminating, as in the opposite side, in a ball, w'. A light open circle, c, of silvered brass, or varnished cardboard, 6 inches in diameter, and divided into eight parts, is fixed centrally round the axle, where it passes through the back cross bar of the framework of the machine, as shown in Fig. 50, which is a back view of the whole apparatus. By means of an index, i, upon the axle, we are enabled in turning the plate to estimate the number of its revolutions.

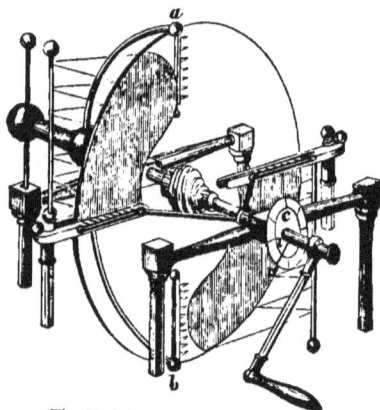

Fig. 50. Back view of Harris's Machine.

80. When the instrument is so arranged as to convert it into a negative machine, the prime conductor is turned upon its insulator into a horizontal position (Fig. 51), and its collecting points placed in contact with the rubbers. The original negative conductor, passing behind the plate, is in this case entirely removed, and its place supplied by a vertical positive conductor, of the form a R b, attached to a wide cylindrical brass ring R, movable with friction upon a fixed socket surrounding the axle. This conductor consists of two light tubular arms, A R, R B, about a quarter of an inch in diameter, each inclining at an angle toward the plate, being held fast on two stout brass pins screwed into the movable ring R. These arms carry collecting points, $a\ b$, as in the original prime

conductor. The branches a R, R b, fit closely upon the pins of support, and may be removed at pleasure. We are enabled, by the circular motion of the ring upon the brass socket surrounding the axle, to adjust the conductor that carries off the electricity of the glass, in an exact vertical position; or in again replacing the original prime conductor in its first position, we can turn down the new positive conductor a R b so as to connect the rubbers, as

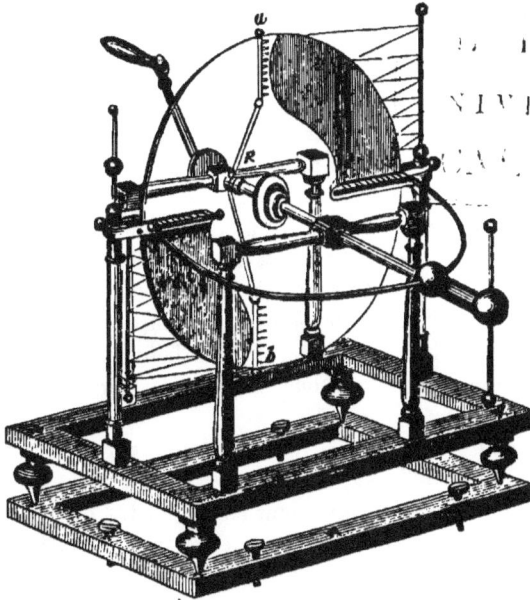

Fig. 51. Harris's Plate Machine for Negative Electricity.

seen in Fig. 50. When acting as a positive conductor, however, it is merely employed to carry off the electricity of the glass, in which case it is connected with the ground; whilst the original prime conductor, acting horizontally, receives the electricity of the rubbers. The machine arranged in this way as a negative machine, is wonderfully efficient, and sends forth a torrent of negative sparks from the horizontal prime conductor; so that both positive and negative electricity are thus obtained.

Such is the construction of the instrument as fitted for positive and negative electricity. It is novel in form, and of convenient and general application.

The base and framework upon which the machine is supported are of seasoned mahogany. The base consists of two open rectangular frames, each 4 feet 6 inches long, and 2 feet wide, the pieces forming the frames being 4 inches wide by 3 inches thick, neatly moulded, and rounded at the angles. The lower of these frames, w, Fig. 51, is set upon castors to admit of an easy movement of the whole instrument, and has six levelling screws through it, in order to give it, when in place, a firm bearing upon the table or floor. The upper frame m q is insulated upon the lower frame w by four stout conical glass feet, 5½ inches long. These are received below in small shallow cavities, and are screwed into the under part of the upper frame, through the medium of brass caps, nuts, and screws.

The glass plate of the machine is supported about 28 inches above the upper frame m q, by means of four cylindrical mahogany pillars, from 1½ inch diameter above to 2½ inches below. They are about 20 inches apart, and are united above by stout cross-pieces, which receive the axle of the plate. The supporting columns and cross-bars are carefully moulded and fashioned, and are securely united to each other, and to the framework, by iron screws and nuts. The centres of the cross-bars, through which the axle passes, are formed ornamentally into cubical blocks 2½ inches wide by 3¼ inches deep: the bars being nicely rounded up to their terminations in other blocks joining the vertical pillars.

81. The diameter of a plate machine may vary, as already observed (73), from 2 to 8 feet, or more. A 3-foot plate, however, is upon the whole the most convenient size. Plates of 18 inches, 2 feet, and 30 inches diameter, are powerful and efficient. The electrical machine constructed by Cuthbertson in the year 1780, for Teyler's Museum at Haarlem, consisted of two circular glass plates, each 65 inches in diameter, fixed in the way already referred to (74), upon the same horizontal axis, parallel to each other, and at a distance apart of 7½ inches. The glass plates were excited by eight rubbers, each 15½ inches long, placed in appropriate frames. The prime conductor was divided into branches supported by three glass pillars, 57 inches long. The branches entered between the plates, and by means of fine points, collected the excited electricity. This machine appears to have been of great power. Two, and occasionally four men were required to work it. When in good action, the following effects were observable:—A sharp steel point presented to the prime conductor drew forth a luminous electrical stream full half an inch in length. The same point fixed to the conductor, so as to project 3 inches from it, emitted similar streams 6 inches long, when a ball of 3 inches in diameter was presented to it; and 2 inches long when another

point was presented instead of the ball. The peculiar creeping sensation, commonly called *spider's web*, was felt on the face of the bystanders often at the distance of 8 feet. A delicately suspended thread, 6 feet in length, was sufficiently attracted by the prime conductor, at the distance of 38 feet, to cause it to deviate sensibly from the perpendicular (5). The electricity of a pointed wire rod, 28 feet distant from the conductor, appeared luminous. A second conductor of similar dimensions being presented, in order to receive the sparks, and a perfect metallic communication being made between the second receiving conductor and the earth, by means of a long brass wire three-eighths of an inch in diameter, it was found that whilst a stream of electricity passed from the prime conductor to the receiving conductor, the brass wire gave small sparks to conducting bodies placed near it. Hence the quantity of electricity evolved from the prime conductor must have been considerable, since a wire one-eighth of an inch in diameter could not perfectly transmit the accumulation to the earth. The sparks between the two conductors were generally 21, but sometimes 24 inches long. The electrical stream between the conductors appeared crooked, and sent forth lateral branches of a large size.

Since that time, plate electrical machines have been constructed from 8 to 10 feet and upwards in diameter, turned by steam machinery, the effects of which have been of a surprising character.

82. The old electricians, with a view of obtaining dense and powerful sparks, were frequently in the habit of employing prime conductors of enormous dimensions (71). In our own time, that is, about the year 1856, Professor Winter, of Vienna, applied to a plate machine of a peculiar construction (the plate was 19 inches in diameter), a circular prime conductor under the form of a large ring, from which he was enabled to obtain unusually long and brilliant sparks, quite equal to the sparks obtained from the machine of Teyler's Museum, at Haarlem. This new form of prime conductor consisted of a ring of polished wood, nicely rounded at its edges, about 1¼ inch thick, and 27 inches in diameter, the ring being constructed in parts, and having a stout metallic wire passing throughout its substance. The ring is upheld by a stout metal rod, 20 inches long, inserted in the prime conductor of the machine, insulated on a strong pillar of glass. Fig. 52 represents Winter's ring. c is the ball of the prime conductor, g the insulating pillar, d the metallic rod, inserted in the conductor that carries the ring. The electricity evolved by the machine accumulates on the ball c of the conductor in the usual way, and is transferred to the

ring R, through its metal support. It is necessary to keep the
ring R at least 2 or 3 feet away from the ceiling of the room.
With an electrical plate of 3½ feet in diameter, Professor Winter
is said to have obtained vivid electrical sparks of nearly 2 feet
in length.   This machine was excited between two circular
rubbers, each 6 inches in diameter, furnished
with collecting points, and consisted of leather
pads stuffed with cotton wool. It appears to
have been powerful.

The advantage of Winter's ring seems to
consist in this, that whilst its rounded edges
tend to retain the accumulated electricity, the
intensity of action is increased by its prominent
circular elongation, and thus a brilliant dense
spark escapes from it.

83. Electrical machines demand especial care
and attention in preparing them for excitation.

Fig. 52. Portion of
Winter's Conductor.

If of glass, it is requisite the glass be clean
and dry; the exciting rubbers should be carefully
covered with electrical amalgam (7), and this should not be
applied beyond the junction of the silk flap with the rubbers,
but quite in the rear of it. All the insulators should be varnished
with a solution of shell-lac in naphtha, dried off by heat; and
it will be found advantageous to interpose a layer of clean, dry,
warm writing-paper between the rubbers and the glass, previously
to turning the plate a few rounds. Indeed the exposure of the
plate to friction with clean, warm, dry paper for a short time, or
during a few revolutions, has a very beneficial effect.

84. When a powerful electrical machine is set in motion in a
dry atmosphere, the rubbers pressing gently on the glass, the
following phenomena will be apparent:—

The conductors being removed, brilliant circular streams of
light, or, as the old electricians styled it, *electrical fire*, attended
by a sharp crackling sound, dart round the plate or cylinder, whilst
luminous corruscations play about its circumference, producing
a very brilliant and beautiful effect. The peculiar creeping sensation, or *spider's web*, is felt on the face and hands, and a vaporous
odour, termed *ozone*, affects the sense of smell. The conductors
being in place, similar phenomena present themselves. Bright
sparks occasionally shoot back over the glass not covered with
the silk flap, to that part of the conductor carrying the collecting
points, and often continue to the rubbers under the flap. If the
knuckle be presented to the glass plate or cylinder beyond the

rubber, a sharp spark will frequently be experienced. When one knuckle is presented to the positive, and the other to the negative conductor, both conductors being insulated, a pungent and unbearable succession of electric sparks is experienced. If the two conductors be united by a metallic wire or rod, all these phenomena vanish. If for the metallic rod a series of small shot be substituted, strung upon a silk line about one-thirtieth of an inch, or even less, apart, a luminous and brilliant effect is produced by the electricity passing from one shot to the other, as if an electrical current existed between the two conductors. Either conductor emits the most powerful spark when the opposite conductor is connected with the ground (41). A similar effect ensues on linking the two conductors together by an iron jack chain, which may be prolonged in festoons by means of insulators.

85. Of the different forms of electrical machine the cylinder is the most simple and convenient, and the least troublesome, as there is only one rubber. In this form, and without any complicated mechanical arrangement, both electricities are at command, for which purpose we have merely to transfer our operations from one side of the machine to the other.

86. The electrical machine is, when well and perfectly constructed, equable and continuous in its action, and may be con-

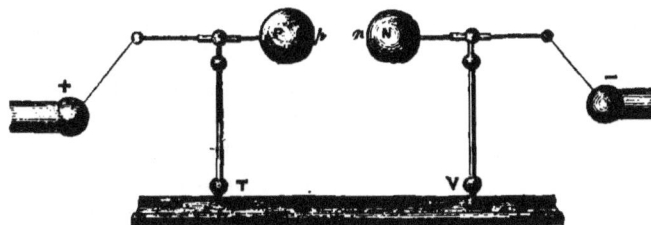

Fig. 53. Opposed Discharging Balls.

sidered to produce at every revolution the same quantity of excited electricity. This is an unquestionable experimental fact, although several electricians, especially in Germany, have entertained an adverse opinion. A very little reflection, however, will suffice to put the fact in evidence.

Exp. 33.—Let p n, Fig. 53, be two light, polished, hollow balls of copper, from 3 to 4 inches diameter, mounted on insulating supports inserted in sliding bases, t v, and set at a given distance apart. Let one of the balls, p, be connected with the positive conductor, and the opposite ball, n, with the negative conductor.

Let the touching points p n of the balls be set at such a distance apart, that on carefully turning the electrical machine a given number of discharges between the balls may take place at each two or three revolutions. By means of a pendulum, or the musical regulator termed a *metronome*, let the time of the revolutions be accurately noted. It will be found that the same number of revolutions in the same time will evolve the same number of discharges between the balls, consequently the same quantity of excited electricity must have been developed at each revolution of the machine. We suppose, of course, the machine to have been carefully constructed, and in good order. The number of discharges between the balls therefore may stand for and represent the quantity of electricity at each given number of revolutions. It has, however, been said by Professor Riess of Berlin, and others, in opposition to this very plain demonstrative fact, that inasmuch as the sparks are commonly observed to decrease in vigour, and also in number, as the charge proceeds, therefore the machine does not produce equal quantities of electricity in equal times, or the same quantity at each revolution. Now it is by no means difficult to perceive the insufficiency of this observation; it is clear that any diminution of the sparks in frequency from the prime conductor, as the charge of a limited surface of accumulation advances, does not depend upon any decreasing power in the machine to produce the increased quantity, but in the decreasing power of the receiving surface to accumulate the quantity produced. Unless the quantity evolved from the machine can be as rapidly taken up as it is produced, it will be, of course, impossible to maintain an infinite succession of equal quantitative sparks between the balls P N, supposing the capacity of the negative ball N to be limited. Directly, however, we connect the ball N with the negative conductor, or with the earth, and set it free from this limitation, we obtain an infinite series of explosive sparks of precisely the same magnitude. Imagine, therefore, the insulated conductor, or other insulated surface charged from the prime conductor of the machine, to have sufficient electrical capacity or extension ; the revolutions of the machine are uniform and equable up to a given limit of charge, and may be taken as an accurate measure of the quantity of excited electricity up to that limit, or nearly so. Let, for example, the limit of charge of a given surface charged to saturation be one hundred revolutions of the plate, we may conclude, without sensible error, that when charged with forty revolutions the given surface would have received twice the quantity of electricity it would have received with twenty revolutions, and such is really the case, as is demonstrable by experiments.

87. *Steam Machine.*—This species of electrical machine is of modern date and construction, and is the result of an accidental observation (P).

Armstrong's machine consists of a steam boiler, A, Fig. 54, insulated on stout pillars of glass. The steam is made to issue from a general steam-pipe through bent iron tubes, *a b c*, terminating in jets of wood, and of which there are a large number. An insulated projecting conductor, N, is placed in connection with the boiler for the convenience of collecting the excited electricity, and a second conductor, P, formed of a metallic case furnished with several rows of points, is placed immediately in front of the jets to receive and carry off the electricity of the steam, and prevent its return upon the boiler, by which the opposite forces would be neutralised. Faraday, who investigated this question with his accustomed tact and penetration, endeavours to show, by a series of masterly experiments, that the electricity thus produced does not depend upon any chemical or other change which may be supposed to arise from evaporation or condensation, but is the

Fig. 54. Steam Electrical Machine.

result of the friction of condensed particles of water whilst being driven by the still issuing steam through the jets, so that, in fact, these particles perform the office of the glass of the common electrical machine (68), and give out vitreous or positive electricity ; the wood jets and pipes act as the rubber, and give out resinous or negative electricity ; the friction of the steam in passing through the jets being the source of electrical power. The electricity produced by this apparatus is enormous in quantity. The sparks from the conductor N, upon an insulated metallic ball, are dense and rapid, presenting frequently the appearance of a continuous flame, and will readily set fire to inflammable matter.

Although the friction theory of the operation of this instrument adopted by Faraday appears consistent with striking facts, it is nevertheless not perfectly clear that the condensation and evaporation of water as steam is altogether without influence in the electrical development.

88. *The Electric Column ; or, Perpetual Electrical Machine.*—

When a series of thin discs of two different metals—zinc and silver, for example, or zinc and copper—are alternated with similar discs of common writing paper in a dry state, the result is a peculiar electro-motive action. The zinc disc of the series exhibits positive, the copper or silver disc negative, electricity ; and the arrange-ment will act powerfully on the gold-leaf electroscope (32). For example, let A B, Fig. 55, be a clean tube of glass, about three-quarters of an inch in diameter, varnished both internally and externally with a thin clear coating of shell-lac dried off by heat. Within this tube is a series of circular discs of very thin zinc and silver, or zinc and copper, alternated with similar discs of dry paper, in the succession of silver, zinc, paper, silver, zinc, paper, and so on. The zinc disc at one extremity of the series exhibits positive, and the terminating silver or copper disc negative, electricity. If the tube A B be capped at each extremity with short caps or rings of brass, A B, communicating with the first and terminating discs, by compressing screw rods, *a b*, passing through the caps at each extremity of the column, one of the caps will affect the electroscope positively ; the other negatively.

Fig. 55.
Dry Pile.

The amount of positive and negative action will of course depend on the extent of the series. A series of a thousand alternations will be extremely sensible to the gold-leaf electroscope, and may be employed with advantage to produce a given divergence of the leaves without violence. With a series of from two to three thousand discs, comparatively powerful electrical effects result.

The most simple way of constructing the electric column is to cut out, by means of a hollow punch, from very thin zinc plates, circular discs, about three-quarters of an inch in diameter. Simi-lar discs should then be prepared in the same way from dry cartridge paper, covered on one side with silver leaf. We then proceed to arrange the discs in succession within the tube, between its capped extremities, in the order of zinc, silvered paper—the silvered side next the zinc ; then again zinc, silvered paper—the silvered side next the zinc—and so on. We have then a series grouped in the order of zinc, silver, paper, zinc, silver, paper, and so on, the zinc being in contact with the silver throughout, and each pair of zinc and silver plates separated by dry paper. One of the caps should be fixed before the discs are introduced into the tube, and each cap should have a compressing screw-rod passing through its centre, in order to press the discs closer together, and effect a good contact with the terminating discs of the column. The extremities of the screw rods should terminate in small metallic balls, these being found convenient for conveying weak electrical charges to the gold-leaf electroscope.

The operation of this species of electrical apparatus appears to depend on a peculiar kind of electro-motion, induced by the contact and separation of dissimilar bodies, and has apparently some relation to the phenomenon of excitation, as previously exemplified (26). We here observe that when the rubber and glass, two dissimilar bodies, are first brought into close contact, and subsequently separated, both evince a state of electrical excitation ; the one being electrified positively, the other negatively : thus leading to the conclusion that electrical excitation is more or less dependent on the contact and separation of dissimilar bodies.

The celebrated experimental electrician Singer, by bringing various insulated bodies into contact and then separating them, succeeded in developing positive and negative electricity. Thus an insulated polished circular plate of zinc, of about 4 inches in diameter, being brought into contact with a well-insulated polished circular plate of copper, of the same diameter, the two plates evince on separation opposite electrical states, which become sensible to the electroscope ; the copper plate being negative, the zinc plate positive. These states are more fully apparent after several contacts of the two plates. Hence it has been inferred that in the operation of the electric column, the associated metals, separated by semi-conducting matter, such as paper, give rise to a propagation of electrical action analogous to electrical excitation ; the zinc acquiring a positive electrical state by contact with the copper or silver, and the interposed paper transferring the positive electricity thus developed from the copper to the next zinc, and so on throughout the series, so that at last all the developed positive electricity becomes accumulated on the terminating zinc disc, and all the negative electricity on the terminating silver or copper disc. Thus the column is, as it were, a spontaneous source of electrical excitation. It will be seen that in this kind of electrical machine there are three associated substances—two metals, and an imperfect conductor—viz., zinc and silver, or copper, and paper ; being so far analagous to the voltaic apparatus, which consists of two metals, and an interposed fluid, and which in the hands of Davy effected such marvellous decompositions. There is, however, this remarkable difference between them, that whilst the voltaic apparatus is celebrated for its electro-chemical power, the electric column has no electro-chemical power at all, but is principally remarkable for its pure electrical effects, resembling common electrical excitation (7).

It was the opinion of Volta that the fluid interposed between the metals of his apparatus serves only to conduct the electricity developed in one pair of metals to the succeeding pair. In a similar way the discs of paper in the electric column may be conceived to break the continuity of the opposite metals, and at the same time transmit the electricity developed by the one pair to the succeeding pair, constituting a species of electro-motion.

89. The pure electrical effects of the column show that the arrangement of a series of zinc and silver, or copper discs, with intermediate paper, is a semi-conducting column, which in its insulated state is positive at one extremity, negative at the other, and neutral in the centre.

Exp. 34.—Connect a column, A B, Fig. 56, of 500 to 1,000

Fig. 56.   Action of Dry Pile on Electroscopes.

alternations, with three delicate gold-leaf electroscopes, A, B, C; one being at one extremity, A; another at the opposite extremity B; and a third at the centre c. The electroscope connected with the copper termination will diverge with negative electricity; that connected with the zinc termination will diverge with positive electricity; whilst the electroscope at the centre c will appear neutral. If either extremity, A or B, of the column be connected with the ground, and thus set free, the leaves of the electroscope connected with that extremity will close; those of the central electroscope c will immediately diverge with the same electricity, whilst the leaves of the electroscope at the remaining insulated extremity will, in accordance with ordinary inductive action, have their original divergence increased (41). This indicates a real electro-motive property in the apparatus by which the zinc extremity accumulates positive electricity, and the copper extremity negative.

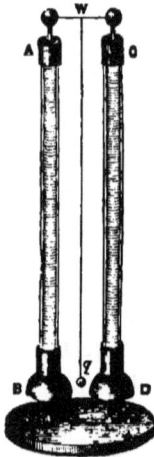

Fig. 57.
Electric Chime.

90. The opposite states of the electric column have given rise to an ingenious instrument termed the *Perpetual* or *Electric Chime*, represented in Fig. 57, in which A B, C D, are two active electric columns, containing about 1,000 groups each, set upon two small

insulated bells, B D, their zinc and copper terminations, A and C, being connected by a small conducting rod, W. A light metallic ball or clapper, q, is suspended between the bells by a fine silk thread, which, if the series composing the column be extensive, will continue to vibrate between the bells (18), keeping up a perpetual chime, and affording some ground for the conclusion that the electro-motive power is perpetual. A series of about 1,500 groups was found by De Luc to sustain a vibration in this way for several years, giving rise to many important meteorological results. If the electric power be really continuous, we have no doubt arrived at perpetual electrical motion. If a sufficiently extensive series be employed, the electrical development appears to be so continuous and of such duration as to go far in support of this deduction.

91. The electric column has been further employed as a delicate means of determining the positive or negative state of a body, through the medium of the gold-leaf electroscope.

Fig. 58 represents a sensitive differential arrangement applicable to this purpose. L is a single gold-leaf electroscope constructed in the usual way (32). The gold leaf is suspended from an insulated metallic rod, between the hemispherical terminations of two conductors, p n, passing through small holes drilled in opposite sides of the glass enclosing the gold leaf. R S is a horizontal insulated electric column, the extremities of which communicate with the rods p n by small conducting wires, and thus place its flattened terminating hemispheres in opposite electrical states (89). Conse-

Fig. 58. Differential Electroscope.

quently, if the electricity affecting the gold leaf through the plate q be positive, the leaf is attracted toward the negative side n; if negative, it is attracted toward the positive side p (27); thus the electrical state of any body affecting the plate q is immediately made evident.

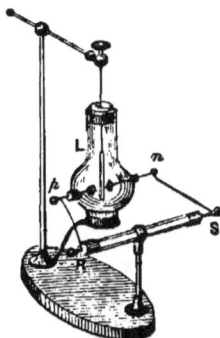

92. The Electrophorus.—If an ideo-electric plate (8), N, Fig. 59, be insulated on a glass support, S, its upper surface being excited by friction, and covered by a metallic plate, P, having an insulating handle, ṁ, the excited electricity does not leave the surface of the ideo-electric plate N to accumulate on the metal cover P, but acts upon the metal cover by simple induction (38), and in such way as to determine the electricity of the metallic plate P upon the excited surface of the ideo-electric plate N.

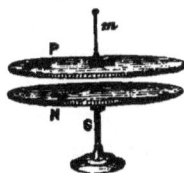

Fig. 59. Single Electrophorus.

Let, for example, the idco-electric plate N be a circular resinous disc—suppose a disc of coarse sealing wax, about half an inch in thickness—its upper surface being negatively excited by rubbing it with dry flannel or soft silk (30). Let P be a circular metallic plate, or cover, of nearly the same dimensions. It may be of wood covered with tinfoil, and about three-eighths of an inch thick, reposing upon the negatively excited surface of the resinous plate. The negative electricity of the excited plate does not leave it, but acts inductively on the metallic cover, determining its positive electricity upon the negatively excited surface of the resinous plate, leaving the metallic cover negatively electrified. If, under these circumstances, we present a conducting ball or the knuckle of the hand to the cover, positive electricity will be given off to it from the knuckle under the form of an electric spark, the opposite electrical forces will combine, and the cover will be again rendered apparently neutral. Let the cover be now raised from off the excited plate by its insulating handle ; it then bears off or carries away the positive spark communicated to it, and which satisfies its previously negative state, so that it will now appear positively charged, and will return to the knuckle or a conducting ball the identical spark it had previously received, and again render the cover neutral. By repeating this process of successive contacts of the excited plate and its cover, and the subsequent withdrawal of the cover, we may bear off and obtain a series of electric sparks to a large amount without a new excitation of the resinous plate. An excited electrical plate and cover thus circumstanced constitutes a species of spontaneous electrical machine, termed an *Electrophorus*, from the Greek words ηλεκτρόν, electricity, and φορεω, to bear or carry off, the metal plate carrying away after each contact an electrical charge. The simple combination of an electric plate having a metallic cover with an insulating handle is all that is absolutely requisite to this result. An electric plate, therefore, having an insulating conducting cover, as represented Fig. 59, may be termed a *single electrophorus*. Any electric substance may be employed for the purpose. A resinous electric plate is, however, generally chosen, resinous bodies being more especially retentive of excited electricity. A resinous plate, consisting of shell-lac, common resin, and Venice turpentine, in equal parts, cast at a temperature of from 230 to 250 degrees, within a metal or wood ring upon a marble slab, or smooth metal surface, is very efficient. [Electrical Manipulation. 94.] A plate of second or inferior sealing-wax is very efficient. A circular plate of glass may be employed with success, especially when consisting of two plates of flat glass joined by means of sealing-wax. [Manipulation. 94.] A plate of this kind is extremely excitable. A plate of brimstone, cast upon glass in a similar way to the preceding, is remarkably efficient. It is, however, liable to crack by changes of temperature.*

93. *The Compound Electrophorus.*—The ordinary method of preparing an electrophorus is to place the electrophorus disc N, Fig. 59, between two metal discs, P and Q, Fig. 60, of somewhat less diameter, one of them, Q, being in connection with the ground, through any convenient conducting stand of support, T, the upper plate P being placed as in the preceding case. Such an arrangement may be termed the *Compound Electrophorus.* The conducting plates may be of light wood, well rounded at the edges and smoothly covered with tinfoil

Let N, Fig. 60, be an electrophorus plate, resting on the metallic plate Q, supported on a light brass pillar and foot, T, and covered by a metallic plate, P, as before. This arrangement is subject to the general laws of induction (41). We have in fact two metallic conducting plates, P and Q, opposed to each other through an intervening dielectric, N. The upper disc P has been termed the *cover,* the under disc Q, in communication with the earth, the *sole.*

Fig. 60. Compound Electrophorus.

This has been found to increase the action of the single electrophorus (92). How it contributes to this result is not immediately apparent, seeing that the evolution of an electrical spark may be obtained without it.

94. In order to explain the action of the compound electrophorus many complicated theoretical views have been advanced. The action of the compound electrophorus, however, admits of a simple and practical solution.

We have seen (41) that an insulated conductor has its capacity for electricity increased by induction when it is opposed to an uninsulated conductor of great superficial extent, or in connection with the ground. In the case of the compound electrophorus, Fig. 60, the cover P is an insulated conductor, the sole Q an uninsulated conductor of unlimited extent, being in connection with the earth, and directly opposed to the cover, through an insulating medium. Hence, by the laws of induction (41), the cover P can receive from the electrophorus plate N, under the

* The best method of preparing electrophorus plates will be found in our chapter on Electrical Manipulation (94).

influence of the sole, a greater charge, under a given intensity, than it could receive without such auxiliary influence. Such appears to be the advantage of the compound electrophorus. An inductive action is set up between the sole and the cover; increasing the power of the cover to bear off electricity from the excited resinous plate.

95. *Electrophorus Manipulation.*—Having raised the cover from off the electrophorus plate (93), so as fairly to expose its surface, we proceed to excite it. Supposing the electrophorus plate to be resinous, this is best effected by gentle friction with woollen stuff, or by dry soft silk, as in whisking a dry soft silk handkerchief across it. Both the woollen stuff and the silk, together with the electrophorus plate, should be warm, dry, and perfectly free from moisture. The woollen stuff may be a piece of warm dry flannel, loosely formed into a roll about a foot long; one end of the roll being held in the hand and the other extremity swung round over the surface of the plate in an oblique direction with a quick turn of the wrist, so as to have repeated contacts with the plate in a way varying between a rub and a blow. This will excite the plate to a considerable extent. The cover P, held by its insulating handle, is now placed on the excited surface. If a conducting communication be made between the cover and the sole by means of the thumb and forefinger, a slight electrical shock will be felt, and a small spark will pass, owing to a combination of the opposite electricities. If the knuckle or a ball be presented to the cover, a strong electrical spark will be elicited in the way already explained (92). If the cover be now raised by its insulating handle from off the excited surface, then, as in the single electrophorus, the same identical spark which had been communicated from the knuckle to the cover will be returned to the knuckle, or any conducting substance presented to the cover. If the electrophorus plate be resinous, we obtain in this way a succession of positive sparks. If the electrophorus plate be vitreous, a succession of what may be termed negative sparks will be obtained in a similar way. The whole operation, as is evident, is an affair of induction.

The electrophorus is a convenient instrument for obtaining sparks of positive or negative electricity, or minute electrical charges, at any required moment. In the case of minute electrical charges, we employ extremely small electric discs, arranged as in the single electrophorus (92). The cover may be of gilt wood. The electrophorus disc may be of the best ordinary sealing-wax. Minute electrophori of this kind will always supply positive or negative electricity, to a greater or lesser amount, for electrical investigation, more especially in the employment of the gold-leaf electroscopes. Instruments varying from 1 inch to 5 inches in diameter, may be advantageously resorted to. An electrophorus from 10 to 20 inches in diameter, if well constructed, has considerable power. The celebrated Volta, Professor of Physics at Pavia, was an original inventor of this instrument; an account of it was published by him in 1775. The Professor of Physics at Stockholm claims, however, to have been the first inventor in August, 1772; as may be seen in the memoirs of the Stockholm Academy. Lichtenberg, in 1777, constructed an electrophorus of gigantic dimensions, and of extraordinary power. His electrophorus plate was 6 Paris feet in diameter. It consisted of common rosin, turpentine, and Burgundy pitch, melted together, and poured whilst in a fluid state upon the surface of a sole (93), up to the given thickness. When cold and solid, the surface was smoothly polished by the usual mechanical processes. The cover of Lichtenberg's electrophorus was well rounded at the edges, had a diameter of 6 Paris feet, and was lifted off and applied to the electrophorus plate by means of a suspension pulley. When this electrophorus was freely excited and in action, sparks are said to have been obtained from it 15 inches long, and of such density and power as scarcely to be endured.

Several gigantic instruments of this kind were also constructed by the Germans; one, 9 feet in diameter, from which sparks are said to have been obtained 13 inches in length, as thick as the little finger.

96. We owe to Professor Phillips a convenient modification of the compound electrophorus. His electrical plate is perforated in three or four points, and short pieces of brass wire inserted in the perforations, so as to be level with its opposite surfaces, and form a communication between the cover and the sole, whenever the cover is applied to the excited plate. The object of this construction is to communicate, by means of the knuckle or other conductor, a supply spark to the insulating cover every time the cover is raised off the surface. In this arrangement the cover is supplied whilst in place through the conducting communication with the sole. The cover may in this way be charged from 50 to 100 times in a minute, merely setting it down and lifting it up as fast as the operator can work. This instrument was about 20 inches in diameter, the cover about 16 inches. It emitted flashing sparks of electricity full 2 inches long.

97. *Electrophorus Electrical Machine.*—Herr Holtz, of Berlin, availing himself of the principle of the electrophorus, has applied

it in the construction of an electrical machine acting not in the
usual way, by simple friction, but by mere induction, he being
under an impression not only that such machines were possible,
but that with a moderate expenditure of force they would exceed
ordinary machines in quantitative effects.

Holtz's induction machine consists of two circular glass discs
(Figs. 61, 62), one of which, A, Fig. 61, is about 16 inches in
diameter, and is mounted on an axis, a, accurately centered, after
the manner of an ordinary plate electrical machine. It is made
to revolve rapidly by means of a multiplying apparatus, at the
distance of about one-eighth of an inch from a stationary plate, B,
Fig. 62, of window-glass, about 2 inches larger in diameter. Both

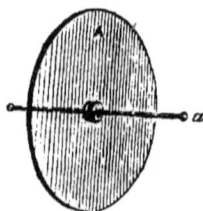

Fig. 61. Movable Disc.   Fig. 62. Fixed Disc.

discs are coated with shell-lac varnish. The fixed plate B has a large
central hole, to allow the axis a of the rotating plate, A, Fig. 61,
to pass freely through and travel clear of it, and admit of the
two plates being set at one-eighth of an inch distant from each
other, or at any other given distance apart. B is also furnished
with two openings or windows, $o\ p$, near which are attached two
surfaces of paper, $m\ n$, with a point from each projecting into the
plane parallel to the openings $o\ p$.

The principle upon which this machine works pre-supposes, as
in the case of the electrophorus, a certain amount of previously
existing electrical excitation in the rotating plate A; which is
effected by any ordinary means, as by holding an excited glass
tube or a stick of resin near the paper points, $o\ p$. On setting
the rotating disc in motion, induced electricity is evolved from
beneath the paper coatings, m, n. Two insulated metal rods, or
conductors, furnished with collecting points, are set before the
rotating plate A, parallel to the axle a, Fig. 61. The points are intended to take up the induced electricity as rapidly as it is evolved,
and give it off under the form of powerful current discharges.

Holtz's machine is mounted on a strong insulating rectangular
frame, with insulating pillars and cross bars (Fig. 63). The striking distance of a spark from this machine is said to vary from

1 to 4 or 5 inches. From three to four discharges are reported to be obtained in one second.* This machine is of ingenious and scientific construction, but is somewhat precarious and tedious in practice. It is necessary to clean it every time it is used, and

Fig. 63. Holtz's Electrical Machine.

to free the rotating disc from the dust which accumulates on it, in consequence of the decomposition of the resinous varnish with which the glass is coated. It is hence questionable whether an ordinary plate machine of Cuthbertson's construction (74) is not more simple and efficient.

98. *The Condenser.*—Analogous to the electrophorus is the electrical condenser, the object of which is to render sensible to the electroscope minute quantities of electricity, which, without its aid, would be inappreciable.

It has been already remarked, with reference to the compound electrophorus, that the increased power derived from the sole is referable to the inductive action between an insulated and an uninsulated conductor (94). Such is, in fact, the principle of the electrical condenser, a contrivance originating in the ingenious experiments of Volta, which led to his memoir " On the advantages of a kind of imperfect insulation." He remarks that the Marquis Bellisoni, having by chance placed his charged electrophorus disc on a table covered with animal skin, and after a short time raising it up, was astonished at obtaining a spark from it. This fact he communicated to Volta, who instituted some further researches bearing on the subject, and eventually proposed the following problems, which he entitled " Electrical Paradoxes."

* A detailed account of this instrument is given in the *Philosophical Magazine* for December, 1865, p. 425.

1. To cause any conductor to retain electricity whether insulated perfectly or imperfectly, or not insulated at all.

2. To accumulate a much greater quantity of electricity on a conductor imperfectly insulated, than can be effected by means of the most perfect insulation.

3. To cause a metallic conductor of small dimensions to retain a charge to a small extent, although repeatedly touched by the finger or a metallic body for twenty or thirty seconds.

4. To avoid a total dissipation of an electrical charge imparted to a conducting body, although the conducting body remains in contact with a neutral conductor for a considerable time.

Volta's experiments go to show that a conductor may better acquire the power of retaining a charge when in contact with an imperfectly insulated body than when perfectly insulated. Thus a conducting plate, M, Fig. 64, suspended by silk lines, and flatly applied to a marble table, N, does not yield up all its electricity, but will, after being removed from the marble, still give a spark, even although the contact with the marble surface be continued for a considerable time. This is

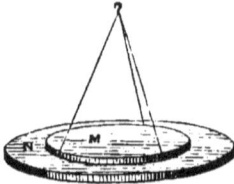

Fig. 64. Volta's Paradox.

what Volta calls a *paradox*. The more extensive the contact, and the better the surfaces apply to each other, the more perfect the experiment. It is only when we touch the table with the edge of the plate that the electricity leaves it. If the electrified plate be insulated in free space in the air, all its electricity disappears in a few minutes, whereas if you place it upon a dry piece of wood or marble the electrical charge remains. Again, if the charged plate M be placed upon a cake of brimstone or resin, still its electricity leaves it, although these bodies may be considered as electrical insulators (15); and the more perfectly the electrified plate M is insulated, the more rapidly will its electricity disappear. Volta makes a great difference between this and the action of the electrophorus (92). The electrophorus disc is never in action unless its electrical surface be first excited ; the excited electricity in this case dissipates but slowly, but remains adherent on the surface as if pasted on it. The case of the charged metallic disc M, just alluded to, is different. If it be made to repose upon a free metallic plate, or upon a plate of wet wood or moist marble, its electricity is soon dissipated.

99. Volta inferred from these results that the circumstances under which a charged conductor can best preserve its charge is not that of perfect insulation, but, on the contrary, a state of such

imperfect insulation that we can scarcely consider it as being insulated at all. Volta further observed that imperfect insulation enables a conductor to absorb or take up electricity more readily than perfect insulation. The effect of a semi-insulating plate upon a metallic plate is to restrain any free electricity communicated to it, and so, by diminishing its intensity (62), enable it to receive under a given electrometer indication a much greater quantity. On separating the metallic plate from its semi-insulating base, the increased charge is immediately apparent (49).

100. The condenser, therefore, consists simply of a semi-conducting disc, M, Fig. 64, reposing on a semi-insulating plate, N, the disc M having an insulating handle, or be suspended by insulating silk lines. If now a very weakly electrified body, charged with an infinitesimal quantity of electricity, 'quite inappreciable by the most delicate electroscope, touch the insulated metallic plate M, the imperfect insulating plate N will enable the metallic plate M so to absorb every particle of electricity in the weakly charged body, as to cause the metallic plate to abstract the whole of it, or nearly so; which otherwise, without such influence, it could not do,—some portion must still always remain. The semi-insulating plate, in fact, increases by induction (41) the electrical capacity of the metallic plate, by which any small amount of electricity imparted to it is at the instant rendered, as it were, insensible to the electroscope. Directly, however, the conducting disc M is raised in an insulated state off the surface of the semi-insulated base on which it rests, the small quantity of charge it had absorbed whilst in contact with the semi-insulating surface N becomes (in accordance with what has been already stated, 98) immediately sensible. Such is the operation of the electrical condenser.

101. Volta gives a list of semi-insulating bodies best adapted to the purpose of a condenser. He prefers marble. But all kinds of marble do not succeed equally well. He thinks Carrara marble the most efficient. Some varieties of alabaster also answer the purpose as semi-insulators. All these substances, however, must be very clean and dry, not only on the surface, but within the substance. Volta tried ivory and other kinds of bone, but not with the same success. He found nothing better than plates of wood previously steeped in oil and well dried by heat. A table covered with white wax, velvet and silk stuff stretched on a table, were found efficient as semi-insulators.

102. The electrical condenser, as more recently constructed, dispenses with a semi-insulating substance altogether, and consists

of two metallic plates, P Q, Fig. 65, one of them, P, being insulated
on a light rod of glass, r, the other, Q, uninsulated, it being fixed
on a brass rod, t, supported on a sliding piece, v. The two plates
are set quite parallel to each other, and are brought, by means of
the slide v, as near together as may be without touching. The uninsulated plate Q supersedes, in this way, the necessity of a marble
or other semi-insulator, there being a small stratum of air interposed between the two plates. When we desire to detect the
presence of an infinitesimal quantity of electricity in a given body,
contact is made with the weakly charged body and with the insulated
plate P, through a small projecting ball p; the minute quantity of
electricity in the weakly charged body becomes immediately
absorbed and condensed by the influence of the approximated
uninsulated plate Q. The plate Q is now withdrawn by means
of the slide v, at a distance from P; the condensed electricity in P
then becomes immediately sensible to a delicate electroscope. The
rod supporting the uninsulated plate Q has occasionally a joint
and stop, w, at its lower extremity, which admits of the plate Q
being turned back from P without moving the slide v. This may
be termed the *single condenser*. If the insulated plate P, Fig. 65,

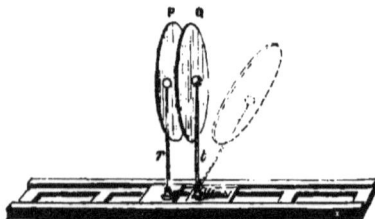

Fig. 65. Single Condenser.   Fig. 66. Double Condenser.

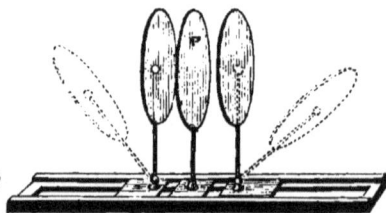

be placed between two uninsulated plates, the plates being separated
by an extremely small interval of air or other dielectric, the
condensing power is greatly increased. This may be called the
double condenser, and is shown in Fig. 66.

103. Cavallo extended the condensing power still further by
transferring the condensed electricity in a first plate to the
insulated plate of a second condenser. The insulated plate of the
second condenser is generally attached to the cap of the double
gold-leaf electroscope (33). The uninsulated plate opposed to the
plate on the electroscope is fixed on a light brass rod, having a
joint and stop attached to a slider connected with the foot on which
the whole is supported, as seen in Fig. 67. The diameter of the
condensing plates in an instrument of large size may vary from 5

to 12 inches: the auxiliary condenser attached to the cap of the gold-leaf electroscope need not exceed 2 inches.

104. Many eminent electricians of past days, not content with this increase of condensing power, endeavoured to arrive at still more delicate arrangements, which, although evincing great ingenuity and wonderful sen-sibility, tend to produce spon-taneously the electricity they were designed to detect, and are hence liable to give rise to equivocal results. We may, for example, refer to Bennett's *Doubler* of electricity, Cavallo's *Multiplier*, Nicholson's *Spin-ning Condenser*, Wilson's *Double Multiplier*, and such like.*

Fig. 67. Cavallo's Condenser.

105. A simple and efficient condenser may be easily constructed by dropping three small spots of sealing-wax on the lower face of an electrophorus cover (93), by which it may be supported at a short distance above the even surface of an ordinary table. We have in this arrangement a metallic plate under the influence of a semi-insulator. If now a weakly-charged body, insensible to a delicate electroscope, be made to touch the metallic plate, and the plate be then raised off the table by its insulating handle, the abstracted electricity immediately becomes sensible directly the plate is applied to the gold-leaf electroscope. Bennett's doubler is said to multiply an infinitesimal quantity of electricity 500,000 times. It is evident, however, that no instrument of this descrip-tion can actually multiply the electrical agency. Its power is necessarily limited to rendering sensible to the electroscope a small quantity of electricity actually existing, which, under ordinary circumstances, is inappreciable.

106. Although the condenser may be considered as an instrument more especially adapted to the rendering very small quantities of electricity sensible to the electroscope otherwise inappreciable, yet the converse of this is not by any means apparent; it does not follow that the condenser renders quantities of electricity, sensibly affecting the electroscope; insensible. It can only diminish the intensity to a certain point. The condensing or inductive power of the semi-insulated plate has a limit, so that the electroscope in

* An account of these instruments will be found in the 77th and 78th volumes of the Phil. Trans., for 1787 and 1788; Cavallo's "Treatise on Electricity;" "Nicholson's Journal," 4to., vol. i. p. 16; and in the article ELECTRICITY, by Sir David Brewster, in the eighth edition of the "Encyclo-pædia Britannica."

communication with an insulated electrified conductor, will always have a certain degree of divergence, the condensing induction of the semi-insulated plate decreases in some inverse ratio of the electricity communicated to the insulated plate, as is, in fact, found by experiment (41).

107. *The Electrical Jar, or Leyden Phial.*—If the plates of the condenser (102), instead of being of sensible thickness, and brought extremely near each other, separated by a thin stratum of air, were to consist of thin metallic leaf, such as tinfoil, and be separated by some solid dielectric, such as glass, we then arrive at an electric combination productive of marvellous effects. For example, let A B, Fig. 68, be a square of common window glass;

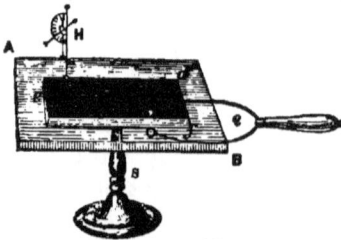

Fig. 68.　Coated Pane.

c d, e f two squares of tin leaf of less size than the glass, smoothly attached to the opposite surfaces of the glass by thin strong paste, leaving a wide insulating margin e d f all round in order to prevent the opposite edges of the tin leaf surfaces from touching each other. We have then what has been termed *a coated pane*; the opposed metallic surfaces c d, e f being termed the *coatings*. With a view of rendering the insulation of the edges of the two opposed coatings c d, e f more complete, the uncoated interval or margin of glass, e d f, separating them, is covered with a thin varnish of shell-lac dissolved in naphtha and dried off by heat. If this coated pane be placed upon a conducting stand, s, one of the coated surfaces, e f, being connected with the ground, we have virtually another form of condenser, the upper plate c d being the insulated or accumulating plate, e f the condensing, or what may be termed the semi-insulating, plate. We may, therefore, in a similar way accumulate or condense upon the insulated surface c d a considerable quantity of electricity not possible to so accumulate without the aid of the lower plate e f. In this arrangement, as is evident, we have an insulated conductor, c d, opposed to a semi-insulated conductor, e f.

108. Thus circumstanced, if we communicate to the upper insulated surface c d a given quantity of electricity, while at the same time a Henley's electrometer, H, is standing on the insulated plate c d, the ball of the electrometer will be repelled to a given distance by the uncondensed electricity on the charged plate, and the two plates, c d, e f, will exhibit opposite electrical states or

forces ; that is to say, if the upper surface or insulated conductor
c d be positive, the opposed surface or uninsulated conductor e f
will be negative when freed from the influence of the positive
coating, by which its negative electricity is held as it were
in abeyance, much in the same way as the cover of the excited
electrophorus emits an electrical spark, directly it is raised off
the surface of the excited resinous plate (92), being then set free
from the inductive action of the excited electric surface.

Exp. 35. Let A B, Fig. 69, represent a coated circular plate of
glass of about a foot or more in diameter ; P N the two coatings,
the whole being mounted on a convenient supporting foot s, as
shown in the figure.   p, n are two delicate reed electroscopes, freely suspended along the opposite coatings P, N.

If, in this arrangement, we connect one of the coatings
N with the ground, by means of a metallic wire, we have,
as in the case of the coated pane (107), an insulated conductor P, opposed to an uninsulated conductor N ; and all
the elements of the condenser (100) realised in a vertical
position. Communicate to the insulated coating P a series
of electrical sparks, suppose positive ; the electroscope *p*

Fig. 69.
Coated Pane.

of the charged side P will diverge, whilst the electroscope *n*, in connection with the uninsulated surface N, will appear
neutral, although the two coatings P, N are really in opposite electrical states ; the coating N being actually negative, its negative
electricity, however, is, as just explained, held in abeyance by the
uncondensed positive electricity on the opposite surface P, operating
through the glass ; hence the negative state is not sensible to the
electroscope *n*. Remove now the connection of the coating N with
the ground, and we have the whole system insulated. Withdraw
from the surface P of accumulation a portion of its charge (which
is easily done by taking a spark from it by means of the knuckle),
the electroscope *p* immediately falls back to the surface P, whilst
the electroscope *n* diverges with the opposite or negative electricity.   If now in a similar way we withdraw a negative spark
from the coating N, we in like manner liberate positive electricity
from the surface P, previously held in abeyance by the inductive
influence of the negative electricity of the coating N, so that
the electroscope *n*, now divergent, falls back in its turn, the electroscope *p* again rising, and thus, by withdrawing a spark, first
from one coating, and then from the opposite coating (which can
always be done so long as the system remains insulated), the
electroscopes *p n* may be kept in alternate reciprocal movement.

Hence, the two coatings P, N are really in opposite electrical

H

states, and under the influence of each other. The coated pane
A B, thus electrically effected, is said to be *charged*.

109. The principles just adverted to involve the theory of *free*
and *condensed*, or *simulated* and *dissimulated* electricity (38), as set
forth by the French philosophers. Faraday calls this theory in
question as conveying erroneous views, if by the terms *free* and
condensed electricity we mean to imply any difference in the
electrical agency itself, the one not being really more free or
condensed than the other. But although Faraday's objections
are just, and consistent with the nature and operation of electrical
force, yet the employment of the terms *free* and *concealed* electricity are not only admissible, but convenient, when limited
by definition, and their meaning and application clearly apprehended. We may, for example, in the case of the condenser,
or the coated pane just alluded to (108), fairly term that portion
of the accumulation not sensible to the electroscope *concealed*
electricity, or *électricité dissimulée*, and that portion operating
more immediately on the electroscope, not so concealed or com-
pensated, as it were, *free* electricity. It is certainly allowable to
draw a distinction between that portion of a charge not sensible
to the electroscope, and that portion by which the indications of
the electroscope are more especially affected.

110. If the electricity communicated to the accumulating surface
(107) (which may be either of the coatings), the opposite coating
being uninsulated, be positive, the pane is said to be *charged
positively;* if negative, the pane is said to be *charged negatively.*
If the opposed metallic coatings of the pane, being charged
either positively or negatively, be joined by a bent conducting-
wire, the opposite electricities unite with a loud brilliant
explosion, resembling in a minor degree the phenomenon of
thunder and lightning. The pane is now said to be *discharged*,
and has been termed by the French philosophers, in consequence
of this explosive effect, a *fulminating square*. If we bring the
finger of one hand in contact with the uninsulated coating, and
a finger of the other hand in contact with the insulated coating of
accumulation, a painful sensation through the arms and across
the breast is experienced, termed the *electric shock.*

111. We may conclude, from the operation of the coated pane
(108), that in charging or discharging a coated surface, the two
opposite electricities or forces are so related, that as much of
either force as we add to or subtract from the one coating, we
subtract from or add to the opposite coating, an important elementary fact, first announced by Franklin.

Exp. 36. Let a coated circular pane, q, Fig. 70, be mounted on convenient supports, and a stand, s, between two small metallic balls $p\,n$, placed at short and equal distances from each coating on either side of it. Let the ball p communicate with the positive conductor P of the electrical machine, and the ball n with the ground by means of a wire or chain. Let the machine be now gently set in motion until the ball p sends off a spark to the adjacent coating. The opposite coating will at the same instant send off a spark to the ball n, and this reciprocal action will continue until the pane q be fully charged (108) ; that is to say, until the positive electricity emitted by the ball p has been held in abeyance by the opposite electricity of the ball n.

We see by this experiment that in charging a coated electric, as much of the one electricity as we communicate to one of the coat-

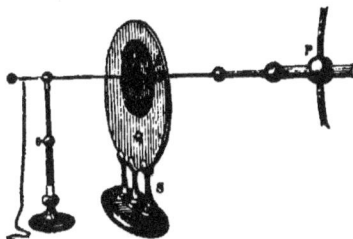

Fig. 70. Coated Pane.

ings, we abstract from the opposite coating, and reciprocally. Suppose, for example, that twenty electrical sparks had been received upon the one coating from the ball p, then twenty equal and similar sparks would be given off by the opposite coating to the ball n, so that the coating opposed to the ball p will be charged positively with twenty sparks, and the coating opposed to the ball n would be charged negatively with twenty sparks. When the two opposed coatings p, n are united by a bent conducting wire in the way just mentioned (110), the whole twenty measures received from the ball p, rush in a dense state of accumulation through the connecting wire upon the coating n from which twenty measures have been abstracted, giving rise to the fulminating effect and electrical shock already adverted to (110). The coated dielectric pane being but another form of condenser (107), we see the necessity of one coating being uninsulated, or in connection with the earth, without which the system cannot be what we have termed *charged*, or in other words, cannot be brought under the dominion of electrical induction. We have, in fact, seen (41) that in opposing two conductors to each other the inductive effect is the greatest possible when one of the conductors is uninsulated or free. In taking alternate sparks from either coating, as in the former experiment, we eventually effect discharge also in successive and alternate portions.

112. The form and extent of the dielectric glass separating the coatings being of little or no consequence to the experiment,

it has been found convenient, instead of a thin flat plate of glass, to employ an ordinary glass jar, coated on its inner and outer surfaces with tinfoil to within a certain distance of its mouth. Such an arrangement has been termed the *electrical jar* or *Leyden phial*, so called from its marvellous electric shock (110), first accidentally noticed by some Dutch philosophers at Leyden, in Holland, whilst endeavouring to electrify water enclosed in a common phial; a discovery which caused much astonishment throughout Europe.

113. The most approved form and dimensions of the electrical jar are represented, Fig. 71. *m b* represent a jar of clear glass of almost any shape or size; it is, however, generally a hollow cylinder. Its height from the base *b*, to the mouth *m*, should be about 20 inches; diameter, *a d*, one-half its height—10 inches; width of mouth, *m*, one-quarter its height—5 inches; height to shoulder, *s*, about a diameter and three-fifths—16 inches; thickness of glass, something less than the one-tenth of an inch. The jar should be rounded and somewhat compressed at the shoulders,

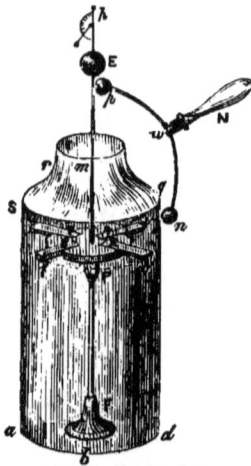

Fig. 71. Leyden Jar.

rising gradually in a flattened curve to the mouth. It is coated internally and externally with tin-foil up to the shoulder s, the coatings being attached to the glass by strong thin paste, and smoothly rubbed down upon its surface. The jar when coated generally exposes about 4 square feet of coating. The insulating, or, as it is termed, the uncoated interval, *q r*, separating the edges of the internal and external coatings, is varnished with a filtered solution of shell-lac dissolved in naphtha.

114. Electricity is communicated to the jar from the electrical machine or other source of electric accumulation, by means of a light metallic tube, E F, termed the *charging rod*, passing through the mouth of the jar to its internal coating. As it is often desirable, especially in employing single jars, to vary the length and extent of the charging rod, it is convenient to have the rod in two separate portions, E P, P F, which consist of two light drawn metallic tubes, sliding with friction one within the other; one of these, P F, being securely held in a wooden foot, F, covered with tinfoil, is a fixture within the jar. The foot F is strongly glued to the internal coating at the base of the jar. This fixed portion of the charging rod passes completely through the foot,

so as to touch the tinfoil beneath, and extends from the foot to a little above the height of the internal coating; where it is further steadied by passing with friction through a similar inverted foot, P. This upper support, P, is furnished with four light cross-arms, t, u, v, w, covered with tin-leaf, inserted in four separate parts, in corresponding grooves, cut for their reception in the upper surface of the inverted foot P, which allows of their being easily placed in position through the mouth of the jar, and extended within it so as to touch the internal coating at opposite points, t, u, v, w, of its circumference. The cross-arms of support are secured by means of long slots cut in them, and small metal screws passing through the slots into the grooves beneath. We are thus enabled to give the cross-arms any required amount of lateral extension, and pressure against the internal coating of the jar. The remaining portion P E of the charging-rod slides freely within the fixed portion, so as to project beyond the mouth of the jar to any required extent, its upper extremity terminating in a charging ball E. Several fine holes are drilled in the projecting portion of the charging-rod for attaching light and finely pointed tubular rods.

The communicated electricity condenses on the inner coating by the inductive influence of the external coating operating through the intervening glass (108); with this view, and in order to satisfy the conditions of the semi-insulated plate of the condenser, the outer coating must be free, the resulting phenomena being precisely the same. In charging a coated jar, as the charge begins to accumulate, it is made evident by means of a Henley's electrometer, h, usually fixed on the ball E at the extremity of the charging-rod. As in the case of the revolutions of the electrical machine (86), there is a limit to the quantity of electricity which the jar can receive. On approaching the maximum or limit of charge, the electrical sparks from the prime conductor begin to subside, and the uncondensed or accumulated electricity (86) flies off in luminous coruscations, and a bright electrical discharge frequently ensues between the coatings over the intervening glass, or uncoated interval. This spark constitutes the phenomenon known as *spontaneous discharge*. In all experimental inquiries with the electrical jar, therefore, it is requisite, as in the case of the revolutions of the electrical plate, to determine the limit of charge (86) of which the jar is susceptible, and not to press the accumulation beyond that limit.

115. If the charging-rod communicate with the external coating by means of a bent wire, w, Fig. 71, an explosion takes place, and the jar is said to be discharged. This bent wire w, or other

conducting communication between the coatings is termed a *circuit.* When mounted on a glass handle, N, with a joint in the centre, and terminating in small balls, *p, n,* the circuit is called a *discharging rod.* The glass handle N shields the hand from any shock incidental to the discharge, whilst the joint at *w* admits of the circuit being extended to any convenient distance between the discharging balls.

116. Such is the tendency of the two electricities developed on the opposite coatings to unite, that absolute contact between the balls of the discharging rod and the opposed coatings is not requisite for effecting the discharge of the jar; it is enough if one ball of the discharging rod, usually the negative, first touch the negative coating, whilst the other ball be brought *near* the charging rod of the jar. In such case a brilliant explosive spark breaks through the interval separating the points of discharge, and the opposite electricities immediately unite with a brilliant explosion (110).

The jar being charged, its inner and outer coatings, as in the case of the fulminating square (108), are in opposite electrical states; that is to say, if one exhibit positive, the other will exhibit negative electricity. Both states may be made evident by placing the jar on an insulating base, in which case we may demonstrate experimentally, as in the case of the coated pane (111), that as much of either force as is accumulated *on,* or abstracted *from* the one coating, is abstracted from, or accumulated on the opposite coating. For example, let each coating in a neutral state of the jar be supposed to contain a quantity of electricity represented by the number 20, that is to say, let each coating be supposed to contain a quantity represented by the number 10. When the jar is fully charged, the whole quantity, 20, is accumulated on one of the coatings; 10 being taken from one coating and added to the other, when the jar discharges through any circuit, the electrical equality between the coatings again obtains, and the quantity 10, taken from the one side, rushes through the discharging circuit (115), and restores the original distribution. The circuit which enables this phenomenon to occur may be of unlimited extent, and may consist of any given kind of substance. If there be many such circuits of discharge, the jar in the act of discharging will affect them all. On applying one of the balls of the discharging-rod first to the outer coating, suppose negative, and then bringing the opposite ball towards the ball E of the charging-rod, suppose positive, the explosive spark of the discharge takes place before the two balls, as already observed (116), actually touch; a phenomenon arising from the attractive force between the opposite electricities, and their tendency to combine (27). Hence if two conducting balls, one connected with the charging-rod, or inner coating of the jar, the other connected with the outer coating, be directly opposed to each other at a given distance, an explosive discharge will ensue between the balls directly the electrical accumulation on the positive coating of the jar has reached a given point, and thus a measure of the charge may be obtained.

117. When a number of electrical jars are joined so as to operate as one whole, we have an *electrical battery*, the construction of which has been usually of a complicated and expensive character, and not always effective. The simplest method of forming an electrical battery is to group about a central jar, c, Fig. 72, upon a square or circular base, A B, covered with tinfoil, a given number of coated jars. All the charging-rods of the group are to be united by means of light conducting wires radiating from the charging-rod of the central jar, so that the whole group may be charged and discharged from the central rod as a single jar. We may in this way, by grouping together nine jars, numbered from 1 to 9 (Fig. 72), easily obtain a battery exposing 36 square feet of coated glass, within a circular space the diameter of which is 30 inches, each jar being similar to the jar already referred to

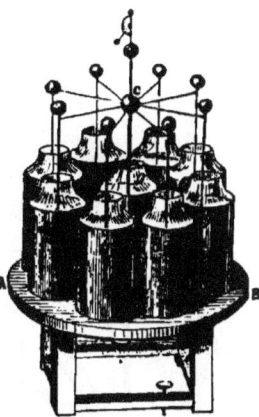

Fig. 72. Electrical Battery.

(113), and exposing 4 square feet of coated glass. By connecting together two or more such groups, a battery of almost any extent may be constructed. The great battery in Teyler's Museum at Haarlem consisted of 135 jars, each jar exposing something less than 1 square foot of coating. The explosion from this battery melted an iron wire 25 feet long and $\frac{1}{50}$th of an inch in diameter.

118. In constructing a battery in the way just described (117), the conductors connecting the charging-rods of the jars should consist of small brass tube, about one-eighth of an inch in diameter, with solid, pointed, sliding wires, projecting from each extremity of the tube, so as to allow of the rod being elongated or contracted to a given distance ; small holes are drilled in the charging-rods for the reception of the pointed connecting-wires. The base A B on which the jars rest is supported on varnished glass columns.

The whole battery, therefore, may be insulated, if required, on withdrawing the conducting communication with the earth. It is evident that by connecting, by means of the discharging-rod (115), the central ball c, in which all the charging-rods terminate, with the tinfoil base on which all the jars rest, we may discharge the whole system as a single jar (Q).

OCCASIONAL MEMORANDA AND EXPLANATORY NOTES.

(O) The cylinder machine may be of any given dimensions, according to the power required. A cylinder 10 inches in diameter and 12 inches in length between the shoulders, gives considerable power in a machine. A cylinder 1 foot in diameter and 13 inches in length between the shoulders, is extremely powerful. A cylinder of 7 inches in diameter and 9 inches in length, with a rubber of 8 inches, is very efficient.

Fig. 73 represents the cylindrical machine as constructed by Franklin, the cylinder P being about 7 inches in diameter and 9 inches in length between the shoulders. Franklin's machine was set up in a firm frame, ff', and was turned by a wheel and band, as shown in the figure.

Fig. 73. Franklin's Electrical Machine.

(P) In 1840, an intelligent workman in charge of a steam-engine at Sighill, near Newcastle, noticed a considerable escape of steam, arising from a leak in the cement about the safety-valve. The engine man being about to adjust the weight of the valve, was surprised by the emission of a powerful spark of electricity, which he found to proceed from the metal work connected with the boiler; as also from the boiler, if he attempted to touch it during the escape of steam, especially if one of his hands were immersed in the vapour. Mr. Armstrong, a scientific gentleman at Newcastle (now Sir William Armstrong), having been informed of this result, lost no time in investigating this new phenomenon.

By means of an insulated brass rod, with a metallic plate at one extremity, and a ball at the other—the latter placed near the boiler, and the plate in the issuing vapour—he obtained sixty or seventy sparks per minute. After a series of interesting inquiries, he succeeded in constructing a new form of electrical machine (87), depending on the excitation of particles of water driven by steam through small orifices.

(Q) The largest jars which the glass-blowers of former days could conveniently make, were 17 inches in height, and were seldom more than 4 inches in diameter. They were thin and very liable to fracture from spontaneous explosion (114). An electrical battery of the old construction was a hazardous and expensive piece of apparatus. It usually consisted of a large number of jars of variable dimensions and thickness, arranged within a rectangular box, Fig. 74, lined with tinfoil. Each jar had a

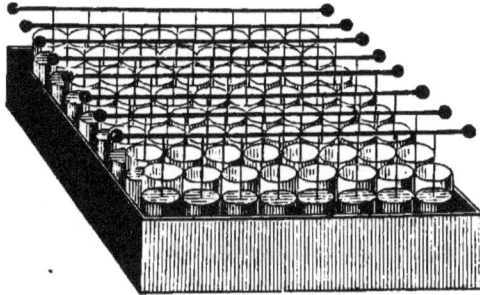

Fig. 74. Electrical Battery.

varnished, wooden cover, with a brass charging-rod, connected by a short brass chain with the internal coating of the jar. The charging rods were all connected together, as represented in Fig. 74, by cross rods, nicely polished and lacquered. From the great number of connecting rods and balls, &c., &c., employed in the construction of an electrical battery of the olden time, this piece of electrical apparatus was necessarily very complicated and costly. The battery employed by Dr. Priestley consisted of sixty-four jars, each jar 10 inches in height, about 3 inches in diameter, and coated to within 1½ inch of the top, exposing 72 square inches, or half a square foot each of coating. The first battery constructed by him consisted of forty-one jars, each 17 inches in height and 3 inches in diameter, coated to within 2 inches of the mouth, each jar exposing a square foot of coating. The whole battery, therefore, exposed 41 square feet, an amount of coated surface which could be very well grouped under an arrangement of 10 jars, in the manner described (117). The jars thus arranged, and grouped together in a box, were very liable to fracture in consequence of spontaneous explosion, by which the whole battery was rendered useless, and consequently involved a costly reconstruction. The simple method described (117) is not only efficient but economical, and easily managed, involving no complication whatever. It is evident, as remarked (118), that by connecting all the charging rods of the jars with the tinfoil lining of the box, Fig. 74, by means of the discharging rod, the battery will be at once discharged as a whole through any given circuit. It is clear that in order to discharge any battery we only require to unite the outer and the inner coatings by a metallic communication.

Let, for example, N, Fig. 74, be a projecting rod or wire passing through the box, so as to touch the inner tinfoil lining upon which each jar rests. If we connect one of the charging balls, by means of the discharging rod, with the projecting wire N proceeding from the tinfoil lining of the battery, the whole system discharges in the manner just observed (118).

CHAPTER IV.

119. *The Unit Jar.*—It was one of Franklin's great facts that in charging the electrical jar, as much of the one electricity as is caused to accumulate on the one coating, the same quantity of the other electricity is developed on the opposite coating (111).

In accordance with this great principle, the author of this work proposed to interpose between the prime conductor and the battery or jar to be charged, a small insulated coated jar, exposing a few square inches only of surface; connecting the inner coating of the small jar with the prime conductor of the machine, and the outer coating with the jar or battery to be charged, or reciprocally, so as to charge the battery with the electricity given off from the opposite coating of the interposed jar. When the small jar becomes charged to a given point or height, an explosion or discharge ensues between the two coatings of the interposed jar, by means of two small exploding balls, and marks the reception by the battery of one charge or unit of the small jar. Hence this piece of electrical apparatus has been termed a *unit measure*, each of its explosive discharges being considered as a unit of charge of a given magnitude, taken in terms of the distance between the two small exploding balls of discharge.

120. The unit jar or measure, Fig. 75, is a small electrical jar about 4 inches long, six-tenths of an inch in diameter, and one-twelfth of an inch thick, exposing about 5 square inches of coated surface, either under the form (113), or straight and open-mouthed. The unit jar is constructed of thin-drawn glass

Fig. 75. Unit Jar.

tube, hermetically sealed at one end, varnished both internally and externally with a thin solution of shell-lac. This unit measure is carefully prepared; the coatings are well rubbed down upon the surface of the glass, and silvered over. The charging-rod *c c'* is a small silvered metallic tube, one-fifth of an inch in diameter, and consists of two portions, movable one within the other, in the manner already described (114). One of these, *c*, is a fixture, within the jar, and projects for about an inch beyond its mouth. The other portion *c'* slides freely within this fixed portion so as to admit of extension or contraction to any required amount. The fixed portion carries upon a sliding piece, *s*, a small exploding ball, *p*, about three-tenths of an inch in diameter; an opposite similar exploding ball, *n*, is fixed on the extremity of a sliding tube moveable within a tubular piece, *t r*, secured by light metallic bands, *m*, *b*, to the outer coating. These exploding balls, *p*, *n*, can, by means of the sliders, be set at any given distance apart as determined by accurate measures or gauges, or a delicate regulating screw, affixed to the tubular piece *t r*, and acting on the ball *n*. The sliding tube of the ball *n*, where it emerges from the tube *t r*, is further graduated into divisions one-twentieth of an inch apart. The base of the jar is capped externally with a brass band, *b*, terminating in a small ball, *a*. The whole is sustained upon a ball and pin attached to the metallic band *m*, so as to admit of turning the jar into any convenient direction.

121. The unit measure is principally employed for measuring the quantity of electricity accumulated on coated surfaces, as in charging the electrical battery, in which case it is either mounted on an independent insulator, and interposed between the machine and battery to be charged, Fig. 76, or is directly sustained on the prime conductor.

In charging a given body, as the charge proceeds, successive explosions between the neutralising balls of the unit jar mark with great precision the number of measures in the battery. The

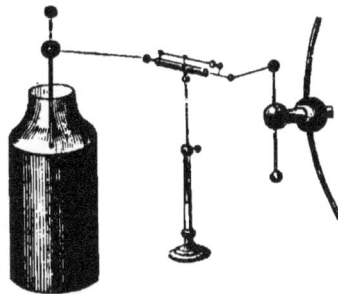

Fig. 76. Position of Unit Jar in charging Leyden Jar.

precise dimensions of the unit jar will of course depend on the requirements of the experimentalist.

122. A small jar exposing about 5 square inches of coated surface will, for general purposes, be found sufficient, although larger units may be occasionally required. If employed to denote small quantities of electricity, unit measures of very small dimensions may be employed, which may be constructed of thin coated

glass tube exposing about half a square inch of surface, or of small thin coated glass plates.

123. Objections have been made to the unit measure similar to those advanced relative to the uniform action of the electrical machine (86), but they are equally inapplicable. It has been urged, for example, that the unit measure does not communicate the same quantity of electricity at each discharge, in consequence of the accumulation in the battery operating upon its outer coating, which interferes with the freedom of accumulation. Such, however, is not the case in the practical application of the instrument. We have already seen (86) that no circumscribed insulated conductor can receive and maintain an unlimited quantity of charge. The limit of accumulative power being determined, every unit discharge within that limit communicates an equal quantity to the jar or battery with which it is connected. This fact is made quite evident by a simple and plain experiment.

Exp. 37. Let the jar A, Fig. 77, be fitted with two exploding balls, P, N; the one, P, fixed on the extremity of a light sliding

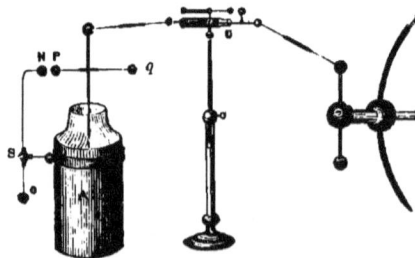

Fig. 77. Self-Discharging Leyden Jar and Unit Jar.

rod, P *q*, passing through the charging-rod of the jar; the other, N, fixed at the extremity of a sliding rod, N o, passing through a slide, s, projecting from a ring on the outer coating. Let the balls be placed and adjusted at successive and equal distances, suppose ·1, ·2, ·3, &c., of an inch apart. Connect the jar A with an interposed unit measure, U, in the ordinary way. The electrical machine being set in motion, note the number of revolutions, and the number of unit explosions required for discharge of the jar at distance ·1, the number required for explosion at distance ·2, and so on. The unit explosions, or the revolutions of the machine (86), are found to be as the distances between the discharging balls P, N, which could not possibly be the case if the unit measure had not been uniform in its action up to the distance three-tenths at least. Let the distance of the discharging-balls P, N be now increased up to a point of distance at which discharge between the balls P, N becomes altogether impeded, or ceases to ensue, in consequence of the jar having reached its limit of accumulation; this immediately determines how far the unit measure may be relied on as a quantitative accumulator.

124. Faraday, who examined this question, arrived at the conclusion that the unit measure is a true indicator of quantity. He remarks:—"Suppose the unit jar has one-tenth the capacity of the large jar or battery to be charged; and that being charged up to its exploding point, it contains ten particles of positive electricity; these ten particles will have then passed into the large jar or battery as a unit accumulation, and none will remain in the unit jar. At this moment the conductor of the machine, the outside of the unit jar, and the charging-rod and ball of the large jar or battery, will be positive to a carrier ball (14). Now although on continuing to turn the machine, the positive state of all these surfaces increases, still the mutual relation of the two exploding balls, and of the inner and outer coatings especially of the unit measure, will be quite the same as before, for no external relation can change their mutual relation, though it may affect the outer coatings both of the charging jar or battery and the unit jar, so that whenever a unit spark does pass between the unit balls, the quantity of electricity passing must be the same because the inductive relation of the coatings to each other through the glass, and the inductive relation of the two exploding balls to each other remain absolutely the same. This is, I think, a rigid consequence of the principles of inductive action."

125. *Lane's Electrometer.*—We have seen (116) that in discharging the electrical jar by means of a discharging rod the explosive spark takes place before the ball of discharge actually touches the positive ball of the jar. A similar result ensues in the case of drawing a spark from the prime conductor of the electrical machine. The celebrated electrician Lane, availing himself of this fact, constructed an instrument for determining the comparative force of an electrical discharge, either under the form of a spark from the electrical machine, or that of an electrical explosion between the opposite coatings of an electrical jar or battery.

126. Lane's instrument (Fig. 78) appears to have been simple and efficient. In this figure M is a pillar about 8 inches high, and as originally constructed is for the most part of baked wood boiled in linseed oil. The pillar is bored through two-thirds of its length for the reception of a brass rod, R, the upper extremity of which carries a brass ball, B, from 2 to 3 inches in diameter. The rod R is movable within the pillar M, and may be set to any given point

of altitude, and finally secured by a stop screw, M. A brass rod,
N H, about a quarter of an inch in diameter, and 6 inches in length,
is cut as a screw with the threads one twenty-fourth of an inch

apart; this rod passes through the
central ball, B, as a nut. One extremity
of this rod carries a well-polished metallic
ball, N, about 1½ inch in diameter, the
other extremity terminates in a milled
head, H, for turning the screw N H. A
graduated rule or scale, S E, divided into
twelve parts, measures the movement of
the screw, and passes out horizontally
under the milled head H; and as the

Fig. 78. Lane's Electro-
meter as originally constructed.

threads of the screw are one twenty-fourth of an inch apart,
each turn of the screw measures the one twenty-fourth of an inch.
A circular plate fixed upon, and movable with the screw rod,
indicates each turn of the screw on the graduated scale S E below.
This plate is divided into twelve parts. The ball N is opposed
to a similar ball, P, projecting from the prime conductor C L of the
electrical machine.

127. The principle of this instrument is simple. The quantity
of electricity in operation will, as Lane remarks, always be
directly as the distance between the balls P, N. If a spark occurs
after four turns of the electrical machine when the balls P and N
are one turn of the screw N H distant, the spark will again occur
after eight turns of the electrical machine, when the balls P and N
are two turns of the screw distant, and so on; that is to say, the
number of revolutions of the machine for each explosion increase
with the distance between the balls P, N.

128. In the simple application of Lane's electrometer, the positive
or exploding ball P is, as just stated, either fixed to the prime conductor, or otherwise insulated and connected with it. By noting the
distance between the exploding balls P, N, together with the number
of explosions in a given time, Lane compared the power of electrical
machines with each other. In applying the instrument to the
explosions of the electrical jar, the ball P is placed in connection
with the knob or positive side of the jar, and the ball N, with the
negative side. The quantity of electricity for each explosion at
a given distance between the balls P, N, is, according to Lane,
proportionate to the extent of coated glass, and the distance between
the exploding balls. Such is Lane's Discharging Electrometer, as
originally constructed.

129. Although the instrument, as at present improved and per-

fected, is deserving of consideration, yet it seems to have been ill-appreciated and not well understood as an instrument of electrical research, electricians having taxed it with defects from which it is certainly exempt. They have complained, for example, of its being liable to premature and irregular discharge, in consequence of particles of dust, or other matter floating in the atmosphere, and accumulating between the discharging balls; also from repeated discharges rendering the surfaces of the discharging balls rough and irregular, and have raised other similar objections to its employment. The author of this work, however, having critically investigated the nature and operation of Lane's instrument, considers these objections, if the electrometer be properly constructed and employed, as unsound and of little moment.

130. The instrument as now revised and perfected by the author, is represented Fig. 79, in which G is a central varnished glass insulating column about 1 inch in diameter, and 14 inches high. This insulating column is fixed in a firm circular foot, F, resting on levelling screws, and steadied by a stout central ring of lead. The column terminates in a strong ebony cap, having a plane surface, about 2 inches in diameter. This insulating column, G, carries a light mahogany stage, S E, upon which are two light varnished insulating glass pillars of support, P' N', about half

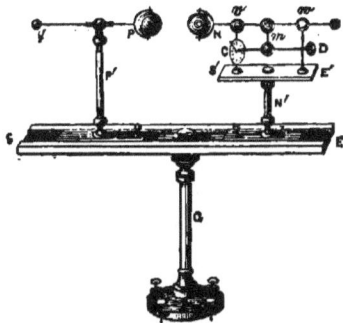

Fig. 79. Lane's Electrometer as Improved by Harris.

an inch in diameter, and from 3 to 6 inches long. These are secured in mahogany sliding bases, a, b, movable on the stage S E. Upon the insulator P' is a brass cap, giving support to a stout horizontal brass rod, P q, about one-eighth of an inch in diameter, and 6 inches long. This rod is movable within a small brass tube, fixed in the upper ball of the cap of the insulator P'; one extremity of this rod carries a finely-polished copper ball, P, about 1½ inch in diameter, easily revolvable within a semicircular, axial, metallic ring. Upon the insulator N' is a lesser varnished mahogany stage S' E', about 6 inches long, and 1 inch wide. Upon this stage are secured two firm vertical metal supports, s', E', between which is the regulating screw C D. The vertical supports s', E' terminate in light metallic balls, v, w, through which slides a plain metal rod, w v, about 10 inches long and one-eighth of an inch in diameter. This rod is acted upon by a central globular nut, m, through which passes a screw-cut rod one-fifth of an inch in diameter, and 5 inches long, the threads of which are one-tenth of an inch apart, and is moveable between the two uprights, s', E'. The movement of the screw is regulated by a milled head, D, fixed on one extremity of the screw-cut rod C D, and by one turn of which the screw is caused to advance or recede one-tenth of an inch. The terminating extremity of the rod $w\ v$ carries a finely-polished copper ball, N, $1\frac{1}{2}$ inch in diameter, similar and directly opposed to the ball P just described; also set in a semicircular axial ring. Balls varying from 1 inch to 5 inches in diameter may be easily and accurately opposed to each other by means of the sliding bases a, b, and caused to present renewed surfaces of explosion by turning them in their axial rings when required. The sliders a, b are steadied in position by tightening screws passing through them. The insulating rod s' carries a graduated circle, divided into ten parts, at the extremity of the screw rod C D, indicating by an index the separation of the balls P, N, to the one-hundredth of an inch.

Lane's Electrometer, Fig. 79, as thus arranged, may be employed to measure the uniform action of an electrical machine (86), in which case the exploding balls P, N, being insulated, are put in communication with the positive and negative conductors through the medium of their respective rods, P q, and $w\ v$. The length and frequency of the sparks may be accurately measured and estimated by means of the opposed balls P, N. In order, however, to obtain a full, round, definite spark for ordinary purposes, the balls should not be less than 4 inches in diameter, nor less than 1 inch apart. Where very minute explosions are required, the balls may be small and near together, but this is seldom needful for ordinary electrical investigation. Gassiot, in applying the principle to small voltaic discharges, constructed a beautiful instrument, by which the distance between two discharging balls may be adjusted to the one-hundredth of an inch, or even less.[1]

131. When Lane's instrument, as thus constructed, is employed for measuring discharges between two balls, P, N, the balls should not exceed $1\frac{1}{2}$ inch in diameter, and seldom be placed more than four-tenths of an inch apart at the utmost. By a little care and attention to the apparatus the exploding balls may be kept free from dust and extraneous matter, and a nicely polished surface of discharge preserved. When employed to measure the explosions of an electrical jar, the ball P is connected with the positive coating through the charging rod, and the ball N with the negative coating (123).

(1) Phil. Trans., 1840.

132. Although this form of the instrument is well adapted for general purposes, the electrometer may be constructed in such way as to be more immediately applicable to the electrical jar, and occasionally constitute a portion of the rod of the jar itself. In this improved electrometer, *a c b*, Fig. 80, is a stout arm of varnished glass rod, about 6 inches in length, and three-tenths of an inch in diameter, bent at *c* to a right angle, each arm, *a c*, *c b*, being 3 inches in length. The arm, *a c*, is cemented at *a* into a metallic socket of support, projecting from a short tubular slider, K. The slider K moves on a tubular rod of brass, *d e*, and with sufficient friction to remain at any required point. The other arm, *c b*, is also cemented at *b* into a socket holder, attached in a similar way to a tubular piece, *h*, within which is a screw movement acting on a short horizontal rod of brass, *f*; this rod carries the polished silver or copper exploding ball N, 1¼ inch in diameter, which is mounted on axial points, as already explained (130). The rod *f*, and the ball N,

Fig. 80. Improved Electrometer.

are acted on by turning the milled head *v*, the degree of motion being measured by an index and graduated circle attached to the tubular piece *h*; one turn of the screw moves the ball N through a distance of one-tenth of an inch; the graduated circular plate being divided into ten parts, we may set the distance to the one-hundredth of an inch. Immediately opposite the ball N is a similar ball, P, set up in like manner on axial points within a semicircular metallic ring, fixed at the extremity of a sliding rod, T P, movable with friction through a short tubular piece on the vertical stem of support *d e*, and terminating in a covering ball of wood. By these several movements the exploding balls P, N may be nicely regulated and adjusted in respect of each other. By the sliding motion at K, the centre of the balls may be accurately placed in the same right line. The horizontal sliding motion of the rod T P enables us to bring the exploding points P, N into contact when the index of the screw motion is at zero. By this motion the exploding points of the balls P, N may be separated by any required interval. Finally, by the axial movement of the balls P, N, we are enabled to renew, as it were, the exploding surfaces, should any abrasion or other defect present itself. The instrument thus constructed is either set upon an independent insulator, I, Fig. 80, and connected with the jar or battery for experiment through the sliding ball T, or terminating ball *d*, or it may

I

be placed immediately on the charging rod of the jar itself, Fig. 81. All the junction edges of the caps, or other joints, are neatly covered by small varnished ebony balls. It being of importance to determine the exact distance of the balls P, N, it is convenient, in addition to the measuring screw within the tubular piece h, Fig. 80, to be prepared with a set of small ivory or other gauges, varying from one-tenth to four-tenths of an inch in thickness; these may be simple oblongs, about 2 inches in length, by 1 inch in width, carefully and accurately measured and prepared, and occasionally inserted between the touching points of the exploding balls for given distances, so as to verify the distance given by the turns of the screw, or for estimating the distance of the balls in the absence of a screw measure.

Fig. 81. Leyden Jar with Attached Electrometer.

133. There are several methods of constructing the screw measure. That represented in Fig. 82, after the manner of what opticians term "the rising sight of a sextant," is neat and compact, the screw part being concealed. We have here three pieces, A, B, C, one within the other; the outer piece A is a hollow cylinder, and fixed; the second piece B moves within the outer cylinder, for which purpose it is drilled, and tapped about two-thirds up its length, constituting a hollow nut about the one-eighth of an inch in diameter; the inner piece C is a small cylindrical piece, having the threads of a screw cut on it to match the hollow cylindrical nut. The inner screw terminates in a projecting square and ball, H, by which the inner screw C may be moved round within the cylindrical nut B, in doing which the piece B, within which the nut is cut, is made

Fig. 82. Screw Measure.

to advance or recede according to the direction in which the screw is moved by the ball H, and it is prevented from turning round in position, within its cylindrical enclosure, by a projecting pin, or by being squared. The movable piece B sends forth a short projection, upon which is fixed an exploding ball, N. By means of a graduated circle and index, s, behind the ball H, the amount of advance or recession of the ball N is accurately measured.

134. We have already observed (129) that electricians in the application of Lane's Electrometer have not resorted to sufficient experimental investigation; they have hence been led to somewhat hasty conclusions concerning it. If we critically consider

the conditions under which the explosive discharges between the balls occur, we immediately detect several sources of error to which it is not unfrequently exposed. It is, for example, quite evident that the exploding balls of the instrument, when in connection with the electrical jar, are really coatings to the air between the balls. The exploding balls of the instrument, when connected with a charged jar or battery, are really coatings of a second charged electric medium; the interposed air, operating between the balls in precisely the same way as the interposed glass between the coatings of the jar. We have, therefore, two coated electrics more or less dependent on each other, charging at the same time the air between the balls, and the glass between the coatings of the jar. Now the dielectric air between the balls being a lesser resisting medium than the dielectric glass between the coatings of the jar, will break down or give way under the accumulation, whilst the glass remains, and thus a circuit of discharge is opened for the quantity actually accumulated at the instant in the jar itself; one of the exploding balls being connected with the charged side of the jar, and the other with the opposite side (128), and thus a relative measure of the accumulation is obtained by noting the distance of discharge between the exploding balls, which allows a given accumulation to take place before discharge occurs, the exploding balls being supposed to be always the same, and always affected in the same manner.

135. *Cuthbertson's Steel Yard Electrometer.* — This instrument was invented by Cuthbertson about the year 1790.

D E, Fig. 83, is a stand of seasoned mahogany, about 18 inches long, 6 inches wide, and an inch thick. A B is a light metallic tubular rod, about 13 inches long, and one quarter of an inch in diameter, terminating in light metallic balls, each about one

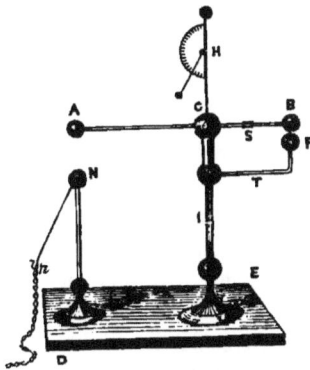
Fig. 83. Steel Yard Electrometer.

inch in diameter. The rod A B is delicately balanced on a knife-edged centre, or axis, the centre of gravity being a little above the centre of motion. The beam and its axis are sustained on a strong glass central rod I, through the medium of a hollow metallic ball, c. The ball c is made up of two hemispheres one immediately over the other. The lower hemisphere is fixed on a brass cap terminating the insulating rod of support

I 2

1.   The upper hemisphere is turned with a groove, and shuts closely down upon the lower hemisphere.   The fulcrum upon which the knife-edged central axis of the beam A B rests is enclosed between the two hemispheres constituting the ball c.   In order to allow of the free ascending or descending motion of the beam A B, the two hemispheres are slit open in opposite directions at each extremity of a diameter of the hollow ball.   One arm, c B, of the axial beam has a slider, s, on it, which, when set at different distances from the centre of motion, loads the arm c B with a proportionate weight, the arm c B being graduated from one grain to 60.   The ball B at the extremity of the graduated arm A B rests upon a similar ball, P, either supported by a light bent insulating arm, T, proceeding from the ball at the extremity of the insulator, I, connected with the central ball c, or sustained, like the ball N, upon an independent insulator.   The ball, N, is immediately under and at a short distance from the ball A, at the opposite extremity of the arm.

136.   In the application of this instrument to a jar or battery, the central ball c is put into metallic communication with the positive or charged side, and the ball N with the negative or uncharged side, through a light metallic connecting wire or small chain.   A Henley's electrometer, H, is usually fixed on the central ball c to denote the progress of the charge.

137.   It is evident from the construction of this instrument that if the stand D E be horizontal and the ball B at one extremity of the axial beam lightly rest upon the ball P, it will remain in that position without the help of the slider s; and if the ball B receives a weak electrical charge, the two balls B, P will repel each other, the balls A, N at the same time exerting on each other a slight attraction in consequence of the weak charge accumulating in the ball A, through the medium of the central ball c and axial beam A B.

The combined operation of these two forces causes the arm c B to ascend, and on account of the centre of gravity being a little above the centre of motion, the ascension will continue till A rest upon N. If the beam be again adjusted horizontally and the slider set to any given weight, it will cause B again to rest upon P with a pressure equal to that weight, so that more electricity must be accumulated before the balls B P again separate, and as the weight on the graduated arm c B is increased or diminished by the slider s, a greater or less degree of repulsion will be required to separate the balls.   It is further evident that when the electrical accumulation in the jar or battery is sufficiently strong, the arm c B rises, and the arm c A falls and descends upon the ball N, when discharge immediately ensues, since the ball A is in communication with the positive side of the battery through the central ball C, and the ball N in communication with the negative side through a light metallic wire or chain, n (110).

138. There seems, however, to have been some little misapprehension on the part of Cuthbertson and the older electricians in their views of the electrical terms *Tension* and *Intensity* (62). They assumed that the quantity of electricity accumulated in a jar or battery will be directly as the force required to separate the balls B, P, and consequently as the position of the slider on the graduated arm. Thus, if the position of the slider s indicated a resistance of 5 grains when the quantity of electricity accumulated was equal to unity or 1, the quantity accumulated would, according to Cuthbertson, be equal to 2 when the slider was set to 10 grains, or double the former resistance. But since it is demonstrable (63) that the intensity of an electrical accumulation is as the square of the quantity, if intensity be taken as a measure of the force, in order to accumulate twice the quantity the slider should be set to 20 grains instead of 10, or four times the resistance instead of twice. The older electricians have evidently substituted what has been termed *electrical tension* (58) of the exploding distances of the electrical spark for its *attractive force*,—a very important distinction, as exemplified in Lane's Electrometer.

Many errors in electrical research have arisen from not being aware of this important law.

139. *The Charging and Discharging Quantitative Electrometer.*— A convenient and ready means of charging and discharging an electrical jar or battery, or other electrical accumulation, with ease, convenience, and safety to the operator, either in the light of day or amidst the darkness of night, has been deemed a desideratum in practical electricity.

Fig. 84 represents and describes an instrument calculated to satisfy the conditions of this important mechanical problem. In this figure A B, is a mahogany elliptical stand or base, sustained on levelling screws, its long diameter being about 15 inches, its lesser diameter about 7 inches, and the thickness three-quarters of an inch.

At each extremity of the long diameter is a strong vertical varnished glass insulator A N, and B P, each three-quarters of an inch in diameter, and about a foot in length. These insulators are set in varnished mahogany feet, each about 2 inches in diameter, firmly screwed into the elliptical base at each of its extremities. The insulating column B P is capped with a stout wooden cap and ball, P; a brass rod, *p q*, about 6 inches long and 1 quarter of an inch in diameter, passes centrally through the ball of the wood cap P, and terminates in a brass ball, *p*, about an inch in diameter. The wood cap of the insulator sends up a vertical brass tube, *l*, about 8 inches long and three-eighths of an inch in

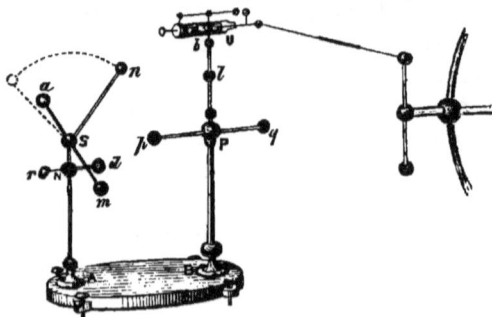

Fig. 84. Charging and discharging quantitative Electrometer.

diameter, having a sliding ball, *l*, on it, which may be adjusted to any convenient altitude.

The lower portion of the tube *l* is a fixture, and proceeds from the ball P of the wood cap, and directly communicates with the transverse rod *p q*, passing through the ball. Within the fixed portion of the tube *l* is a second light tube terminating in a brass ball, *b*, sustaining a small unit measure, U (120), communicating by a light tubular connection with the conductor of the electrical machine. The outer coating *n* of the unit measure U communicates with the jar or battery to be charged through the sliding ball *l*, which has holes drilled in it for the reception of small connecting rods. The insulator A N, at the opposite extremity of the elliptical base, is capped with brass, supporting a metallic ball, s, within which is a strong joint movable with friction, and acted upon by a varnished glass rod, *a m*, about 5 inches long and one-quarter of an inch in diameter, firmly cemented in a short brass socket passing through the flanges of the joint, and terminating in small, varnished mahogany balls, so that by means of the glass rod *a m* and axial tube, passing through the flanges of the joint, we may cause the joint to turn in either direction, according to the direction of the force applied to the terminating balls at the extremity of the axis. A stout brass rod, s *n*—about 6 inches long and a quarter of an inch in diameter, carrying a brass ball, *n*, about three-quarters of an inch in diameter—is firmly screwed into the centre piece or flange of the joint, acted upon by the axial insulating rod *a m* above mentioned, so that by turning the axis *a m* in either direction, the ball *n* may be raised or depressed to any required extent. The ball N on the brass cap on the insulator A N carries a short stout brass rod, $r\ d$, in which is inserted the termination of the circuit leading to the outside of the jar or battery.

140. It is evident by this arrangement that since the ball p is in direct communication with the positive coating of the battery, through the connection of the sliding ball and tube l, with the outer coating of the unit measure U, and the ball n, in connection with the circuit leading to the outside of the battery, directly we cause the ball n to descend upon the ball p, we discharge the whole accumulation by putting the inside and outside of the battery in direct metallic communication with each other, through any given circuit we choose to establish (110), and this may be effected either in the broad light of day, or in the darkness of night, with ease and safety to the operator, by simply turning the insulating axis $a\ m$.

141. *The Quantity Jar and Transfer Measure.*—The object of these instruments is to communicate to insulated conducting, or other bodies, given measured quantities of electricity, for which purpose we require a sort of reservoir of force of given magnitude, from which we may abstract, at any instant, accumulated electricity, and by means of a carrier (14) deposit the quantity abstracted on a given surface. The reservoir of force may be termed a *Quantity Jar.*

This jar, as prepared and arranged, is represented in Fig. 85. J is an ordinary electrical jar of the form already described (113). It is one foot high, 6 inches diameter, and ·1 of an inch in thickness, and exposes about 200 square inches, or about 1¼ square foot, of coating, the height of the coating being about 9½ inches. It has an open mouth of about 3 inches diameter. The jar is carefully varnished, and fitted with a charging rod, its coatings being well and smoothly rubbed down on the glass. The jar, thus prepared, is placed upon a circular varnished glass salver, S, within a light open ring of varnished wood, fitted to the salver so as to retain the jar in a central position. The jar is finally covered by a light bell-glass screen, E, resting in a concentric groove in the wood-ring, within which the jar is placed. The bell-glass screen has an open mouth of sufficient diameter to allow the charging-rod and ball, P, to project

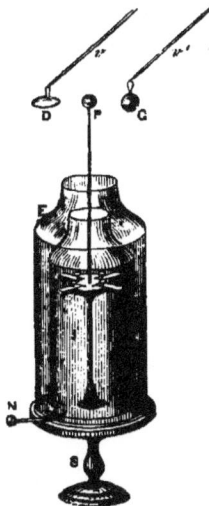

Fig. 85. Quantity Jar.

through and beyond it to any required extent ; this bell-glass envelope E exceeds the diameter of the jar by about 2 inches, and is varnished, inside and out, from its mouth to about an inch below the height of the outer coating of the enclosed jar. A wide band of tinfoil is pasted diametrically across the surface of the glass plate of the salver, upon which the jar rests, and is continued over the rim of the salver, and terminates in the under surface. A similar short band of tinfoil is pasted around the wooden ring at opposite extremities of a coincident diameter, so as to have contact with the tinfoil band traversing the surface of the glass beneath. A communication is thus ensured with the outer coating of the jar when required, and for which purpose there is a small wire and ball, N, projecting from the band of foil where it passes over the edge of the wooden ring. The jar thus circumstanced is charged to a given intensity, through a unit measure, in the usual way (121), communication being at the same time made with the outer coating of the jar and the ground by means of the small projecting ball N.

142. A jar, when charged, insulated, and covered up in this way, retains its charge for a long time, and will communicate a succession of precisely equal charges to an insulated carrier (14) when brought into contact with the metallic ball at the extremity of the charging rod of the jar, communication being at the same time made between the outer coating of the quantity jar and the ground, when applying the carrier.

It is desirable to employ for experiment two quantity jars thus arranged and prepared, one for positive and the other for negative electricity, or for two jars combined when required. The positive jar may have a small band of red paper, about a quarter of an inch wide, pasted over the edge of the outer coating, in the way recommended by Mr. Singer. The negative jar may be furnished with a similar band of blue paper. These bands, whilst they serve to restrain spontaneous explosion, denote at the same time the electricity with which each jar is charged.

143. *The Transfer Measure.*—This instrument is to be employed, in conjunction with the quantity jar, for the purpose of abstracting from it given measured quantities of electricity, and communicating the abstracted quantity to a given insulated surface (14). The transfer measure consists of a light, thin, circular metallic disc, D, or globe, G, Fig. 85, having an eye of fine silk gut firmly secured to it, by which the plate or globe may be taken up on a long slender rod of varnished glass or vulcanite, $v\,v'$.

If the measure be a plate, it is requisite to secure the suspension eye effectually, for which purpose a small loop is tied cen-

trally in a fine thread of varnished silk-gut; two fine holes are
drilled obliquely upward, on each side of the transfer plate, from
the lower edge of the circumference at each extremity of a
diameter, so as to extend from the edge of the circumference to
a little within the surface: the extremities of the suspension gut
are then passed outward through these holes to the edge of the
plate, in order to bring the loop in the centre. The gut is secured
by fine wooden pegs, nipped off fair with the circumference. The
vulcanite rod or holder is tapered, rounded off, and slit open by
a thin saw cut for the reception of and for holding fast the suspension loop of the plate. When a light rod of varnished glass is
employed, the glass is bent a little upward at one extremity to
prevent the gut loop from falling off it.

The suspension loop is easily secured in a globular transfer by
drilling a fine hole in it, inserting the extremities of a simple loop
in the hole, and finally securing it by a small wooden peg, nipped
off fair with the surface. A complete insulation is essential to the
perfect operation of the transfer measure. The transfer measure
thus prepared is charged to saturation, or to a given intensity by
contact with the charging ball of the quantity jar.

144. The practical application of the quantity jar and transfer
measure is as follows. The jar being placed on its insulating
salver, Fig. 85, and covered up (the ball P of its charging-rod
projecting clear of the mouth of its bell-glass screen), it is charged
to a given intensity, through a unit measure, mounted on a long
insulator (121). The jar being charged, is now withdrawn
from the unit measure and electrical machine, when it is ready
for experiment, and the application of the transfer plate, which,
after contact with the charging ball as just mentioned, is transferred to the subject of experiment.

145. It has been shown in the author's paper in the Philosophical Transactions for June, 1864, that the quantity of electricity
which circular planes, or globes, can receive up to a given point
of intensity, or saturation, is as their diameters. It is easy, therefore, to calculate a series of circular planes, or globular transfers
of definite values, taking the circular or globular inch as unity,
calling the circular plate of an inch diameter, after Cavendish, a
circular inch of electricity; and a globe of an inch diameter, a
globular inch of electricity; the plate or globe being, as just
observed, charged to saturation, or up to a given intensity.

In the following table are given the quantities of electricity in
particles, or units of charge, contained on circular plates, or globes
of given diameters varying from ·25 to 2 inches, together with

the respective intensities. In this table the circular inch, that
is, a circular plate of an inch in diameter, and one-fifth of an inch
thick, is taken as unity, and supposed to contain 100 particles, or
units of charge.

Diameters, or Units of Charge.	CIRCLE.		GLOBE.	
	Particles.	Intensity.	Particles.	Intensity.
0·25	25	0·062	35	0·124
0·50	50	0·250	70	0·500
0·75	75	0·560	105	1·120
1·00.	100	1·000	140	2·000
1·25	125	1·560	175	3·120
1·40	140	1·960	196	3·920
1·50	150	2·250	210	4·500
1·60	160	2·560	224	5·120
1·75	175	3·060	245	6·120
2·00	200	4·000	280	8·000

146. Although the foregoing table will be found useful in the
course of experimental inquiry, it is seldom requisite for general
purposes to employ transfer measures exceeding the globular
or circular inch. For estimating small quantities of electricity,
however, to be communicated to the gold leaf, or other very
sensitive electroscopes, very small transfer plates or globes of ·25
of an inch diameter, or even less, may be resorted to.

Perfect insulation of the transfer-measure being necessary, care
must be taken in perfecting the insulating power of the vulcanite
or glass rod, by which the plate or globe is to be suspended.
These insulations should be as slender as possible, nicely varnished, and be made very dry.

147. *Hydrostatic Electrometer.*—In this instrument (Fig. 86),
the force of attraction between a charged disc, P, and a neutral
disc, N, in connection with the earth, is hydrostatically counterpoised by a cylinder of wood, c, accurately weighted, and partially
immersed in a vessel of water. The neutral disc N and its balance
are freely suspended by two fine lines, *l l'*, in opposition to each
other, over the circumference of a light brass balance-wheel, w,
2·4 inches in diameter, weighing about 160 grains. The line *l*,
suspending the disc N, is of fine gold thread, by which its free
electrical state is always preserved. The thread of suspension
passes through a small hole drilled through the circumference of
the wheel at the extremity of a vertical diameter, and is held there
by a knot tied in it ; from this it passes over the circumference,

and terminates in a small hook, to which the suspension lines of
the plate N are appended ; the plate itself is hung by three lines

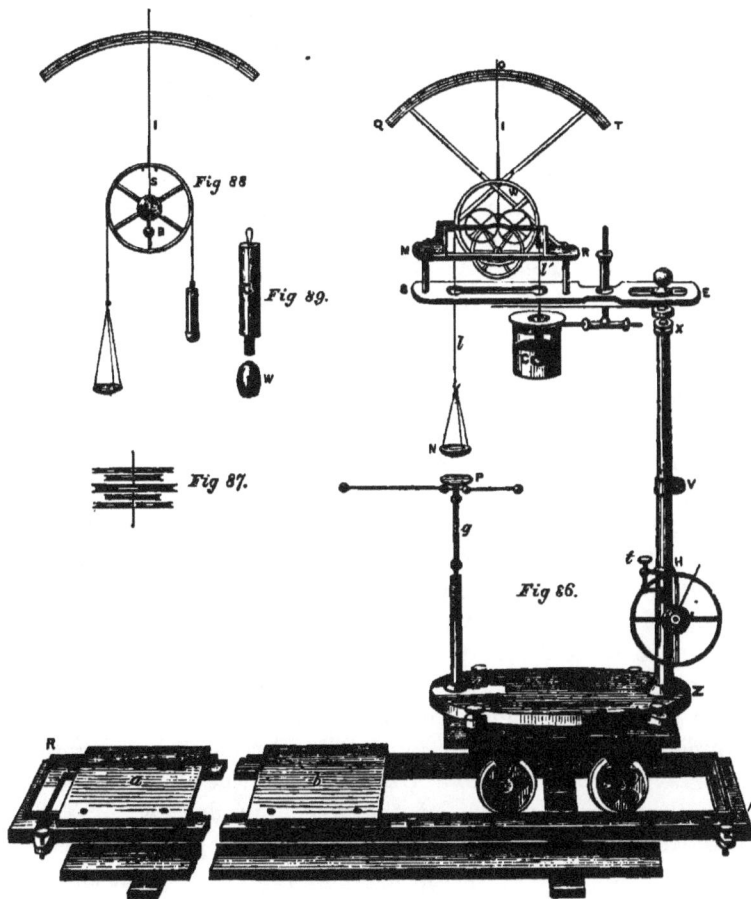

Fig. 86. Harris's Hydrostatic Electrometer.
Figs. 87, 88, and 89 show some of the separate parts enlarged.

of gold thread attached to opposite points of its upper surface, and
converging to a centre after the manner of a scale pan.   The line
l, suspending the counterpoise c, is of silk thread, and is fixed in
a similar way through a hole in the circumference of the wheel,
close to the former; from this hole it passes over the circumference, and also terminates in a hook sustaining the counterpoise.
The circumference of the wheel is slightly grooved for the reception of the lines of suspension.   The attracting discs, P N, are of

light wood, smoothly gilt, about 1½ inch in diameter and ·1 of an inch thick.   These dimensions, however, may be varied.

The wheel, W, is mounted on a steel axis, the extremities of which are hardened and turned down to fine pivots; these pivots rest centrally on the circumferences of two light friction wheels (Fig. 87) of 1½ inch in diameter, each weighing about 40 grains or less.   The axis of the large wheel is prevented from falling to either side by four smaller, or check wheels, each 1 inch in diameter, and weighing 20 grains.   These are placed two on each side of the terminating pivots of the axis of the large central wheel.   The last-named six wheels are delicately poised, and set on hard steel axes centred in jewels or in hammered brass, similar to the wheel-work of chronometers, so that the large wheel has perfect freedom of motion, and is influenced by the slightest force between the attracting plates P, N.

Whatever this force be, it is speedily counterpoised, hydro-statically, within certain limits, by the elevation or depression of the cylindrical float C above or below the water level.   The force thus counterpoised is measured by an index, I (Fig. 86), of light straw reed 7 inches long, weighing 1½ grain, attached to a radius of fine steel wire, S (Fig. 88), passing through the circumference of the wheel W.   The weight of the index I is counterpoised by a small brass ball, B, movable with screw motion on a similar steel wire, continued on from the boss to an opposite point in the circumference, so that the two steel wires constitute a diameter of the wheel.   The index moves over a graduated quadrant, Q T (Fig. 86), of about 7 inches radius, accurately divided into 45 degrees on each side of its centre, O; indicating the movement of the wheel, in degrees, in either direction, O Q, O T, the centre of the arc being zero.   Each 5 degrees of the arc corresponds with the $\frac{1}{40}$nd part of the circum-ference of the wheel, so that the plate N ascends or descends about one-tenth of an inch for every 5 degrees of the arc.

The arc, Q T, is composed of cardboard about three-fourths of an inch wide, and is attached to a light open quadrant of support, con-structed of thin slips of wood fixed immediately behind the wheel, and its index upon a brass plate, M R, carrying the wheelwork: this brass plate is 4 inches long and 1½ inch wide, and is, with the wheelwork and quadrant of measure, supported upon two small pillars of brass 1½ inch high and one-fourth of an inch in diameter, screwed into a brass stage or platform, S E, 11 inches long, and from 1½ to 2 inches wide, where it receives the small brass pillars. The stage is secured at its narrow extremity upon the cap of a vertical column, x z, by means of a nut and projecting screw passing through a slot in the stage, so that the stage may be adjusted to project from the column to any required distance, not greater than 8 inches. The hydrostatic counterpoise c, Fig. 89, is about 2 inches in length, and ·3 of an inch in diameter; but may be greater or less than this, if required. A stout brass wire is screwed centrally into its lower extremity, leaving a short projection of about one-eighth of an inch, upon which is the thread of a screw for the reception of a small cylindrical or globular weight, w, sufficient to counterpoise the suspended disc, N, and cause the float to sink freely in the water when not held in equilibrio by an opposite force. The upper extremity has a conical cavity sunk in it for the reception of fine shot, or other small weights, which may be required to occasionally regulate its position in the water. Should the equilibrium, in certain cases, be not readily obtained on the side of the suspended electrometer disc, N, it is effected by placing a thin metallic ring upon its upper surface (Fig. 86). The water vessel carrying the hydrostatic counterpoise c, is about 2½ inches in diameter, and 3 inches deep, being an ordinary inverted cupping-glass. It is closed in by a light wood cover, having a round hole through its centre for the passage of the counterpoise float, and is held in a ring of brass at the extremity of an horizontal rod, acted on by a vertical screw motion, contained in a short brass cylinder fixed in the platform carrying the wheelwork. The water vessel c, by this screw motion, may be raised or depressed to a certain extent, so as to cause the index i to move in either direction, o q, o t. The stage s e, sustaining the electrometer wheels and other portions of the apparatus, is, as just stated, secured upon the cap of a vertical column, x z, fixed on an elliptical base, y z, 1 foot long by about 7 inches wide, set on four levelling screws. The lower portion of this column consists of two brass tubes, u h, u z, each about 5 inches long and 1 inch in diameter, movable one within the other by a rack and pinion at u. This portion of the column, when fully elongated by the rack and pinion, is about 9 inches in length; it carries, by means of a tube v h, a wooden column, v x, about 1 inch in diameter and 1 foot long, terminating in the cap x; the lower end of this wooden column is fixed in a brass tube 2¼ inches long, movable by a second rack and pinion, v, within the tube v h, which is screwed upon the upper tube, u h, of the lower rackwork, so that the wooden part of the supporting pillar can be raised or depressed alone.

The attracting disc p, immediately opposed to the suspended plate

N, rests on a small circular plate of brass three-fourths of an inch in diameter, screwed upon the brass cap of an insulated rod of glass, g, 10 inches long by one-fourth of an inch in diameter. This insulating rod is movable with friction, through fine cork, within a graduated slider, movable within a tube fixed upon a sliding piece, Y, at the extremity of the elliptical base. The insulating support g, of the attracting plate P, being movable within the graduated slider, may be regulated to any convenient height, giving a clear insulation of about 8 inches. A short piece of small brass tube, about 1 inch in length and one-tenth of an inch in diameter, is fixed transversely through the cap of the insulator, immediately under the brass plate supporting the attracting disc P, for the occasional reception of connecting wires, or of a transverse movable rod terminating in small brass balls; and by which an electrical jar, or any other insulated body, may be conveniently connected with the attracting plate P of the electrometer. In order to estimate the elevation or depression of the suspended disc N, by means of the rack T, carrying the column U V, a vertical scale of measure of about 4 inches in length divided into twentieths of an inch, is applied to the upper brass tube, so as to measure the movement, up or down, of the wooden column carrying the electrometer wheels. The zero line of this scale coincides with the edge of the lower tube when the two tubes are coincident, and there is a small tangent screw, t, by which the movable zero line may be made perfectly coincident with the fixed line. One turn of the pinion of the rack raises the wooden column carrying the wheelwork $\frac{2}{3}$ths of an inch. In addition to this scale of measure, a light open circle of 4 inches diameter, and divided into 64 parts, surrounds the pinion of the rack. The axle of the pinion carries a light index movable upon it with friction, so that the index may be set to zero of the circle, when the small vertical scale of the rack is accurately adjusted to zero. We may in this way estimate the vertical movement of the rack to the $\frac{1}{100}$th of an inch, or very nearly; for since the graduated circle surrounding the pinion is divided into 64 parts, one division corresponds to the $\frac{1}{64}$th part of $\frac{2}{3}$ths of an inch, equal to ·0101 ; that is, the $\frac{1}{100}$th of an inch very nearly. The suspended disc N of the electrometer, therefore, will be raised or depressed $\frac{1}{100}$th of an inch for each degree of the circle, according to the direction in which the milled head U of the pinion of the rack is turned.

The levelling screws on which the elliptical base Y Z, carrying the whole of the electrometer apparatus, is fixed, rest on the platform of a small railway carriage, movable on rails, R, R', so as to transfer the instrument from one point to another on the line of the rail with the least possible disturbance. The platform is 10 inches long and 7½ inches wide. The rails are of brass, and are set 5 inches apart upon an open mahogany frame about 5 feet long by 7½ inches wide. The frame is sustained on levelling screws, supported at its opposite extremities on two circular tables fixed on central columns, which being movable in stout brass tubes fixed on levelling stands, can be adjusted to any convenient height. There are two light travelling platforms, a, b, movable on the rails, 1 foot long by 7 inches wide; these are for the support of insulated or other bodies to be placed in connection with the electrometer in the most convenient way. Beside these travelling stages there is a light platform stage at the side of the rails 3 feet long and 7 inches wide, resting on two projecting bars attached by clamp screws and nuts to the under-part of the open rect-angular frame upon which the rails are based. This platform is for the purpose of placing insulated or other bodies in convenient positions at the side of the electrometer.

The bars supporting the platform being movable, may be caused to project on the other side of the rails, so as to transfer the platform to the other side of the electrometer, or admit an addi-tion of a second platform on the opposite side of the rails, for the convenience of supporting on insulators extensive conducting surfaces in any required direction.

148. It may be observed that unless the wheelwork is delicately mounted, and accurately set in place, a small index error of a degree or a degree and a half may arise upon the measurement of extremely small forces. This, however, is easily detected by a slight oscillation communicated to the index I, by which it will be immediately seen whether the pivot of the great central wheel is liable to shift its position either to the right or to the left. There is no difficulty in making a small allowance for this error, should it occur.

149. This electrometer is peculiarly adapted to the measurement of electrical force, which can be estimated in terms of an admitted standard of weight. If, for example, with a given hydrostatic counterpoise, 1 grain added to either side moves the index 5 degrees in either direction, 2 grains should move it 10 degrees, and so on. We may, therefore, estimate degrees in grains weight. Thus, if we had a registered force between the attracting plates, P, N, of 15 degrees, that would, in this case, be equivalent to a force of 3 grains.

Another advantage in using this instrument is, that the expe-

riment requires only a short time for its accomplishment; so short
a time, indeed, as to supersede the necessity of any kind of calculation for electrical dissipation. When the electrometer insulations
are perfectly dried off, and the air dry, the index will often remain
unmoved for at least ten minutes, and sometimes longer; whilst
any experiment with the instrument may be concluded in less
than three minutes, and occasionally in less than one minute.

150. The mechanical conditions of the electrometer being fully
apprehended, the experimental manipulation will be as follows :—
The index being accurately set at zero of its arc, either by pouring
a certain quantity of fluid into the water-vessel, or by a small movement of the screw motion that supports it, the accuracy of its indication is determined by placing small weights of a grain each, or
less, either upon the suspended plate N, or in the conical cavity
of its counterpoise.   If 1 grain move the index 5 degrees in
either direction, 2 grains should, as just stated (149), move it
10 degrees, and so on, which it will not fail to do if the instrument be carefully constructed, so that the index error, if any, will
be small. The hydrostatic counterpoise c must, however, be previously immersed for a short time in water, so that no part of it
remain dry; for if it were so, it would interfere with its correct
position in the fluid.   When in perfect order the index may be
caused to oscillate by a slight movement of the wheel, in either
direction, with perfect freedom, and will invariably return to the
zero of its arc when the oscillation ceases.

151. In the application of the electrometer to experiment,—as
for example, to the determination of a charged electrical jar, or
to that of an insulated circular plate, or any other charged surface,
—we first determine the distance at which the force between the
attracting plates is to operate.   This distance is measured in the
following way :—The index of the arc being carefully adjusted to
zero, and the vertical scale and the index of the circle of the lower
rackwork also adjusted to zero, the surfaces of the attracting plates
P N are brought into contact by the elevation or depression of
the slide supporting the lower disc, or by a more accurate adjustment, by means of the upper rack-work v operating on the wooden
column supporting the wheel-work of the apparatus.   By means
of this rack the suspended plate may be lowered or raised to the
last degree of precision.   The plates P N being thus in contact,
and the index of the arc, together with that of the scale of
measure of the rack, adjusted to zero, we now proceed to raise the
column supporting the wheel-work and the suspended plate N
through a space equal to the required distance between the plates

P N,—suppose, for example, half an inch. We know, therefore, that when the index of the arc is at zero, the two plates P N are ·5 of an inch apart.

152. The following experiments may serve as illustrations of the operation of this instrument.

Exp. 38.—Let it be required, for example, to investigate the law of electrical attractive force, between the circular planes P, N, as regards quantity. Connect the electrometer with a hollow spherical conductor of 5 inches diameter, having an open projecting circular mouth of 3 inches diameter, so as to freely admit a transfer measure of 1 circular or globular inch (145). Proceed now to transfer to the interior of this sphere a given quantity of electricity, suppose 5 globular inches; the discs of the electrometer will attract each other, and the index will move forward in the direction o Q, Fig. 86. (To prevent the suspended disc from descending beyond a certain limit, a small stop-pin is inserted in one of the radial arms of the arc on the side o Q of the index.) The water vessel is now lowered by the screw motion, until the index is again brought back to zero of the arc. We are now assured that whatever be the force between the attracting plates, it is operating at the given distance of half an inch. To determine the amount of this force, we discharge the electrical accumulation. The hydrostatic counterpoise now descends in the fluid, and the index recedes in the direction o T, until an equilibrium again ensues between the counterpoise and the suspended disc, showing the amount of force previously in operation. Let the force thus shown be, for example, 5 degrees, being supposed to be equivalent to a force of 1 grain (149). Charge the electrometer, through the hollow sphere, with 10 globular measures, or twice the former quantity; the hydrostatic counterpoise will again be raised, and the index move forward in the direction o Q, until impeded by the stop-pin, against which it rests. We now further lower the water vessel, until the index again rests at zero; the force between the plates, whatever it be in this case, is still operating at distance ·5. We determine the amount of this force as before, by discharging the electrical accumulation. A hydrostatic balance again ensues between the suspended disc and its counterpoise float, so that the index will have again receded in the direction o T; showing the amount of force in degrees due to 10 circular inches, or twice the former quantity, and which, in this case, will be 20 degrees, or four times the former force. Thus, whilst the quantities of electricity are as 1 : 2, the forces are as 1 : 4, or as the squares of the quantities. The electrometer, therefore, together with the method of experiment, is in strict accordance with an admitted and well-known law of statical electricity, and is so far entitled to our confidence (R).

Exp. 39.—Let it be now required to investigate the law of attractive force as regards distance. Raise the vertical pillar another half inch, which will double the distance between the attracting plates when the index is at zero of the arc. Repeat the charge of 10 globular inches, which by the last experiment shows a force of 20 degrees at distance ·5. The index will again advance in the direction o q. Bring the index again to zero, which is effected by operating on the water vessel as in the former case. We now, as before, discharge the accumulation.

The hydrostatic counterpoise will again descend in the fluid, and the index again recede in the direction o т, indicating the force between the plates at 1 inch, or double the former distance, which will be now 5 degrees, or only one-fourth the former force. In this case we see that whilst the distances are as 1 : 2, the forces are as 4 : 1, the quantity of electricity being constant; that is to say, the force of attraction between the electrometer discs is in an inverse ratio of the squares of the distances. The electrometer, therefore, together with the method of experiment, is in strict accordance with well-established laws of electrical action. The same laws may be observed with increased quantities at increased distances. It is requisite, however, to observe that to get a greater range of distance or force, we must take such distances or forces as come within the grasp of the instrument. We could scarcely, for example, obtain a staple equilibrium of 45 degrees at the short distance of ·5 of an inch only between the plates, which would be the force due to 15 globular inches measured as before. To exemplify the law of quantity, therefore, for several terms of the series, it is requisite either to take an increased distance or a less force.

153. *The Scale-beam Electrometer.*—In this instrument the common scale-beam is adapted to the purpose of measuring the attractive or repulsive forces between bodies of various kinds and forms, charged with given quantities of electricity, and placed at given distances from each other. The Scale-beam Electrometer is represented in Fig. 90. L M is a delicate scale-beam, 10 inches in length, weighing about 300 grains, and will readily turn with one-tenth of a grain when loaded with 100 grains. It is hung in the usual way from a convenient support, w. From one arm of the beam is suspended, by three fine gold threads, a light circular gilt plate, N, or other body, accurately

balanced by an equivalent counterpoise, in a scale-pan, s, suspended from the opposite arm M. Being accurately balanced, the attractive force between the suspended plate N and a plate, P, placed immediately under it, is estimated by grain weights placed in a second scale pan, T, hung by three silk lines from the opposite arm M of the beam L M. The scale-pan s is attached beneath the scale-pan T by its three lines of suspension, which are continued on for the purpose.

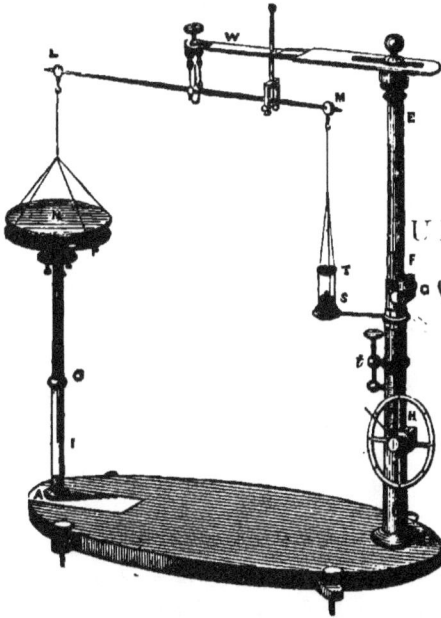

Fig. 90. Scale-beam Electrometer.

The attracting plate P, immediately under the suspended disc N, is supported on a varnished insulating rod, I, 6 inches long; and three-quarters of an inch in diameter.· In order to insure the parallelism of the two plates P N, the attracting body P rests on four small levelling screws, passing through a brass plate, supported on the insulator I, by means of a tube, sliding with friction through another tube, proceeding from a ball, C, at the extremity of the insulator, and by which the altitude of the plate P may be adjusted, and electrical charge communicated to it.

The insulator I, with its attracting plate P, is fixed in a sliding foot, A, forming part of an elliptical base, A B, 16 inches long, its width across being 9 inches, and its thickness 1 inch. By

K 2

means of the movable foot A, the insulating support I carrying the attracting plate P may be adjusted to any given point.
The opposite extremity B of the elliptical base carries a compound
column of support, B E, variable in length from 20 to 30 inches.
This column carries a light rectangular projecting brass plate of
support W, about 10 inches long, half an inch wide, and one-eighth
of an inch thick. This plate supports the delicate scale-beam L M,
and its appendages.

In order to steady the beam and give it occasional support, there
is a small open rectangle of brass, attached to a rod, made to
slide within a tube screwed into the plate of support W. The
beam passes through this rectangle, so that by means of a cross-
pin, and small holes in the sides of the rectangle, the beam may
be steadied in any position, and its motion, by a due adjustment of
the sliding rod, restrained.

The compound column B E supporting the scale-beam M N, and
its suspended bodies, N, S, consists of different parts, E, F, G, H,
and B.

The upper part E is a cylinder of extremely light wood
(Chap. V., 35), about a foot in length, and a full inch in diameter.

The cylinder is inserted and fixed in a brass socket, F, three
inches in length, at the side of which is a delicate rack, acted on
by a pinion fixed at the upper extremity of a brass tube, G, within
which the socket carrying the pillar slides with friction. By
means of this rack and pinion the portion E may be adjusted
to any given altitude.

The part G is screwed on the upper extremity of a following
stout brass tube, H, at the side of which is a rack similar to that
of the tube F, acted on by a pinion fixed at the upper extremity
of a final brass tube, B, screwed into the elliptical base A B at the
extremity B. By means of this rack and pinion the total height
of the compound column B E may be varied.

In order to estimate any change which may be effected in the
altitude of the column B E by means of the lower rack, the
pinion moving the rack is furnished with a light open graduated
circle, divided into 64 parts, having a delicate index movable
with friction about the milled head. One revolution of the pinion,
as measured by the index on the graduated circle, raises the
pillar B E $\frac{18}{30}$ths of an inch; and since the circle is divided
into 64 parts, one division raises the pillar the one-hundreth of an
inch, or very nearly. Besides this circular measure there is a
small movable scale, graduated into twentieths of an inch, attached
to the lower tube H of the column, and acted upon by a tan-

gent screw, *t*, so as to accurately adjust the zero-point of the scale to the upper edge of the lowest tube B of the column. In order to steady the scale-pan s with its counterpoise, there is a light tube, sliding with friction on the tube G, carrying a small projecting circular plate on which the scale-pan rests.

154. It is easy to comprehend by the following instructive experiments, the nature and action of this instrument.

Exp. 40. Let it be required, for example, to determine under given conditions the attractive force between two equal spheres, P, N, Fig. 91, of a given diameter, suppose 2 inches, each weighing

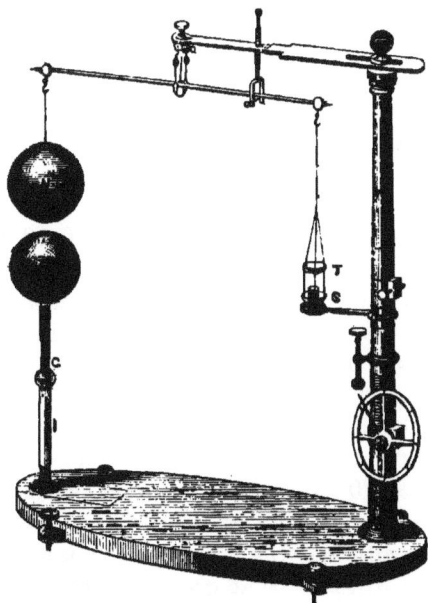

Fig. 91. Electrical attraction of spheres.

150 grains. The sphere N is delicately suspended from the beam immediately over the sphere P by a fine, double, gold thread; the attracting sphere P resting on the conducting tube Q, at the extremity of the insulating rod I. The sphere N being accurately balanced in the way already described (153), the attracting sphere P immediately opposed to it is then put into communication with an electrical jar of given dimensions, exposing suppose about three square feet of coated surface. The jar is then charged from the prime conductor of the electrical machine through a unit measure (121). When the accumulation amounts to a given number of units or measures of charge, the two spheres

attract, and the sphere N descends upon the sphere P. The force
of this attraction is estimated by grain weights placed in the
scale-pan T. Thus the electrical attractive force between the
spheres with a given quantity of electricity is estimated in
a known standard of weight. Suppose, for example, at the instant 10 units have been accumulated, the sphere N has descended
on the sphere P, and raised a weight of 12 grains placed in the
scale-pan T. In this case the attraction between the spheres P, N,
at a given distance from each other, and measured between their
near points, may be considered as amounting to 12 grains. It
is however necessary to accurately determine this distance, which
is arrived at by first bringing the touching, or near, points of the
two spheres into contact, and causing the spheres to rest freely
on each other, by means of the upper rack and pinion. They are
then separated by a given distance, as measured by the vertical
scale or the graduated circle of the lower rack. When the distance
of the surfaces is for example, ·5 of an inch, the quantity of
charge being 10 units, suppose the attractive force between the
two spheres to be equivalent to a weight of 12 grains in the
scale pan T.

Exp. 41. Let the diameters of the spheres remain the same as
at first, that is to say, 2 inches, the quantity of electricity (that
is to say, 10 units), and the distance being also the same. Then
as we have seen, (Exp. 40), the force of attraction is 12 grains
placed in the scale-pan T.

Let the quantity of electricity be now doubled, that is to say
20 units instead of 10; the attractive force, as measured by the
number of grains placed in the scale-pan T, will in this case be four
times as great as with half the quantity; that is to say, the weight
in the scale pan T will now be 48 grains instead of 12, as in the
former case.

Thus whilst the quantities are as 2 : 1 the forces are as 4 : 1, or
as the squares of the quantities.

Exp. 42. Increase the distance between the plates to 1 inch,
or double the former distance, the quantity of electricity remaining
as in Exp. 40—that is to say, 10 units, which is equivalent to a
weight of 12 grains; the attractive force decreases to 3 grains, or
one-fourth the former weight. We have in this case (as determined by this instrument,) the distances as 1 : 2, and the attractive
forces as 4 : 1, in which case the *attractive force varies in an inverse
ratio of the square of the distance*—a long sought and important
problem.

Exp. 43. Let the diameters of the two spheres P, N, be now doubled, that is to say, 4 inches instead of 2, as at first. The areas of the attracting surfaces opposed to each other will, in this case, be four times as great, the surfaces of spheres being as the squares of their diameters. Let the suspended sphere N of 4 inches diameter, and which weighs about 350 grains, be balanced by an equivalent weight placed in the scale-pan S. Let the number of unit accumulations remain as in the preceding experiment (42), that is to say, 10 units, the distance between the near points of the hemispheres being adjusted as before (Exp. 40); that is to say, to ·5 of an inch. In this case the attractive forces between the spheres P, N, of double diameter, will now be only one-fourth as great as between those of half the diameter; that is to say, the attractive forces will be inversely as the areas of the opposed spheres, or as the squares of their diameters. Hence the attractive force between the spheres of double diameter will now only raise 3 grains, placed in the scale-pan T, instead of 12. In this way the attractive forces between spheres, plates, and other bodies opposed to each other under certain conditions, may be accurately weighed, and estimated (S).

155. *Harris's Two-threaded, or Bifilar Balance.*—There are no departments of science which call more for the perfection of quantitative measurement, and a clear perception of what we really measure, than those of Electricity and Magnetism. If we except the valuable researches of Professor Robison, of the French physicist, Coulombe, the recent investigations of Dr. Faraday, and the experiments of the electrician, Brook, of Norwich, we can scarcely be said to possess, in common electricity at least, any connected series of experiments with electrical repulsive forces, carrying with them a rigid numerical value. In the various inquiries which philosophers have instituted into the elementary laws of electricity, it has been their endeavour to perfect our methods of electrical research, whether relating to the quantity of electricity in action, intensity, inductive power, or other elements requiring an exact numerical value, and by operating with large statical forces, both attractive and repulsive, to avoid many sources of error inseparable from the employment of minute and almost insensible agencies. Coulombe's balance of torsion, and other delicate instruments of repulsion, have been employed with this view. The Bifilar, or Double-threaded Balance (invented by the author in the year 1831, and described in the Transactions of the Royal Society for the year 1836) has been resorted to with much advantage as a measure of repulsive force in electrical and magnetic researches.

156. This instrument is constructed on the following principles. If a delicate electrical needle, M N (Fig. 92), of moderate weight, and about ten inches in length, be suspended from two fixed points a a' by two similar filaments of unspun silk a b, a' b', without torsion or twist, placed parallel to each other at equal distances from the axial line c c', it is evident that its position of rest will be horizontal, and in a plane passing through the needle. Whenever, therefore, we turn the suspended needle about the imaginary axis c c', the lines of suspension a b, a' b' tend to cross each other, or become deflected from the vertical, so that the distance c c' will be shortened. We have hence a reactive force derived from the weight of the suspended needle, which becomes imparted, as it were, to the threads of suspension

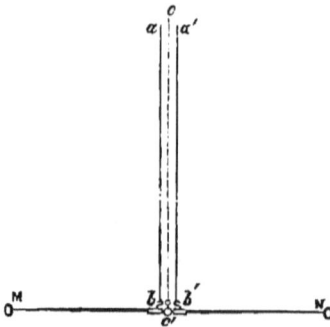

Fig. 92. Bifilar Suspension of Needle.

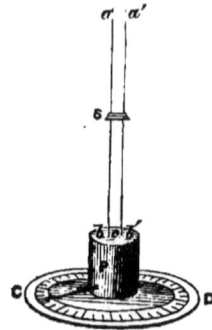

Fig. 93. Bifilar Suspension of Cylinder.

a b, a' b', since the centre of gravity of the mass will again tend to rest in its previous position, and the suspended needle will be similarly circumstanced to that of a body falling down a very small circular arc. If, therefore, the needle be freely abandoned to this reactive force, a vibratory oscillation of the needle ensues, by observing which we may determine, by the formulæ for oscillating bodies, the reactive force of the bifilar threads producing the oscillations.

157. With the view of determining this reactive force, a solid cylinder of wood, P (Fig. 93), 2 inches high, and 2 inches diameter, is suspended by two parallel filaments of unspun silk, a b, a' b', inserted in the points b b' in a diameter of the upper surface of the cylinder P at equal distances from its centre, c, and kept apart by light, small, intermediate cross stays of cork, s. An index of light reed projects from the cylinder for the purpose of allowing observations to be made on a graduated circular card, C D, as to the duration and extent of the oscillations produced by the crossing of

the threads on each other. The threads $a\ b$, $a'\ b'$ (Fig. 93), are sustained in a convenient light mahogany frame (Fig. 94), mounted on a solid base, A A'. The suspension threads $a\ b$, $a'\ b'$ pass through fine holes in a movable cross-bar, $r\ r'$, Fig. 94, and are joined above to a strong piece of silk thread continued through holes in a second bar, $d\ e$, and are finally wound round regulating cylindrical winders at d and e. By these means the respective lengths and distances of the threads of oscillation $a\ b$, $a'\ b'$, are regulated, and the cylindrical weight P caused to hang parallel with the plane of the graduated card c D. By varying the situation of the bar $r\ r'$, the length of the threads of oscillation may be changed, and by a succession of fine holes in the cylindrical weight P, corresponding to similar holes through the bar $r\ r'$, we may vary their distances apart.

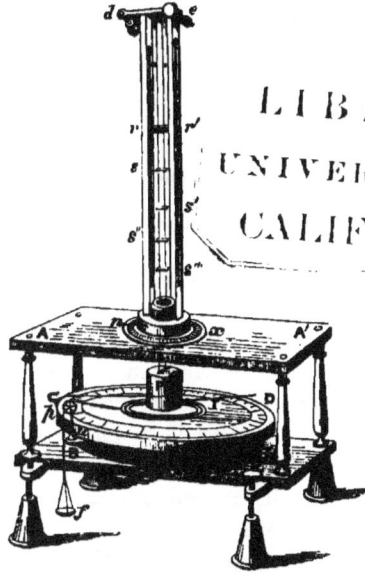

Fig. 94. Bifilar Balance.

158. By these mechanical arrangements, and the oscillations of the cylinder, it is shown that the reactive force imparted to the bifilar threads of suspension in turning the cylinder upon its vertical axis, and allowing it to oscillate, is directly as the distances between the threads, inversely as their lengths, directly proportionate to the weight of the suspended body, and as the angle of twist or torsion of the bifilar threads upon each other; results similar to those arrived at by the celebrated Coulombe in determining the reactive force of the fine wire employed by him in his beautiful instrument the Balance of Torsion (166).

159. Having obtained a given length and distance between the threads, the centre of the cylinder P (Fig. 94) is caused to hang immediately over the centre of the graduated circle c D, as shown by contact with the finely-pointed extremity of a short cylindrical rod, passing from beneath upwards through friction corks in the central block B B'. Small stays s, s', s'', s''', of light reed or cork, are inserted between the threads at given distances, in order to prevent them from closing upon each other.

The index I turned aside to an angle of 60°, being the suspended body P allowed to oscillate freely. In order to preserve
a very free oscillation, the fine central point is depressed from
beneath the base and block B B'.

By carefully noting the rate of oscillation, the following results
are immediately arrived at :—

The time of an oscillation is as the square root of the length of
the threads of suspension, divided by their distance apart, and
is altogether independent of the weight of the oscillating body.

The oscillations are isochronous at all angles.

From these results we may, by the general formula for oscillating
bodies $n = \dfrac{P \; \pi^2 \; a}{2 \; g \; T^2}$—employed by Coulombe, in his experiments
on the torsion of wires, easily deduce the laws of the reactive
force imparted to the threads.

In this formula n is the force, in terms of a unit of weight
$=1$ grain, which applied perpendicularly at the extremity of
a lever of a unit of length $= 1$ inch will resist the reactive force
imparted to the threads, when the suspended body has been turned
about its axis through an arc of 60°, whose chord is equal to the
unit of length, $=$ radius, $= 1$ inch. P is the weight of the
cylinder in terms of the unit of weight, $= 1$ grain ; π the ratio
of the circumference to the diameter of the circle ; a the radius of
the cylinder or circle $= 1$ inch ; g the force of gravity, $= 386$
inches, equal to the distance through which a body falls in free
space in a second of time ; T the time of an oscillation in terms
of a unit of time $= 1$ second.

In applying this formula, it is easy to perceive that the value of
n will vary with the squares of the distances between the threads
of oscillation divided by their lengths, and that, contrary to the
law of torsion, as deduced by Coulombe, is as the weight of the
cylinder P, hence we have $n \propto \dfrac{P \; d^2}{l}$; and since the oscillations
of the cylinder are, as just stated, isochronous at all angles,
we may conclude that n is also proportional to the angle of
deflection of the threads.

160. These results are verified mechanically and experimentally
in the following manner :—

A weight placed in a light paper scale-pan, f (Fig. 94), and
carefully counterpoised, is caused to act tangentially to the
circumference of the cylinder P, by means of a slender filament of unspun silk, $f \; p \; P$, passed round the cylinder and
led over an extremely delicate watch-pulley p. This pulley is

attached to a slide and socket, *u*, fixed to a circular rim, *u* D, movable about the interior block carrying the graduated card c D, hence the line *f p* P may be exactly set at right angles to the radius of deflection in all positions of the index I, and the precise weight determined requisite to balance the reactive force of the threads at any given angle, or otherwise, by turning the whole frame of suspension *d s'''*, in a circular socket formed in the transverse piece A A' through any number of degrees, as measured by a graduated card and index *n x*, and preserving always the index I of the oscillating cylinder P at zero, we arrive at the reactive force of the threads of suspension at any required angle.

The results of a series of experiments conducted in this way, completely verified the above deductions (158), the weight requisite to maintain the index at an angle of 60° being as the weight of the cylinder P multiplied into the squares of the distances between the filaments of the suspension threads, divided by their length. It was also found to vary with the angle of repulsion, as determined by Coulombe.

161. The following tables, abridged from a greater number of experiments than it is desirable to mention here, afford a sufficiently practical evidence of the truth of these results. In these tables the unit of weight is 1 grain, the unit of length 1 inch, the unit of time 1 second.

TABLE I.

Showing the rate of oscillation with different lengths and distances of the threads.

Weight of Suspended Body, 960 grains.   Angle of Oscillation 45°.

Length.	Distance.	Oscillations in 60''.	Time of ten Oscillations by observation.	Time of one Oscillation.
6	0.25	28·50	21	2·1
24		14·25	42	4·2
6		46	13·1	1·31
12	0·4	32·50	18·5	1·85
24		23	26·2	2·62
24	0·8	46	13·1	1·31

Similar results were obtained when the angle of oscillation was increased to 180° and upwards, as also when the weight of the cylinder P was varied from 960 to 480, and 240 grains respectively, the radius being in each constant.

The rate of oscillation was taken with a valuable chronometer, by which portions of time so little as the one-sixtieth part of a second could be well estimated.

TABLE II.

Showing experimentally the weight in grains requisite to resist the reactive force of the threads at an angle of 60°, their length and distance apart being varied, as also the altitude of the cylinder P.

Length.	Distance.	Weight in grains on a lever of 1 inch.		
		= 960 grains.	P=480 grains.	P=240 grains.
6	} 0·25 {	2·675	1·325	0·66
24		0·67+	0·325	0·15
6		7·+	3·525	1·75
12	} 0·4 {	3·55	1·775	0·885
24		1·75	0·875	0·425
24	0·8	6·85	3·425	1·750

The smaller weights employed in these experiments could not be considered as mathematically exact; they were, however, sufficiently accurate for the purposes required. They consisted of 10ths of grains, 20ths, 40ths, and 100ths. The numerical values in the above table are those which resulted from the position of the index, so far as these small weights could determine; and it will be seen that the approximation to the values deducible from the preceding Table I., by means of the formula $n = \dfrac{P \pi^2 a^2}{2 \, gl}$, are as near as could be expected from such an experiment.

TABLE III.

Showing the weight in grains by calculation and experiment, required to balance the reactive force of the threads at various angles of deflection, from 0 to 300°, the threads being 24 inches in length, and ·25 apart, and prevented from collapsing by small stays of cork (159), inserted between them at equal distances.

Weight of Cylinder P=960 grains.

Angle of Deflection	10	20	30	60	90	100	120	150	180	200	240	270	300
Force by Formula	·115	·23	·34	·69	1·03	1·15	1·38	1·72	2·07	2·3	2·76	3·1	3·45
Force by Experiment	·11+	·22	·34−	·67+	1·	1·1	1·35	1·7	2·	2·25	2·725	3·+	3·425

The pulley p (Fig. 94) employed in these experiments is extremely delicate. The small scale f in which the weights were placed, weighed the ·1 of a grain, and was suspended by a filament of a thread of the silkworm ; that part of it passing over the pulley being particularly slender and flexible. The ·01 of a grain was in this way rendered very sensible on the index. The approximations in this table are sufficiently close to show that the reactive force of the threads is as the angle of repulsion or deflection of the needle ; the threads being prevented from collapsing by intermediate stays.

162. Fig. 95 represents the bifilar balance in its most perfect form. A B is a flat, polished, mahogany base, about 16 inches square and an inch thick. A, B, C, D are four light grooved mahogany pillars, each 10 inches high and half an inch in diameter, screwed into the angles of the base. The grooved pillars A, B, C, D receive four glass panes, constituting, together with the pillars, a sort of light glass cage, within which is suspended by bifilar threads, $a\,b$, $a'\,b'$, (157), a light insulating needle M N (156). Each extremity of the needle carries a small circular gilt disc, M N, four-tenths of an inch in diameter, attached to the needle by an insulating thread of shell-lac. A stout vertical metallic axis, H, passing with friction through a central cork, firmly fixed in

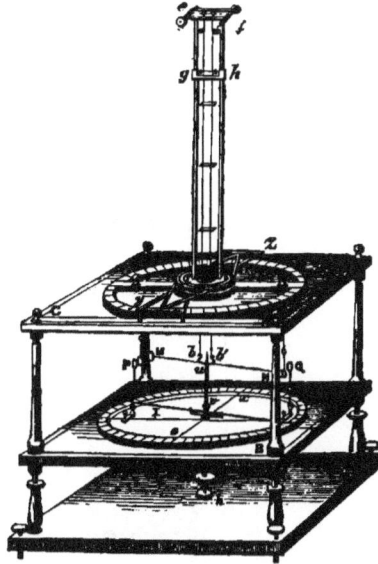

Fig. 95. Bifilar Balance (improved form).

the base A B, carries a circular disc or movable base of light wood, $o\,x$, 10 inches in diameter, and a quarter of an inch thick, firmly united, centrally, to the extremity of the axis H, so that by turning a milled head, h, beneath at the lower extremity of the axis, the circular base $o\,x$ can be moved round. To facilitate this movement upon its square base of support, through which the central axis of motion passes, very fine rollers are delicately inserted in the under surface of the movable base $o\,x$. Two small gilt discs, P Q, similar to those at the extremity of the needle, are fixed on delicate sliding insulators (Chap. I., Note E), sustained in small feet inserted in opposite points

of a diameter of the movable base o x, immediately opposite
and nearly to the discs M, N of the needle, so as to admit of
one of the discs M or N, at the extremity of the needle, coming
very nearly to and parallel with one of the discs P or Q at the
extremity of the sliding insulator. The bifilar threads a b and
a' b' are sustained in a convenient, light, open, metallic frame, e t',
such as already described (157). This light frame-work is constructed of two small, parallel, vertical, metallic tubes a t, a' t',
14 inches long, fixed in a firm circular metallic base, movable,
centrally, in a socket within a light elevated cross-piece, y x.
Two light metallic bars e f, g h, attached to sliders movable on
the vertical tubes a t, a' t', unite the vertical side-pieces of the
frame e t' with each other. The bifilar threads a b, a' b', pass
through fine holes in thin metallic plates, connected with the
bar g h. Thus, by means of the fine holes in the cross-bar g h,
and the sliders on the side-pieces a t and a' t', the lengths of the
bifilar threads of suspension and their distances apart may be regulated to any required extent, in the way already explained (157).
The suspension threads a b, a' b', are attached to short lengths of
stout silk thread, passing round cylindrical winders in the bar e f
above, and terminate in the needle suspension below. The needle
is suspended by means of very fine wire hooks b b' (Fig. 96),

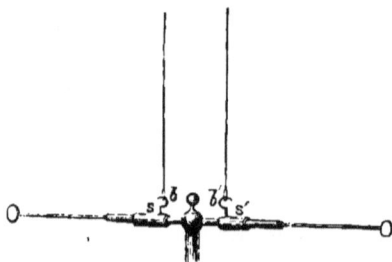

united to short sliding-pieces
of small silvered brass tube.
These sliders are movable
upon a short length of brass
tube carrying the needle, and
by which it is sustained. A
descending vertical tubular
rod, u v (Fig. 95), 6 inches
long, and one-eighth of an
inch in diameter, is attached

Fig. 96. Mode of Suspending Needle.

to the small tube b b' carrying the needle, and extends to the
centre of the under surface of the movable circular base o x, where
it is steadied by a fine needle-point, in which it terminates, and
corresponds with a fine hole drilled in the termination of an
interior axis h, passing through the axis H of the movable base,
in such a way as to admit of the two axes moving one within the
other, after the manner of the hands of a watch.

Thus, whilst the needle is accurately centred, it is prevented
from acquiring any undue lateral motion. The axis h carries a
forked lever for arresting and regulating the position of the
needle through a light index, I, of reed at the lower extremity of

the vertical tubular rod *u v* proceeding from the central tube *b b'* of the needle.

A fixed graduated circle surrounds the movable base *o x*, within which the circular base moves. This circle is divided into 360 degrees. The long index I at the lower extremity of the descending tubular rod *u v* attached to the centre of the needle, moves over the graduated circle surrounding the movable base *o x*, and indicates in degrees the angular deflection of the needle by the repulsive force between the discs M or N and P or Q when brought near each other and electrically charged with the least possible quantity. The instrument indicates an extremely small charge communicated to the opposed discs.

By turning the circular basis on which the supporting frame *e t'* of the bifilar threads rest, we are enabled to deflect the needle from its position of rest any number of degrees. Hence the number of degrees required to reverse the torsion is a measure of the repulsion—that is to say, a measure of the turn of the bifilar threads upon each other. To effect this with precision, a graduated circle, divided into 360 degrees, is fixed on the upper surface of the circular base *t t'*, together with a fixed line or index to denote the zero point. By this arrangement the instrument with the needle is under perfect command, and the index I easily adjusted to the zero points of the graduated circle *o x*.

As it is often convenient to pass the fixed insulated discs P, Q downward through the cover C D of the cage, the insulators, carrying the discs, slide through fine cork, closely fitted in capacious holes in one of the bars of the cover movable under the bridge *y z*, so as to admit of the fixed discs descending within the cage; and thus one of the fixed discs P or Q, and one of the discs M or N of the needle may be placed parallel and very nearly to each other by turning the movable circle *y z*, or movable base *o x*. We may in this way operate upon the repelling discs either above, through the cover of the cage, or beneath, by means of the movable circular base *o x*, the distance between the repelling discs being measured in degrees of a circle of 1 foot in diameter divided into 360 parts. The insulators carrying the fixed discs P, Q, consist of strong vulcanite or other insulating rods, about 6 inches long, and the sixteenth of an inch in diameter.

163. One or two simple experimental illustrations will be sufficient in explanation of the operation of this instrument—the Two-threaded or Bifilar Balance.

Exp. 44. The required lengths and distances of the bifilar threads being duly adjusted, together with the horizontal position

of the needle, by means of the arrangements just described (162),
bring one of the insulated discs P or Q parallel and very nearly to
one of the discs M or N of the needle, which may be done either
by the movable circle y z of the cage above, with which the fixed
discs P and Q are very frequently connected, or by means of the
movable circular base below, in which the fixed discs P and Q are
commonly inserted. Charge either of the fixed discs P or Q with
a given quantity of electricity by means of a transfer plate and
quantity jar (144), suppose 1 circular inch (145). The disc M or
N of the needle will be immediately repelled by the charged disc P or
Q to a given distance (27), suppose 10 degrees, as indicated by the
index I, previously set to zero of the circle o x.   .

In this case the indicated arc of 10 degrees is very little different from its chord, so that the distance of the fixed disc P or Q
and the disc of the needle M or N may be correctly represented by
10 degrees. Hence we have distance of repulsion 10 degrees, and
repulsion 10 degrees, quantity of electricity being 1 circular inch.

Exp. 45. The quantity of electricity remaining the same—that
is, 1 circular inch—bring the two repulsive discs M or N and P or Q
within 5 degrees of each other, or half the former distance. The
repulsion, as indicated by the index I, will rapidly increase as the
distance between the discs decreases; so that the index I—when
the discs are one-half the previous distance apart, that is to say,
5 degrees instead of 10—will indicate a repulsion of 40 degrees,
or four times the former.

Thus, whilst the distances are as $1 : \frac{1}{2}$, the repulsions, as measured
in degrees, are as $1^2 : 2^2$, that is, as $1 : 4$, clearly showing, as in
previous instances of attraction (154), that the repulsion varies in
an inverse duplicate ratio of the distances between the repelling
surfaces. We arrive at the actual reactive force of the threads upon
the needle by some such means as those referred to (162), as for
example, by turning the movable basis (on which the frame supporting the bifilar threads rests) in a direction contrary to the
twist of the threads, and observing how many degrees we
require to turn the circle in a reverse direction to bring the
needle back to its primitive position.

Exp. 46. The fixed disc P or Q, and one of the discs M or N
of the needle being previously set parallel and very nearly to
each other, as in the former experiment indicating a force of
10 degrees, charge the fixed disc P or Q with 2 circular inches,
or twice the former quantity, by means of the transfer plate and
quantity jar (144). The index I will now indicate 20 degrees, or
twice the former repulsion of 10 degrees.

Thus the distances of repulsion are in this case directly as the quantities of electricity in action.

164. We find by these experiments: 1st. That electrical repulsion varies in an inverse duplicate ratio of the distances between the repelling surfaces. 2nd. That with a double quantity the repulsion extends to a double distance, and is directly as the quantity.

165. It may perhaps be as well to observe that in Exp. 45, whilst the quantity remains constant, the distances vary; and in Exp. 46, whilst the distance remains constant, the quantities vary (T).

166. *The Balance of Torsion.*—If a cylindrical weight, w (Fig. 97), be suspended by a fine, elastic, metallic wire, or silk line, *w e*, from a fixed point, *w*, so that the axis of suspension be immediately under the fixed point, and the cylindrical weight and wire be turned round a little upon its point of suspension, the wire will be somewhat twisted, and will have acquired an elastic force, which causes it, together with the suspended weight, to tend to regain its previous position, and if set free, the weight together with the wire will assume a state of vibration or oscillation.

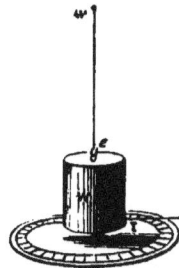

Fig. 97. Coulombe's Torsion Apparatus.

The elastic force set up in the wire *w e* has been termed *the force of torsion*, or *twist*. By carefully observing the duration of a certain number of oscillations of the suspended weight and wire, we may determine, by means of the formula for oscillating bodies, given by Coulombe in his celebrated memoir on the force of torsion, in the Proceedings of the Royal Academy of Sciences of Paris, the reactive force producing the oscillations.

The simple apparatus employed by Coulombe is represented in Fig. 97. The weight w, suspended by the metallic wire or silk line *w e*, carries an index, I, which marks on a graduated card, I, the duration and extent of the oscillations. By varying the weight of the suspended body w, the length of the wire of suspension, its tension and thickness, and the kind of metal of which it consists, we may determine the laws of the reactive force of the torsion of the wire, in reference to the elements upon which the laws of torsion depend.

167. To determine the reactive force of a wire thus subjected to torsion, different weights are suspended from a fixed point by metallic wires of various kinds, lengths, and thicknesses.

The following are the general results of Coulombe's physical investigations :—

The reactive force of the twist or torsion of a wire is as its length, thickness, and tension.

The force of torsion of a wire is as the angle of torsion—that is to say, the angle at which the wire is twisted or turned upon itself.

Upon these elements of force Coulombe constructed his beautiful *Balance of Torsion* for estimating small physical forces, more especially those of electricity and magnetism.

168. The balance of torsion (Fig. 98) consists of a light needle, *m n*, 6 inches long, suspended by a fine wire, *w e*, within a cylindrical cage, A B, 12 inches in diameter, and 12 inches in height ; it is covered by a circular plate of glass, A C, 13 inches in diameter. The centre of the glass plate is pierced by a hole, H, 2 inches in diameter, to receive a vertical glass tube, T H, through which passes, centrally, the fine metallic wire, *w e*. The needle consists of three separate parts, *m, e, n*— a central part, *e*, terminating in two stiff threads of gum-lac, *m, n*.

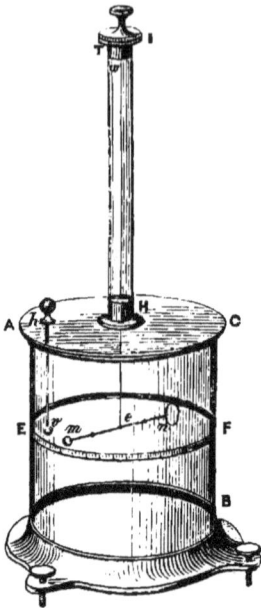

At one of the gum-lac extremities, *m*, is a light pith ball, about half an inch in diameter, weighing about a grain and a half. The opposite gum-lac extremity, *n*, carries a circular gilt paper disc, about 1½ inch in diameter, saturated with turpentine, which, whilst it tends to counterpoise the pith-ball *m*, restrains the oscillation of the needle.

Fig. 98. Balance of Torsion.

Directly an electrified body approaches the pith-ball *m* at the gum-lac extremity of the needle *m n* similarly electrified, the needle is repelled, and turns the wire suspending the needle so as to impress upon it, from its fixed point, torsion or twist. It is the reactive force of this torsion which was employed by Coulombe to measure the force of the repulsion of the needle.

The torsion wire *w e*, carrying the needle *m n*, is subjected to the strain or tension of a weight appended to it, just less than sufficient to break it. The circular glass plate A C closing the cage, A B, has another hole, *h*, through it, about an inch in diameter for the convenience of passing any particular substance into the cage. A second pith-ball, v, at the extremity of a stiff thread of gum-lac, similarly electrified to the ball m of the needle, is passed through the hole h into the cage A B so as to act upon and be exactly opposite the ball m of the needle.

The upper extremity of the glass tube T H receives, after the manner of a telescopic joint, a short, polished, metallic tube, about 2 inches in length, on which is fixed a circular silvered plate, I.

The edge of this circular plate is divided and graduated into 360 degrees, having a zero point, from which the divisions of the circle are estimated.

The torsion wire $w\,e$ is suspended through the medium of delicate pincers, fixed in the under surface of the plate I, closing the upper extremity of the tube T H. A graduated circle, E F, also divided into 360 degrees, surrounds the cage A B at the level of the needle. The position of the graduated circle is such that its zero commences from the point opposite the fixed pith-ball v. Since the tube T, terminating the tube T H and carrying the plate I, turns with a little friction within the tube T H, the zero of the graduated plate I on the tube may be brought into any required position relatively to the divisions of the graduated circle E F, surrounding the cage A B below ; and as the pincers holding the torsion wire $w\,e$ proceed from the under surface of the plate I, the pincers holding the wire $w\,e$ and the wire itself, together with the suspended needle, must also turn with it. Thus, the wire $w\,e$ becomes subjected to torsion to a given extent.

169. If the centre of the pith-ball m of the needle be supposed to correspond precisely with the zero of the circle E F surrounding the cage before the introduction of the pith-ball v ; and if the centre of the fixed pith-ball v, when it is introduced, be made to take the place of the centre of the pith-ball m of the needle, it will displace the ball m, and therefore turn it aside through a space equal to the sum of the radii of the two balls. If either of the two balls thus in contact be electrified, the electricity will be shared between them, and they will repel each other (27), but as the ball v is fixed, the repulsion will only take effect upon the ball m, which will separate from v, and cause the needle $m\ n$ to move from v ; m will continue thus to depart from v, twisting the wire as it turns, until the reaction of the torsion of the wire balances the repulsive force, m will then remain at rest at a distance from v, indicated by the division of the circle E F, surrounding the cage, opposite to which it stands. The number of that division may be taken to represent the angle of torsion of the suspending wire,

L 2

and this angle is always proportional to the reactive force of the torsion, or the force with which the torsion wire $w\ e$ endeavours to recover its position of rest.

In this manner the force of repulsion at different distances may be measured and observed.

170. Such was the elasticity of the wire employed by Coulombe that after having been subjected to the torsion of many circles it accurately returned to its first position.

By turning the plate I, suspending the torsion wire $w\ e$ and the needle $m\ n$, the ball m of the needle may be brought opposite the zero of the graduated circle E F, when it will be immediately under the hole h, through which the pith-ball v and its gum-lac insulator is passed into the cage A B, so as to be directly opposed to the pith-ball m of the needle.

171. The extreme degree of sensibility of Coulombe's Balance of Torsion, and the infinitesimal quantities it is capable of measuring may be collected from the consideration of the dimensions of the torsion wire employed. In the experiments of Coulombe (174), by which the laws of electrical attraction and repulsion were investigated, the suspending wire was 28 inches in length, and was so fine that one foot of its length weighed only the one-sixteenth part of a grain. This, however, was found too fragile for actual experiment. The radius of a circle described by the ball m of the needle was four inches. The elasticity of the wire employed by Coulombe was so perfect that when turned through one entire revolution, or 360°, its reactive force amounted to no more than the 340th part of a grain, and since Coulombe found that the reactive force of torsion is proportional to the angle of torsion (166), the force corresponding to the motion of the ball m through one degree of the circle was only the 122,400th part of a grain; thus dividing a single grain into 122,400 parts, and rendering each part distinctly visible. Coulombe, however, as we have already remarked, found it convenient to employ wires of less sensibility. The sensibility of a torsion wire was found by Coulombe to increase with the length of the wire, all other things remaining the same, and to increase inversely as the fourth power of its diameter, the length being constant; so that the sensibility of the balance of torsion is in the direct proportion of the length of the torsion wire, and in the inverse proportion of the fourth power of its diameter. If, for example, the length of the wire be doubled, the sensibility of the instrument will be doubled; and if the diameter of the wire be reduced in the proportion of two to one, its length being the same, the sensibility of the instrument will be sixteen times as great;

that is to say, the reactive force will be one-sixteenth greater with a wire of one-half the diameter.

172. To communicate electricity to the fixed ball v of the balance we employ a transfer of given dimensions (145), according to the quantity of electricity we wish to impart, the transfer being introduced into the cage through the hole h in the cover.

Such is the sensibility of the torsion balance, that if, after having excited a stick of sealing-wax and weakly electrified the ball m of the needle, negatively, the excited wax be held at a distance of 3 feet from the electrified ball m, the ball will be repelled 90 degrees.

173. In the forces which are manifest between small particles of matter, the first circumstance to be noticed is the fact that the energy of the particles on each other is augmented in some proportion as the distance between them is decreased. The analogies suggested by physical forces generally, the intensities of which increase with the diminution of distance, and more especially the law of gravitation, by which the energy of that force increases inversely as the square of the distance between the particles, naturally leads to the question :—according to what law does the force of electrical attraction or repulsion increase or decrease as the distance between the attracting or repelling bodies decreases or increases ?

If the nature of electricity were perfectly known, this law could be deduced by general reasoning, so that the manner in which electrified bodies would comport themselves in any position in which they might be respectively placed could be certainly foretold. But the physical principles from which electricity arises not being known, the investigation of its laws and the establishment of a just theory respecting its nature depends upon our knowledge of observed phenomena. To determine, therefore, the law of electrical action, it is necessary to submit electrified bodies to their mutual attractions and repulsions at different distances, to measure the actual amount of that attraction and repulsion at those distances, and, by comparing the results of such measurement with the distances themselves, to discover the dependence of one upon the other. This was effected by Coulombe, by the aid of his Balance of Torsion.

174. The following are elegant experimental illustrations— originating with Coulombe in his early investigations in electricity—of the application of his torsion balance to physical research.

Exp. 47. The centre of the pith-ball m, Fig. 98, corresponding precisely with the zero of the circle E F, surrounding the cage, and the ball r being passed into the cage through the hole h, so as to be in contact with the ball m when at zero of the circle E F,—charge the balls m, r, by means of a transfer plate of given dimensions (145), with a quantity sufficient to repel the needle 36°.

In this case the repulsion will be 36° and the torsion force 36°.

Turn the plate I, closing the tube T H, 126° back against the direction of the already existing torsion ; the effect of this will be to impress an additional torsion of 126° on the wire of suspension $w\,e$ and to bring the balls m, r, within 18° of each other, or half the former distance of 36°.

We now have for the two torsions, 36 and 144. In this case the distances of repulsion are as 36° : 18°, that is as 2 : 1; whilst the forces of repulsion corresponding to those distances are as 36 : 144, that is as 1 : 4; or, in other words, the electrical forces are in an inverse ratio of the squares of the distances.

Exp. 48. Turn the plate I farther back against the repulsion 441°, this will impress on the torsion wire $w\,e$ an additional torsion of 441°; This will bring the balls m, r, within $8\frac{1}{2}$° or 9° of each other, or half the former distance.

Comparing these three experiments, we have for the distances of repulsion 36, 18, and 9, and the forces corresponding to those distances 36, 144, and 576. The forces, therefore, still continue to vary in the inverse ratio of the squares of the distances, being in accordance with the great general law of electrical action (U).

175. *The Thermo-Electrometer.*—It was long since observed by Mr. Children, that the heat evolved by metallic bodies in transmitting an electrical charge is in some inverse ratio of their conducting power,—a principle generally admitted, not only as a reasonable deduction, but also as being established by a great variety of facts.

176. The effect of the electrical discharge on metallic bodies is to raise their temperature to a greater or less degree, and in many instances to render such bodies red-hot, and to dissipate them in melted globules.

The fusion of wire has been hence frequently resorted to by the older electricians (Van Marum, Cuthbertson, Singer, and others) as a measure of electrical discharge. This method, however, has often been found tedious and uncertain in its results, and, in many instances, quite inapplicable to refined inquiry. Thus Cuthbertson, in comparing the results of some experiments made by himself on the fusion of wires by electricity with those of similar experiments by Dr. Van Marum, observes :—

" The doctor might, perhaps, have been been led into a mistake in the following manner. He might not have been aware of the different degrees of ignition caused by electric discharges, but only judged of the force, by the wires being converted into balls, by which great mistakes will frequently arise; for if a wire be taken 18 inches long, and of such a diameter that when a jar or battery is charged to such a height as just to cause it to run into balls, much shorter lengths of that same sort of wire may be subjected to the same force, and still only be converted into balls. If only 7 inches were taken, nothing but balls will appear; the only difference will be that the balls will be smaller and dispersed to a greater distance, which might be easily overlooked. If 6 inches of the same sort of wire be taken it will be converted into balls and flocculi, or brown oxide of iron; so that, to be accurate in this point, the lowest degree of ignition must be had, which is known when the charge has passed; the wire will be red-hot the whole length, and afterwards fall into balls "*—a rather precarious condition.

177. The Thermo-Electrometer, whilst it avoids all destruction of the metal, indicates, at the same time, the comparative heating effect of the electrical discharge, and admits of an accurate estimate of its force.

178. This instrument (Fig. 99) is of very simple construction, being little more than an air-thermometer, having a fine metallic wire, or wires, passed air-tight through its ball. A glass tube, *a b c*, the interior diameter of which is regular and somewhat less than one-tenth of an inch, has one of its extremities, *b c*, bent upwards and outwards for about 2 inches, and is united by welding to an elliptical glass cup or reservoir, R, about 2½ inches in length, 2 inches in width and depth; this reservoir contains a small quantity of coloured spirit, such as is employed in the spirit thermometer. The vertical part of the tube is sustained by a graduated scale, *a b*, 2 feet in length, divided into 150 degrees, each degree being

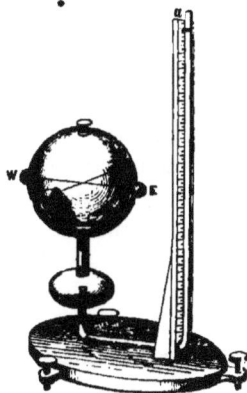

Fig. 99. Thermo-Electrometer.

one-tenth of an inch. This scale is fixed upon a convenient base, *b c*, and the point on it at the level of the coloured spirit in the tube is marked zero. Upon the reservoir R is screwed a glass ball of 5 inches diameter, having a metallic wire, W E, or wires, passed air-tight across its centre.

* Cuthbertson's "Practical Electricity," pp. 185—186.

The method of fixing the wire is easy.   Two flanches of brass, *f*, *f'* (Fig. 100), with shoulders and projecting screws, are cemented in and over the holes drilled through the glass ball for the passage of the wire; the wire is passed directly through the ball by means of corresponding small holes drilled through the flanches, and, being gently put on the stretch, is secured by fine wooden or metallic pegs, by which the wire is slightly compressed in the hole.   When metallic pegs are used, a small groove should be cut in them for the passage of the wire. Both the pegs and extremities of the wire project a little for the convenience of removal, and thus wires of various kinds and of different diameters may be easily substituted.   The whole is finally rendered air-tight by means of small brass balls, B, B', flattened on one surface, and screwed on the flanches against leather collars, so as to render the ball air-tight.   Beside these flanches, the ball is also furnished with a valvular projection, v, attached in a similar way to its upper part, which being rendered air-tight by a screw and leather can be occasionally opened, so as to form a communication with the external air, and thus the level of the coloured spirit in the reservoir R may at all times be adjusted to zero of its scale.

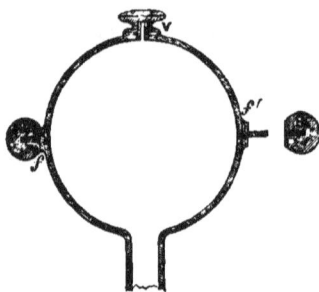

Fig. 100. Mode of Fixing Wire.

179. For the convenience of passing more than one wire through the centre of the glass ball, there are additional holes, furnished with brass flanches cemented to the glass, which are supplied with covering balls in the way above described. We may thus employ several wires of different metals, and cause an electrical charge to pass through either of them, without interfering with the wires already in the ball; and thus obtain a Thermo-Electrometer of different degrees of sensibility, according to the metallic wire employed.   Under these circumstances, when an electrical charge of sufficient force is passed through the wire or wires in the ball, the relative degree of heat evolved is made evident by the ascent of the fluid along the graduated scale *a b*, Fig. 99.

180. The application of the Thermo-Electrometer in promoting electrical investigation is easily seen by a few, simple, direct, and efficient experiments.

Exp. 49. Pass successively through the ball B of the Thermo-Electrometer (Fig. 101) fine wires of different metals; firmly fix them in position, as already described (178). The wires should be drawn down through successive holes in a hardened steel plate to about the one-eightieth of an inch in diameter, and consist of silver, copper, gold, zinc, platinum, iron, tin, and lead.

Charge an electrical jar, A (Fig. 101), exposing about 4 square feet of coated surface (113), with a given number of units (119), through a unit measure, U, and having completed the required metallic communications, discharge the given accumulation by means of a charging and discharging quantitative electrometer E (139),

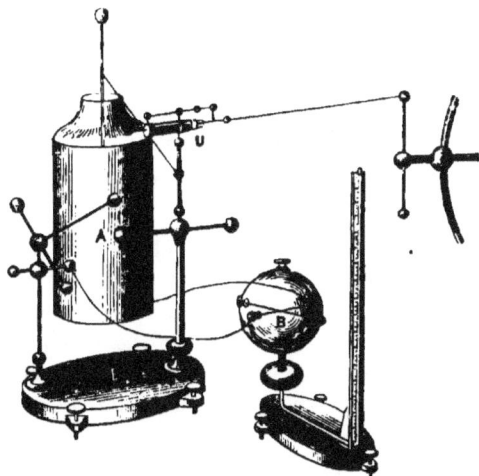

Fig. 101. Mode of Using the Thermo-Electrometer.

through the particular wire under examination. Observe the heating effect of each metal, by the ascent of the fluid along the scale. We find for copper, 6°; for silver, 6°; for gold, 9°; for zinc, 18°; for platinum, 30°; for iron, 30°; for tin, 36°; for lead, 72°; from which it appears that *silver* and *copper* are the *best* conductors, as being the least heated by a given electrical discharge; *lead* is the *worst* conductor, as being the most heated by the same electrical discharge, whilst *iron* and *platinum* are between the two.

181. The susceptibility of fine wires, of different metals and of given diameters, to become heated to a greater or less extent by the electrical discharge enables us to convert the Thermo-Electrometer into an ordinary electrometer of different degrees of sensibility. Thus (180), whilst a given electrical charge passed through a fine wire of copper of given diameter in the ball, causes the fluid to ascend along its scale only 6°, the same charge will, when passed through a platinum wire of the same diameter enclosed in the ball, raise the fluid to five times the height, or 30°,

and when passed through a wire of lead of similar dimensions will raise the fluid to no less than twelve times the height, or 72° ; thus affording a wide range of observation between copper, platinum, and lead.

The metal best adapted to the purposes of an ordinary electrometer is platinum ; it is easily acted on, and not liable to oxidisation. The wire in the ball may vary in diameter from the one-fiftieth to the one-hundred-and-fiftieth of an inch, according to the requirements of the experiment. This method of estimating electrometer effects, by means of the electrical discharge through metallic wires, is convenient, and susceptible of much greater accuracy than the ordinary means by fusion (176).

182. Fig. 99 represents a Thermo-Electrometer with two wires, silver and platinum, each one-hundredth of an inch in diameter, passed air-tight through its ball, by which electrometer effects are readily determined.

Exp. 50. Charge the electrical jar A (Fig. 101), exposing about 4 square feet of coated surface, with 10 units through the medium of the unit measure U. Having completed the requisite metallic communications, through the charging and discharging quantitative electrometer, discharge the accumulation through the platinum wire. Observe the ascent of the fluid along the scale— suppose 10 degrees.

Exp. 51. Accumulate 20 units, or double the former quantity of electricity, on the given surface of 4 square feet, and discharge the accumulation through the same wire. The fluid will now be observed to ascend along the scale of the instrument four times the former height, or 40 degrees ; by which we perceive that the heat evolved by a given electrical discharge is as the *square of the quantity of electricity.*

OCCASIONAL MEMORANDA AND EXPLANATORY NOTES.

(R) The error arising from the small quantity of charge left upon the transfer-plate or globe after each deposit may in most cases be neglected, or otherwise calculated or avoided altogether. Suppose the relative diameters of a transfer-plate or globe, and a given circular plate or globe to be as 1 : 12 ; in that case the charge would be twelve times as great upon the larger plate or globe as upon the smaller plate or globe. To put in evidence how small a quantity would in this case be left upon the transfer after contact with the given surface, we may observe that the charge of which the larger plate is susceptible is at least twelve times that of the smaller plate, and therefore after the contact the charge of the smaller disc will be shared between the two, the larger one consequently taking eleven-twelfths of the whole charge, leaving only a residual of one-twelfth part of the transfer charge on the circular inch, which (145) would be considerably less than could be sustained by a disc of one-tenth of an inch in diameter. There will of course be a slight increase after each transfer, but for the few transfers generally employed the error arising therefrom would be of no moment. It is easy to estimate the amount of the residual on the transfer by placing a hollow sphere, having an open mouth (57), in connection with a delicate electrometer, and after each contact of the transfer measure with the given insulated conducting plate, deposit the residual within the hollow sphere. We know in this case that all the residual will be taken up by the hollow sphere; the electrometer will, consequently, indicate the total sensible quantity of electricity which the small transfer plate has carried away after any given number of transfers, the distance between the electrometer attracting discs remaining constant.

This kind of experiment will be quite sufficient for estimating the greatest error which could possibly arise from the small quantity carried off by the transfer measure after a given number of deposits.

(S) Light circular plates, hollow spheres, and other light bodies for experimental investigation by means of the Scale-beam Electrometer should be constructed of a very light wood, termed white cedar lath wood (Chap. V., 35), and should be constructed in the way referred to Chap. III., 53. A circular plate of cedar lath wood, 7 inches in diameter and three-tenths of an inch thick, when carefully prepared in the lathe, and lightly varnished, was found to weigh only 700 grains.

A hollow sphere of 5 inches diameter and three-twentieths of an inch thick, constructed of this wood, was found to weigh less than 24 drachms, or 3 ounces; and, consequently, well adapted to the delicate Scale-beam Electrometer.

(T) The Rev. Dr. Lloyd, Professor of Mathematics in the Dublin University, appears to have arrived at a different result from that of Coulombe. Instead of the reactive force of torsion being, according to Coulombe, as the angle of torsion, it appears, by Dr. Lloyd's magnetic investigations, the reactive force is as the sine of the angle, and not as the angle itself. This difference, however, with Coulombe's result has probably arisen from the circumstance that Dr. Lloyd's experimental investigations depended upon bifilar threads, not sustained at given distances apart by cross-stays (159), but by bifilar threads which in turning were allowed to collapse upon each other, and which would greatly influence the result of the mathematical analysis of the force.

Fig. 102. Bifilar Suspension.

Let, for example, 2 r = distance between the threads = $m\ m'$ = $n\ n'$; P P' = length of suspended needle = 2 a; angle of inclination of the thread at the point n to the vertical = i; horizontal angle through which P P' has moved, owing to the repulsive force under examination = θ; tensions of threads, $m\ n$ and $m'\ n'$ (which are not precisely the same) = T and T' respectively; weight of the suspended needle P P' = W; horizontal force disturbing P P', and acting at right angles, = f.

Then taking moments about n in the horizontal plane containing P P', we

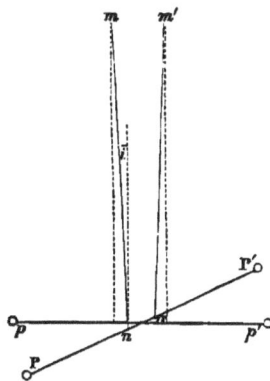

have (τ' sin. i) 2 $r = f (a + r)$ . . . . (1). Similarly, taking moments about n' we have (τ sin. i) 2 $r = f (a - r)$ . . . . (2).

Resolving the forces of the system vertically, we have ($\tau + \tau'$) cos. $i = w$; $\therefore \tau + \tau' = \dfrac{w}{\cos. i}$; but $\dfrac{\tau'}{\tau} = \dfrac{a + r}{a - r}$, dividing (1) by (2); $\therefore \dfrac{\tau'}{\tau + \tau'} = \dfrac{a + r}{2 a}$; and $\tau' = \dfrac{a + r}{2 a} (\tau + \tau') = \dfrac{a + r}{2 a} \dfrac{w}{\cos. i}$.

From (1) $f = \dfrac{2 r \sin. i}{a + r} \tau' = \dfrac{2 r \sin. i}{a + r} \times \dfrac{a + r}{2 a} \dfrac{w}{\cos. i} = \dfrac{r}{a} w \tan. i$.

Now perpendicular distance of point of suspension above the index needle P P$'$ = l cos. $i = p$ say, l being the length of each thread in the bifilar balance, and $r \theta = p$ tan. i accurately = l cos. $i \times$ tan. i $\therefore$ tan. $i = \dfrac{r \theta}{l} \times$ sec. i. But i being a very small angle sec. i is nearly = 1, and $\therefore$ tan. i may be considered $= \dfrac{r \theta}{l}$, this being too small by only a small quantity.

$\therefore f = \dfrac{r}{a} w$, tan. 1 becomes $-\dfrac{r}{a} w \dfrac{r \theta}{l} = \dfrac{w r^2}{a l} \theta$, and since $\dfrac{w r^2}{a l}$ is constant, depending upon the dimensions and weight of index needle P P$'$, length of threads and distance between them, it appears that $f \propto \theta$; that is, the force deflecting P P$'$ varies directly as the angle of deflection when f acts at right angles to P P$'$. If the force be considered as acting in the direction of the chord of the arc in which P moves, then the force f so acting, $= \dfrac{w r^2}{a l} \times \dfrac{\theta}{\cos. \frac{\theta}{2}}$, in which case $f \propto \dfrac{\theta}{\cos. \frac{\theta}{2}}$. But since, in the practical application of this instrument the arc between the repelling disc and the repelled end P of the needle is less as the force under consideration increases, cos. $\dfrac{\theta}{2}$ will differ very little from unity; and we may, consequently, without any sensible error, consider that $f \propto \theta$, or as the horizontal angle through which the needle has moved owing to the repulsive force, which is the same result as that arrived at by Coulombe.

(U) It may appear difficult to understand how, by turning the plate I in the reverse direction of the torsion, we thereby increase the torsion force of the suspending wire. It will however appear, on a little reflection, that whilst the suspending wire is liable to torsion from one extremity in one direction, it is acted upon by torsion on the other extremity in the opposite direction; that is to say, it is liable to twist by the action of the plate I above in one direction, and by the resisting repulsive force below in the opposite direction. It is, in fact, in the same condition as that of a rope twisted at both ends in opposite directions.

CHAPTER V.

The most common articles required by the practical electrician may be thus enumerated :—

(1.) A moderately sized turning lathe, with turning tools, drills, and other requisites.

(2.) A small work-bench of solid oak, fitted with parallel and common vices, supported on strong iron levelling-screws, so as to give it a firm and steady position.

This bench may be about 4 feet long, 18 inches wide, and about 3 feet high. It should have two drawers, one at each end, immediately under its table surface, and two or three capacious drawers in front. The drawers at the extremities are for small - tools, and other selected articles, such as fine broaches, and piercers of various kinds ; delicate pliers, nippers, forceps, and the like.

(3.) Saws of various kinds and sizes, for wood and metal.

(4.) Planes of various kinds, small and large ; spokeshaves.

(5.) Screwdrivers, chisels, gouges, gimblets, bradawls.

(6.) Files and rasps of different sorts and sizes, flat, round, half-round, and square.

Slender round rasps and files, usually termed rat's-tail, are useful in completing roughly perforated holes.

(7.) Pincers, pliers, hand-vices, cutting nippers.

(8.) Callipers of various dimensions for taking measurements.

(9.) Cutting knives and scissors.

(10.) Large and small drills, with drill-bow and stock handles to fix and work in a vice.

(11.) A set of punches from a quarter of an inch in diameter upwards, including such as are used for gun wadding : a square or circular mass of lead for punching on.

(12.) One or two portable anvils mounted on wooden blocks.

(13.) Hammers for fine and coarse purposes.

(14.) A stock, with bits of various sizes.

(15.) Oil stones, pressure oil-cans of a conical form for dropping oil.

(16.) A diamond for scratching, or cutting glass.

(17.) One or two tobacco-pipe stems for blowing a stream of hot air on any substance through their heated extremities.

(18.) A set of heating-irons (Fig. 103), for warming glass rods, and other insulators. These should be inserted in handles, and should be of various forms and dimensions.

The globular iron *a* is very convenient for drying the interior of glass jars.

Irons having curvilinear or other surfaces, such as *b*, *c*, *d*, *e*, are convenient for drying off cylindrical or other insulators. Two curvilinear irons, such as *b*, *c*, or *d*, applied against each other may be made to surround glass rods of various sizes.

Heating-irons of large mass, mounted upon a foot as represented at *f*, are applicable in varnishing glass insulators of various forms and sizes; being fixed in a wooden holder they are easily

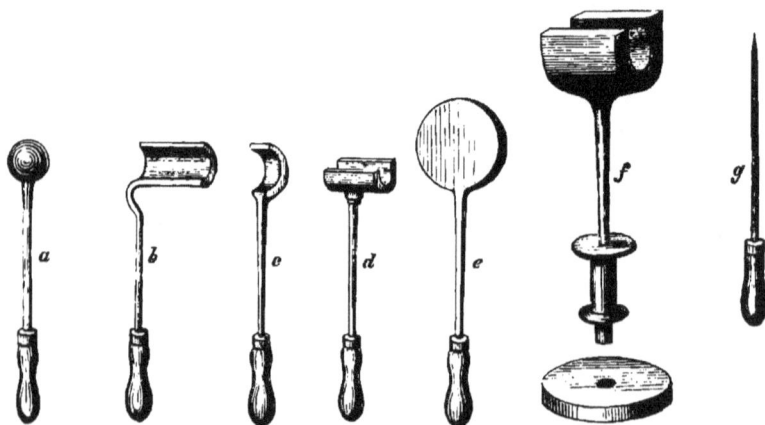

Fig. 103. Heating Irons.

heated in the fire, and subsequently supported by the wooden holder on a stout wooden foot. The glass to be varnished is held over them.

(19.) A set of burning-irons in handles *g* (Fig. 103), varying from the size of a common steel knitting-needle to half an inch in diameter, and from 6 inches to 1 foot or more in length. Such irons, together with the rasps (6), are useful in perforating holes through corks and other substances.

(20.) Earthenware pipkins of various sizes, chemical crucibles, and such like.

(21.) T-squares, straight-edges, and graduated rulers, with plumb-lines attached.

(22.) Case of drawing instruments.

(23.) One or two right-angled triangles for drawing.

(24.) Spring compasses and dividers, coarse and fine.

(25.) A square mahogany board set on levelling-screws, for obtaining a level surface when necessary.

(26.) One or two spirit-levels.

(27.) A dividing-board for graduating card-paper, or other circles. This consists (Fig. 104) of a graduated brass circle, from 1 foot to 18 inches diameter, inlaid, centrally, in a square mahogany board. The meridian circle of a globe may be employed for this purpose. A central point is screwed into the board, against which rests the notch of a straight metallic rule, so adapted as to be, when in place, a prolongation of the radius of the brass circle. The cardboard to be divided having circles swept ˉupon it by accurate compasses, is pinned down to the board, the centre of the circle to be divided being in the centre of the brass circle. The radius ruler is moved round over the graduations of the brass circle, and the degrees marked off by a drawing-pen on the card circle within.

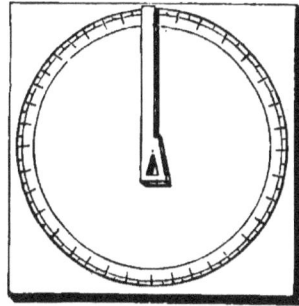

Fig. 104. Dividing Board.

(28.) A good double-barrelled air-pump, with receivers of various sizes.

(29.) Balances with scales. A balance to weigh from 1 ounce to 3 pounds, and which will turn with a few grains when loaded with its greatest weight.

(30.) A balance for more refined purposes, which will weigh from half a grain to 3 ounces, and will turn with half a grain when loaded with 1000 grains.

(31.) A balance for still more refined purposes, which will weigh from the tenth of a grain to 1 ounce, and will turn with one-tenth of a grain when loaded with 500 grains.

The first two may be suspended from a lever arm projecting from a brass standard rod, so that by a fine silk cord passing over a pulley to a lever foot beneath, the beam may be raised, and the scale-pans lifted off the platform upon which they rest.

The brass standard in these cases is usually fixed on a neat mahogany rectangular box, having a drawer for the reception of the necessary weights.

The second, or finer balance, should be enclosed in a small glass case, with front doors, or a sliding front plate of glass—which can be elevated to any height.

The third, or very refined balance, should be mounted in a way similar to the fine chemical balances, and for which it is to be a substitute for delicate purposes. This should be enclosed in a glass case similar to the former. A chemical balance of the highest quality and construction is not called for in ordinary electrical investigations.

(32.) *Weights.*—There are several kinds of British weights in use, such as the avoirdupoise weight, troy weight, and apothecaries' or chemists' weights.

The different divisions of avoirdupoise weight are tons, hundreds, quarters, pounds, ounces, and drachms. In this weight 16 drachms make 1 ounce = 960 grains, and 16 ounces 1 pound = 15,360 grains.

The divisions of troy weight are pounds, ounces, pennyweights, and grains. In this weight 24 grains make 1 pennyweight, 20 pennyweights 1 ounce = 480 grains, and 12 ounces 1 pound = 5,760 grains.

In apothecaries' or chemists' weight the divisions are pounds, ounces, drachms, scruples, and grains. In this kind of weight 20 grains make 1 scruple, 3 scruples 1 drachm = 60 grains, 8 drachms 1 ounce = 480 grains, 12 ounces 1 pound = 5,760 grains.

These different denominations of weight are apt to lead to much confusion, so that it becomes desirable, for the sake of simplicity, to resort to a unit standard of weight, of which a grain is to be considered the unit. Every denomination of weight, therefore, is easily reducible to a certain number of grains. Thus we find by the chemist's weight that 1 drachm contains 60 grains, being 3 scruples; and if by avoirdupoise weight there are 16 drachms in an ounce, an avoirdupoise ounce must be 60 × 16 = 960 grains, and if there are 16 ounces in the pound, an avoirdupoise pound must be 960 × 16 = 15,360 grains. In apothecaries' weight the number of grains in the ounce is 480, being only half the number of those in the avoirdupoise ounce.

The different species of weights are usually made in sets, so as to fit one within the other, especially the larger weights—pounds and ounces. Small weights, such as grain weights, are made of thin metal, such as sheet brass, silver, gold, or platinum, with small stamped points on them indicating the number of grains, No. 10 being marked with a cross; or they are made of fine platinum wire, being a single piece to indicate 1 grain, and bent pieces, indicating by the number of turns or bends their weight in grains: these are generally from 1 to 5 (Fig. 105). Four weights—of 1, 2, 3, and 4 grains—are sufficient to weigh 10 grains; and similar weights of 10 grains will conveniently weigh up to 100 grains. Fractions of grains are generally made of small platinum wire, bent in the way just described; so that with 1, 2, 3,

Fig. 105. Grain Weights.

and 4-tenths of a grain we may take any fraction of a grain in tenths. It is perhaps needless to remark that these weights should be preserved in small pill-boxes or cases, and placed in the drawers of their respective balances. The beams of the balances should be carefully preserved and frequently cleaned.

Besides these balances and weights, measures of capacity are requisite, such as quarts, pints, half-pints, graduated glasses of various kinds, one or two tubular measures, divided into cubic inches.

(33.) A small collection of chemicals, such as alkalies, acids, rectified spirit of wine, naphtha.

Pestles and mortars of porcelain and iron.

Glass, porcelain, and metallic funnels, small and large.

One or two spirit-lamps; cotton for wicks.

A table bellows blowpipe, for melting, bending, or blowing glass, with lamp and other requisite apparatus.

Blowpipe for the mouth.

Other kinds of blowpipes for exposing bodies to a high temperature will be found useful.

(34.) A few iron ladles of various sizes with lips. It would be convenient to have a very large iron ladle for applying heat, after the manner of a sand bath. The simplest form, however, of a sand bath, is a small crock, such as used for melting pitch, filled with sand and placed over the fire.

(35.) A few logs of a very light kind of wood termed white cedar lath-wood, imported from Nova Scotia, will be found useful in constructing electrical apparatus, more especially where levity is an object. It is durable, and little liable to warp or change in shape. Its levity is remarkable—a cubic inch of it does not weigh above 98 grains. A circular plate of 2·8 inches diameter and one-tenth of an inch thick was found to weigh only 100 grains. A circular plate of 7 inches diameter and three-tenths thick was found to weigh 700 grains only. A hollow sphere,

M

of 2 inches diameter and about one-tenth in thickness, was found
to weigh only 150 grains (154). A hollow sphere, of nearly 5 inches
diameter and about three-twentieths in thickness, was found to
weigh 600 grains.   This wood is sightly in appearance when
got up in a lathe, varnished and polished, and is useful for delicate electrical purposes.

(36.) A gilder's cushion for gold-leaf, with cutting-knife, palette-
knife, wooden forceps, together with a wide gilder's brush and a
hare's foot, for brushing away loose gold.

A gilder's cushion consists of a rectangular board, about 1 foot
long, 6 inches wide, and three-quarters of an inch thick, upon
which is laid several pieces of flannel, or cloth, of the same
dimensions.   The whole is covered with calf-skin leather, such
as is used by bookbinders, the rough side upwards, secured to
the sides and ends of the board by fine tacks, the superfluous
edges of the leather being cut away to the size of the board.   The
cushion is bound round at the edges with narrow silk ribbon
secured over the tacks by glue or paste.   When thus finished the
leather surface should be hard and fair, and very little elastic.   It
is usually rubbed over with a little prepared chalk, or very fine
whitening, which is well rubbed into the leather, and the superfluous particles brushed away.

(37.) A stout wooden bowl, suspended by three strong hempen

lines, containing a heavy iron shot from thirty to
forty pounds in weight, for pulverising (Fig. 106).
The lines of suspension are attached to the bowl
by means of small iron hooks and eyes screwed
to the bowl.   The suspension lines converge to a
central hook after the manner of a scale-pan, by
which the whole is hung up to a beam, so as to
take the weight of the shot and bowl.   The substance to be pulverised being broken into small
fragments, is put into the bowl with the shot.
The shot, by a dexterous circular impulse given
to the bowl externally by the hands, is caused to
spin rapidly round in it, by which the matter,
falling continually between the sides of the bowl
and the shot, is speedily reduced to a fine powder.
The bowl has a light cover to prevent any dust
from flying about.   This is a convenient and useful piece of apparatus for pulverising substances,
such as resin, lac, sealing-wax, and such like.

Fig. 106. Pulverising
Bowl.

If the student be anything of a mechanic, and construct his own apparatus, he will find all these things essential, either in whole or part, so that, with the occasional assistance of clever workmen, he will have little difficulty in carrying out any experimental research in which he may be engaged.

(38.) Glass in almost every form is essential, as rods of glass, glass tubes, and plates of various dimensions. Glass rods (sometimes called *glass cane*), vary from the eighth of an inch to 1 inch in diameter, and from 1 to 2 feet in length. Tubes of glass from one quarter of an inch to 1 inch in diameter, and upward.

Mica or Muscovy talc, in bulk and in thin plates. Tourmaline.

(39.) Resinous substances of various kinds, such as amber, lac, brimstone, common resin, beeswax, turpentine, sealing-wax (red and black), and other resinous bodies. Lac and brimstone are useful in electrical excitation; as are also gutta percha, caoutchouc, and other gums; also vulcanite and ebonite—new combinations of these substances with brimstone—and other electrics, are important, all these being good insulators.

(40.) Silk, being an electrical substance, should be at hand in all its forms—such as soft silk stuff; plain white silk, as employed for fine silk handkerchiefs; soft white and other silk; old silk handkerchiefs; silk ribbon of different widths; silk thread; long filaments of unspun thread from the silk-worm; common silk-worm gut, such as used in fishing, and which will be found an excellent insulator; silk, oiled on one side only, so as to give a somewhat rough surface on the other; oiled silk generally. Where a firm silk surface is required, plain white or other silk may be strained over plates of wood with intermediate layers of thick cloth or flannel.

(41.) Dry and baked wood, more especially mahogany,—which becomes insulating after immersion in hot oil.

(42.) Animal furs and skins of various kinds, such as cat-skin, hare-skin, tanned with the fur on. Soft new flannel and woollen stuffs, fine Basil leather; skins of chamois leather, which are admirable for cleaning glass.

(43.) The substance termed spongio-piline—a combination of sponge and wool—is firm, elastic, and useful to the practical electrician. The best exciter for rough glass tubes, brimstone, and other negative electrics, is soft, new, dry flannel. Cushions constructed of these substances will be very useful.

(44.) Light substances of various kinds, such as straw reeds, used in bonnet-making; also other kinds of reeds and light tubular vegetable substances generally; quills of small birds, and other quills; fine downy feathers; bog down; very fine cotton wool; corks of all descriptions—fine bottle corks, bungs, and phial corks of all sizes. Elder-pith, pith of the sola plant, and such like. Gold and silver leaf; Dutch metal leaf; gold-beaters' skin.

(45.) Fine threads of various kinds; fibres of unspun silk from the silkworm, just mentioned (40), and of very fine flax; fine lengths of human hair; spiders' web wherever it can be collected (Chap. I, note A). All these are frequently required for the suspension of light bodies requiring easy motion.

(46.) Balls of various kinds and sizes; balls of smooth fine cork; light balls, made from the pith of elder, or the sola plant, varying from one-tenth to half an inch in diameter; a cylinder of the pith of the sola plant, 10 inches long by $\frac{1}{10}$ths diameter, was found to weigh 100 grains, so that a cubic inch of it would weigh about 10 grains. A ball of the sola plant of half an inch diameter weighs only about 1 grain.

A ball of elder pith, three-tenths of an inch diameter, was found to weigh one-fifth of a grain.

A ball of fine cork seven-tenths of an inch diameter was found to weigh 7 grains. Small wooden balls of various sizes, neatly turned; light, thin, hollow metallic balls, of different sizes, generally of sheet brass or copper.

(47.) Drawn metallic tubes, of various sizes, from one-eighth to half an inch in internal diameter, and from 1 foot to 18 inches or 2 feet in length. These tubes are usually of brass or copper, sometimes silvered, especially the smaller ones, and slide easily one within the other. Solid brass rods and wires, of various diameters and sizes; small rings of fine iron, copper, brass, and other wires.

(48.) Stability being essential to the success of many important inquiries in electricity, it is desirable to be provided with some rings of lead of different dimensions, from 2 to 5 inches in diameter, and from half an inch to 2 inches thick, and having a round hole in the centre from half an inch to an inch in diameter, or more.

(49.) Small lumps of pure zinc and tin; mercury. These are the ingredients in electrical amalgams (94).

(50.) *Aurum Musivum*, already mentioned, Chap. I. (7), also termed *Mosaic gold*, is efficient as an amalgam for the rubber of the electrical machine (100).

(51.) Circular and other foot pieces nicely got up in a lathe. These should be of various diameters and thicknesses, and turned with as few ornaments and projections as possible, they should have a central socket or hole to receive and uphold any vertical rod or glass support which may be required. A wide deep groove should be turned in the under surface of the foot to receive a ring of lead which should be run into it with a view to stability. Whilst in the lathe the foot should be nicely varnished in the way described (70).

(52.) *Preparation of electrical balls.*—Hollow, metallic balls for electrical purposes are usually prepared by casting two half spheres as thinly as possible, and then soldering them together. The junction is neatly effected by previous preparation in a lathe, so that the edges may exactly apply one over the other. They are then finished off and lacquered in the way described (69); one of the hemispheres is usually left a little thicker at its apex to receive a hole and tap for a screw. The joint should not be exactly in the centre of the ball, for which purpose one of the hemispherical castings should be somewhat larger than the other, so that we may drill small holes, if necessary, in the equator of the ball without meddling with the joint. We may obtain at the ironmonger's shops light, three-quarter, hollow spheres in brass pressed up by machinery, and having an iron screw in the centre, generally employed as supports for articles of furniture. These are convenient in constructing hollow metallic balls for electrical apparatus. The iron screw, with the attached part of the hollow ball, being removed, the deficient portion of the sphere is supplied by a brass casting from a mould accurately turned to the required segment. This segment is nicely fitted by means of a lathe, and soldered to the remaining part of the ball, the whole being turned off so as to be completely spherical. In this way we have three-fourths of the ball without a joint. For the convenience of tapping, the added segment should be thicker at its apex than the rest of the ball.

Another method of producing light hollow metallic balls, is by beating up two hemispheres of sheet copper in hemispherical moulds, and then uniting them by soldering as before. When united they are carefully finished off in the lathe, and lacquered. Very beautiful electrical balls may be obtained in this way, and, if very thin sheet copper be employed, are extremely light.

(53.) *Wooden balls.* These are convenient for many purposes, and may be of any kind of wood; boxwood, ebony, and mahogany are three kinds of wood often employed for solid balls intended to cover the extremities of brass or glass rods, and may be from a quarter of an inch to an inch in diameter. They must of course be got up in a lathe, and should be nicely varnished whilst in it.

Light, hollow balls of wood are best made of the white cedar lath-wood already described (35). A block of the wood is first roughed out in the lathe of a spherical form, and then cut by a saw into two hemispheres; each of these hemispheres is then hollowed out in the lathe to given dimensions. The two hemispheres are then firmly united by means of a rabbet joint, and glue; after which, the rough sphere is again returned to the lathe, and turned down to a fair spherical surface, and of such a thickness and dimensions as to leave the hollow sphere required. Very light spheres of this description are important in the measurement of electrical attractive forces by delicate balances to which they are usually suspended. (Chap. IV., 154).

Hollow wooden balls may be formed of other woods, the weight being diminished by removing the interior portions as in the preceding case.

(54.) *Cork balls*, being valuable in electrical experimental research, should be got up in a lathe from very select pieces of cork of the finest texture, but great care is requisite in turning them.

In preparing cork and other soft balls, the following method may be occasionally resorted to. A hemispherical cup is to be turned in a piece of freestone, common brick, or very hard wood, and being adapted to the lathe, so that the cup may revolve centrally, the cork—previously roughed into a spherical form—is held in the revolving cup, and turned round in it until it becomes fair, and smooth, and rubbed down into a sphere of the same diameter as the hemispherical cup.

A lathe is not absolutely necessary for this purpose; the cork being neatly fashioned by eye into the form of a sphere with the assistance of a sharp pen-knife, is perfected by turning the ball round in all directions by hand in the hemispherical cup. If a hard wood be used for the hemispherical cup a small quantity of emery, or other abrasive powder, rubbed into the cup will greatly shorten the process.

Light substances of soft delicate texture, such as the pith of elder or the pith of the sola plant, may be readily formed into balls in a similar way; that is to say, by first fashioning the mass by means of the hand and a sharp pen-knife into a spherical form, and then turning it round in a hemispherical cup of common house-brick or freestone.

(55.) *Piercing holes through corks.*—The employment of cork in the construction of electrical apparatus is frequently required, cork being a most valuable substance, especially when of fine grain and compressible. We frequently require cork plugs with holes through them for the admission of glass rods or tubes. The best method for piercing holes through cork is the following :—Having selected a piece of thin-drawn brass tube of the required size, file one of its extremities to a cutting edge, and with this bore centrally from each end of the cork. We shall speedily get through it, leaving a clean hole in the cork. The brass tube in leaving will have taken out a cylindrical core, which will be found remaining in the tube, and may be usefully employed as a plug of small size. If the thread of a fine screw be cut upon the boring extremity of the brass tube, the operation will be greatly facilitated. It enables us, as it were, to screw the tube through the cork.

The common method of piercing corks is by means of a red hot iron ; or, having first bored a small hole through the cork with a gimblet or some other boring tool, we may proceed to enlarge it with a rat-tail or other round file (6). The first method, however, is the best, and it will be found convenient to prepare a series of boring tubes in the way above-mentioned, varying from 3 to 6 inches in length, and from a quarter of an inch to 1 inch in interior diameter.

Very light balls may be obtained from the gall-nut. These are often found in the forest, penetrated by an insect, and hollow within. When turned truly in the lathe, so as to be perfect spheres, they form extremely light balls.

(56.) *Metallic conducting rods of various lengths and sizes.*—These are convenient for uniting detached insulated bodies, as, for example, electrical jars, or in connecting them with the conductor of the machine, and such like. They are constructed of a central tube, having two pointed sliding wires passing with friction into it from each end, so as to admit of adjusting the length of the conducting-rod to any extent required. These conducting-rods may vary from 4 inches to 2 feet in length, having central tubes of from one-tenth to a quarter of an inch in diameter. A great number of these connecting-rods should be prepared of various lengths and sizes.

(57.) *Varnishing, Varnishes, and Cements.*—The art of covering glass insulators and other pieces of electrical apparatus with a film of dry, hard, non-conducting varnish, impervious to moisture, is an important process in practical electricity. The liability of insulators, more especially glass, to condense moisture on their surfaces, is fatal to its perfect insulation. It may be proper here to observe that any kind of varnish employed for obviating this source of error must be free from all clamminess, so that nothing may adhere to the varnished surface. Hence, any preparation containing balsam, oil, and the like, is inadmissible. Varnishes for electrical purposes should dry speedily, and solidify into a hard, thin, transparent film.

(58.) The great basis of varnish for electrical apparatus is a resinous gum called " lac," usually obtained in commerce under the three following forms—" stick-lac," " seed " or " gum-lac," and " shell-lac " (note V).

Seed-lac and shell-lac are the forms usually resorted to in preparing electrical varnishes. Shell-lac is best adapted for glass. Seed-lac is commonly used in preparing varnish, commonly termed *lacquer*, for guarding metallic substances against oxidation.

(59.) *Shell-lac and seed-lac varnishes* are thus prepared :—

Formula 1.—Put 1 ounce of the shell-lac of commerce into a wide-mouthed 8·ounce phial, containing 5 ounces of well-rectified naphtha.

Close the bottle with a cork, and let it stand in a warm atmosphere until perfectly dissolved. Shake the mixture frequently, and pass the fluid through a paper filter, such as is used by chemists. Add rectified naphtha to the solution from time to time in such quantities as will enable it to percolate freely through the filter. The filter must be changed when necessary for a new one. We obtain in this way a thin, transparent fluid, perfectly clear, and of a light colour. This must be preserved in a clean phial, well stopped. The quantity of rectified naphtha to be employed must be regulated by the fluidity of the original solution, so that it may be easily filtered. The naphtha must be very highly rectified, otherwise we do not get a clear solution ; it will appear muddy, and not settle finely, nor will it filter.

The kind of lac best adapted to electrical varnish is the common, coarse lac employed by hatters and in other trades ; it is rather of a brownish cast. The very refined preparations of shell-lac sold by the chemists do not always produce a clear solution ; but frequently one of a thick, muddy character, which does not filter clearly.

We may occasionally substitute, for naphtha, highly rectified methylated spirit.

(60.) *Varnish or lacquer for metallic surfaces, such as polished brass, copper, and other metals.*

Formula 2.—Dissolve 1 ounce of bruised seed-lac in 6 ounces of rectified naphtha.

Keep it, as before, in a warm atmosphere, in a closed 8-ounce phial, until the solution is complete. Shake the mixture from time to time. Pass the solution, as in the last case, through a filter, and preserve the clarified solution in a clean phial, well stopped. This is the best and most simple form of lacquer for metallic electrical apparatus.

(61.) *Varnish for Wood.*—Both seed-lac and shell-lac are employed as the bases of varnishes for wood. The lac is usually dissolved in highly proof-rectified spirit, or rectified naphtha. A very small portion of Venice turpentine is sometimes added to the solution to render the varnish less brittle.

Formula 3.—Dissolve 1½ ounce of shell-lac in 5 ounces of rectified naphtha, or highly rectified spirit, as in the preceding formula. The solution being complete, add 1 scruple of Venice turpentine, or about a drachm of common resin. When perfectly dissolved, shake the mixture, filter, and strain off.

(62.) There is a kind of varnish, called *quick-drying cabinet copal,* very valuable for wood apparatus, and which dries with a fine glossy surface.

(63.) *French Polish.*—The celebrated French polish is a solution of shell-lac in naphtha, sometimes combined with gums.

Formula 4.—Gum-copal, and gum-arabic, of each 1 drachm. Reduce these to a fine powder, and sift them through a muslin sieve. Put the powders into a capacious, wide-mouthed bottle, together with half an ounce of fine shell-lac coarsely powdered. Add half a pint of rectified naphtha.

Close the mouth of the bottle and expose the mixture to a gentle heat for several days, occasionally shaking it. The simplest method of heating is to fill a common iron saucepan with fine sand ; immerse the bottle in the sand, and set it on a stove or a hot plate until the whole is sufficiently warm, when it may be put aside for a time. The heat must not be great. When the solution is complete, which may be easily perceived by shaking the bottle, filter and strain off.

(64.) *Electrical Varnishes. Mode of Application.*—In the application of electrical varnishes we must carry on the operation in a warm, dry atmosphere, the presence of moisture being fatal to success. If moisture be present, the varnished surface, instead of appearing transparent and clear, assumes a dirty, opaque appearance, and becomes covered with irregular striæ and scoriæ, which not only injure the value of the varnish, but spoil its appearance. Besides this precaution, it is requisite to get the surface to be varnished perfectly clean and free from grease. The best way of effecting this is to wash the article to be varnished in a weak solution of soda, and subsequently in warm clean water, after which dry and wipe off with clean linen free from lint; finally, expose it to a gentle heat, so that the surface of the applied varnish may not chill.

The bodies to be varnished may be exposed to the heat of a common fire in any convenient way. A common plate-warmer, such as used for domestic purposes, having a door in its back which can be thrown open when required, is admirably adapted to varnishing purposes. It should have one or two round holes in its sides for the reception of glass-rods, tubes, or other articles. These may be passed into and through it, and exposed to heat. The varnish is to be applied by camel-hair brushes of various descriptions, flat and round; and it would be convenient to prepare a few with double holders, one straight, the other oblique (Fig. 107), so that by means of a curved handle we may apply them readily to an interior concave glass surface.

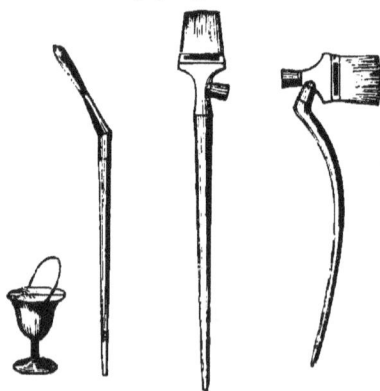

Fig. 107. Varnish Brushes and Cup.

Previously to applying the varnish, a portion of it should be poured out in a well-dried and warm conical cup, similar to an egg-cup (Fig. 107), having a ring, with a small wire strained across it, fitted to its mouth, so that, when the brush has been dipped into the varnish, the superfluous fluid is readily disengaged by drawing the brush across the wire. The cup, previously to receiving the varnish, may be gently heated before a fire, in order to free it from all moisture, and render the varnish as fluid as possible. It is convenient to have several metallic rings of various diameters fitted on different sized varnish cups.

(65.) *Application to Glass-rods, Tubes, and Glass generally.*—The rod or tube being duly prepared in the way above described (64), and its temperature raised to a gentle heat, is held over a concave red-hot iron, such as already alluded to (18). The brush having been dipped in the varnish, and freed from its superfluous fluid, is swept lightly along the rod from end to end, turning the glass round over the hot iron during this process, and taking care not to go over the same place twice. The varnished surface should then be exposed to the heat of the iron from end to end, turning it round until the varnish is nearly dry; it should then be removed into the warm atmosphere of the fire, where it should remain until completely hard, which will be in a very short time. The operator must be cautious not to employ the varnish except in a dilute and thin state, so as to obtain an attenuated transparent coating upon the glass. If a first application does not appear to be sufficient, let it be varnished a second time.

(66.) *Application to Glass Tubes.*—It is not always easy to varnish the interior surface of a glass tube, in consequence of the difficulty in applying the brush.

If the tube be sufficiently large, a full round brush, with a long handle slightly curved forward close to the brush, may be employed. As soon as the varnish has been applied, a red-hot iron wire terminating in a little round knob, and held by a convenient handle, as shown at *a*, Fig. 103, is to be carefully passed into and through the tube, so as to harden the surface by heat, and dissipate all moisture; or a stream of heated air may be passed through the tube by holding it perpendicularly over a bright fire. The tube should then be placed in the drying oven as before. If the tube be so small as not to admit of the entrance of the brush, the following method may be employed : Enclose a small lump of fine cotton-wool in a piece of very fine thin muslin, and secure it so as to form a kind of ball a little exceeding the diameter of the tube ; pass a piece of stout twine through the tube, and attach this ball to the end of it; moisten the muslin surface freely with the varnish so as to make it sufficiently wet, and then by means of the cord draw the ball rapidly through the tube, the tube having been previously heated. Let a current of warm air be passed through the tube immediately after. If this first application be not sufficient, repeat the operation.

(67.) *Application to Glass Jars.*—The uncoated interval of the electrical jar—that is to say, that portion of the glass which is left exposed to the air—requires to be covered with a thin coating of lac varnish both inside and out. This operation is a delicate and sometimes a difficult one, more especially as concerns the interior surface of the jar. In large jars varying from 8 to 14 inches in diameter, having wide open mouths, the course to be pursued is as follows : The coating of the jar having been completed up to the height required, place the jar horizontally with its mouth towards a common fire, turning it round from time to time until the glass becomes fairly warm. This will be better effected through the back of the warming-oven already described (64), the mouth of the jar being within the oven. Finally, the red-hot heating-iron *a*, Fig. 103, may be passed carefully into the jar through its mouth, taking care to avoid contact with the glass. In this way the exposed glass becomes readily and evenly warmed, both inside and out.

The interior surface of the glass jar should be first varnished, for which purpose an assistant should hold the jar obliquely over a red-hot drying-iron (18), while a wide camel-hair brush, mounted on a curvilinear handle (Fig. 107), and moistened with varnish, is being introduced into the mouth of the jar, and rapidly swept round its interior surface until the whole of the surface has been lightly covered. Immediately this is done the globular drying-iron (18) should be heated, and carefully introduced into the jar, and moved round and about it until the coat of varnish has been well dried off. The jar should then be exposed to a gentle heat by holding it in its first position over the red-hot drying-iron. The outer surface should next be varnished, which is a comparitively simple process. The jar is to be placed upright before the fire. A wide brush, moistened with shell-lac varnish, is then rapidly swept round the exterior surface, after which the varnished surface is to be exposed to a moderate degree of heat, which is best effected by a heated iron, moved gradually round it at a moderate distance, until the varnish appears perfectly dried off, and becomes clear and transparent.

Small electrical jars are more easily handled, and placed in any position than large ones, but the mode of treatment is the same, only the brushes and heated irons must be adapted to the size and dimensions of the jar. Where the jars are very small, the mouths not exceeding three-eighths of an inch in diameter, it will be necessary to employ small camel-hair pencils at the end of curved handles, in order to varnish the interior surface. By this method, and careful manipulation, a hard transparent surface, very repellent of moisture, may be obtained. Some judgment and practice will still be necessary to enable the operator to arrive at the precise temperature to which the glass should be exposed, and the regulation of the distance from the sources of heat. The temperature of the glass previously to varnishing should be moderate —just sufficient to enable the varnish to run freely over the surface without becoming chilled. On the other hand, the glass must not be so hot as to drive the varnish into vapour, and cause the lac to solidify in streaks.

(68.) *Application of Varnish to Plates of Glass.*—Glass plates are to be treated as in the preceding cases, the glass being first gently warmed, and the varnish quickly applied by means of a very wide camel-hair brush, rapidly swept over the surface. Two very thin coats are generally requisite, the brush being carried for the first coat in one direction, and for the second coat in a transverse direction. It is not easy to apply the varnish to a glass plate so as to give it a uniformly transparent appearance; but it may be done if proper precautions be taken. Plates of mica and other electrics requiring varnish are to be treated in a similar manner, taking care in all cases to operate in a dry, warm atmosphere, and avoid exposure to moisture.

(69.) *Application of Varnishes to metallic surfaces.*—The kind of varnish best adapted to metallic surfaces is given in Formula 2 (60); but the method of application is the same as before, with similar precautions. The surface must be quite clean, free from grease, and be well got up, nicely polished, and otherwise prepared. For this purpose the surface is to be washed in a weak solution of soda and then in hot water. The process known as *lacquering* is simple, but requires care and practice. A thin coat of varnish should be laid on, the thinner the better, so long as the metal is covered. A great point for appearance sake is the previous preparation of the surface. The beautiful appearance of philosophical instruments in opticians' shops mainly depends on this. For ordinary practical purposes, however, it will be sufficient for the electrician, if his apparatus be prepared in a less costly manner. The bodies to be varnished must be exposed to a moderate degree of heat, and the varnish applied to them by fine camel-hair brushes. Well polished copper and brass should, after varnishing, be slightly rubbed over with a clean rag or flannel moistened with finely drawn linseed oil.

(70.) *Varnishing Wood Surfaces.*—The formula No. 3 (61) is a good varnish for wooden apparatus generally. It is to be applied by means of wide camel's-hair, or other brushes, as in the former instances. The piece to be varnished should be well got up, and gently warmed before varnishing, after which it should be placed in a warm atmosphere. The surface should be finally rubbed over and polished off with a clean rag, moistened with finely drawn linseed oil. Wooden pillars, and apparatus which admit of being turned in a lathe, are best varnished before removal from it; being finally polished off with a clean oiled rag while in motion. The most perfect varnish for flat surfaces is *French Polish* (63). The polishing process is somewhat tedious; there are two or three different methods. The following is very effectual. Roll up a piece of thick flannel, cloth, or listing into a cylindrical wad. Moisten one end of this wad freely with the varnish given in formula No. 4, or any other kind of lac-varnish. Formulæ Nos. 3 and 4 answer very well. The wad wetted with the varnish, is to be covered with clean linen, moistened with good drying linseed oil; the surface to be polished is now rubbed with this wad, first in the direction of the fibres of the wood, and then in small circles, completing a small portion at a time, until the surface is covered. Finally the surface is to be rubbed in a similar way with a clean cushion moistened with alcohol, by which the polish is perfected. This is termed *spiriting*. There are two other methods of applying the polish. A cylindrical wad is to be made as before, faced up with some folds of fine linen. A little linseed oil is applied by the finger to that part of the wad intended for polishing. A small quantity of the varnish is then allowed to drop out from the mouth of the bottle containing it, upon the oiled portion of the rag. The surface to be polished is rubbed with this as in the former case, repeating the application of the varnish and oil when necessary, that is, so soon as the wad is rubbed dry. This operation requires some patience and practice. The surface, as in the previous case, is to be rubbed first in the direction of the fibres of the wood, and then in small circles. This is the English method. The French first moisten the cushion or wad with the varnish and a little olive oil, and rub the wood in the direction of the fibres, repeating the operation several times until the wad becomes dry. The wood is then rubbed with a cushion moistened with olive oil and a little tripoli. The finishing lustre is arrived at by friction with a soft skin, such as chamois leather. The process of varnishing by French polish being tedious, and in many instances not required, it will be sufficient for all ordinary purposes to apply the polish, Formula 4, by a wide camel-hair brush in the usual way. When dry the surface is polished off with a clean rag moistened with quick drying linseed oil.

There are several other varnishes applicable to wood; those before enumerated, however, will be sufficient for general purposes. The following, nevertheless, may be occasionally employed with advantage :—

Formula 5.—Copal, coarsely powdered . . . . . 3½ ounces
　　　　　　Camphor . . . . 10 grains.
　　　　　　Highly rectified methylated spirit . . 28 ounces.

These ingredients should be put into a clean bottle, and digested in a sand heat, until the solution is complete. The heat should be such as to cause bubbles to rise from the bottom of the bottle. Put the solution by for a few days to settle, and pass the supernatent clear liquor through a paper filter. We have then a fine bright solution, which is to be applied by means of a brush in the usual way. This varnish does not dry so quickly as lac-varnish, but when dry has a very bright appearance, especially after being rubbed over with an oiled cloth. The residue may be treated with more rectified naphtha, and a second produce obtained.

Formula 6.—Seed-lac . . . . 3 drachms.
Gum mastic . . . . 2 ,,
Venice turpentine . . . . 1 ,,
Highly rectified methylated spirit . . 2 ounces.

Digest carefully in a sand heat, and proceed as before.

It may be as well to observe that in the application of varnishes to wood by the brush in the usual way, the grain of the wood, however smooth the surface, is liable to become raised, giving it a coarse appearance. If we wish to obtain a very smooth varnished surface, we must get rid of this by rubbing down with fine glass-paper until a new and perfectly smooth surface is obtained. This done, a second coat is to be applied as before. To avoid the absorption of the varnish by the wood, a coating of thin isinglass is sometimes first applied and, previously to varnishing, rubbed smooth when dry. Lac being rendered soluble in water by means of borax, we may thus obtain a species of lac varnish of a simple and most useful kind.

Formula 7.—Shell-lac . . . . 1 ounce.
Borax . . . . ¼th ,,
Water . . . . 6 ounces.

Put these ingredients into a clean wide-mouthed bottle, and raise the temperature to nearly a boiling heat by means of a sand bath. In a short time a clear solution of the lac and borax is obtained. Set it by to cool, and then pass the solution through a paper filter. This is a useful varnish, equal to many spirit varnishes. It is valuable as a vehicle for water colour and is the basis of Hindoo ink. When once dry, water has no effect upon it.

(71.) *Varnishing Paper.*—It is often requisite to protect paper surfaces against dirt and moisture, more especially graduated paper

scales, circles of cardboard, and such like, as also to render a dry
paper surface more or less insulating.

Formula 8.—Dissolve 1 ounce of Canada balsam in 2 ounces of spirits
of turpentine. Put these ingredients into a clean,
wide-mouthed bottle, and digest at a gentle heat.

Before the solution is quite cold, pass it through a paper filter
into a clean bottle. A heated iron (18) should be held over the
filter in order to preserve the solution in a sufficiently liquid state.

Mode of Application.—The paper to be varnished must first be
prepared with a coating of clear, thin, isinglass size, and well
dried. The varnish being poured out into a small cup with a
cross-wire (64), should be exposed to a gentle heat, and then laid
on rather quickly, taking care to sweep the brush in one direction
only, and not to touch the same part twice. If one coat be not
sufficient, repeat it.

(72.) The following is a fine varnish for paper:—

Formula 9.—Gum anime and gum sandaric, of each . ½ ounce.
Gum mastic . . . . ½ ,,
Highly rectified methylated spirit . . 8 ounces.

Reduce the gums to a very fine powder, put them into a clean
capacious bottle with the methylated spirit, and digest in a gentle
sand-heat, keeping the bottle closed. When the solution is complete, pass it through a paper filter.

Mode of Application.—The paper to be varnished must first have
a thin coating of isinglass size. As this varnish dries quickly,
care must be taken in laying it on to rapidly pass the brush over
the surface and not to touch the same part twice.

Isinglass size is obtained by dissolving small strips of isinglass
in boiling water. The solution should be thin, and filtered through
paper whilst hot. As the solution solidifies in cooling, it will be
necessary to keep it in a liquid state by standing the vessel containing it in boiling hot water, as in the case of common glue.

(73.) Amber has been occasionally employed as a basis for
electrical varnishes. The preparation of Amber Varnish is difficult, tedious, and often dangerous. It is best procured from the
varnish maker.

Amber, when fused and dissolved in oil of turpentine, constitutes
a brilliant electrical varnish. It dries speedily, does not crack,
and is very durable and hard when dry ; it resists boiling water
and friction, withstands the stains of ink or coloured liquids, and
maintains its splendour for any length of time.

The difficulty, however, of preparing and even obtaining good amber varnish is such as to greatly discourage its use, more especially as the lac varnishes previously described (59, &c.), are efficient.

(74). *Application of Varnish and Oil to Silk.*—Fine oiled or varnished silk is of great value in electrical experiments, and is much used in the construction of the electrical machine. The silk, however, for this purpose should be oiled or varnished on one side only.

The following method of varnishing silk for electrical apparatus is very effectual and simple. A stout, open, rectangular frame of wood, similar to a picture-frame, is prepared, of any required dimensions—say 4 feet long by 15 inches wide in the clear. The silk is to be first strained upon this open frame, securing its edges to the wood by means of fine tacks. The silk being thus strained is to be brushed over on one surface with a mixture of boiled oil and turpentine, in the following proportions :—

Formula 10.—Boiled oil . . . . 6 ounces.
Clear spirits of turpentine . 2 ,,

Let these ingredients be well stirred together in a clean vessel; then, with a fine painter's tool dipped in the mixture, and the superfluous fluid allowed to run off over a wire (64), brush over one surface of the silk from end to end, and set it by to dry. When dry, give it a second coat, if necessary. The mixture of oil and turpentine should not be very thickly laid on. Silk may also be varnished over with the Canada balsam varnish (71) in a similar manner. For the purpose of the electrical machine the thinnest kind of silk, called *Persian*, should be employed. Thicker kinds of silk strongly varnished are requisite for other purposes.

(75.) *Electrical Cements.*—Cements of different kinds are requisite in electrical manipulation, for attaching brass to glass, or glass to wood, &c. For cementing glass rods into wood or brass caps, and such like, there is no cement equal to the best sealing-wax, which consists of shell-lac, boiled Venice turpentine, and vermilion, or some other colouring matter. A coarser kind of sealing-wax may be occasionally employed, for which the following is a good formula :—

Formula 11.—Shell-lac . . . . 2 ounces.
Powdered resin, of the finest quality . 1 ounce.
Fine Venice turpentine . . . . 1 ,,
As much vermilion, or other colouring matter, as required.

Mode of Preparation.—First dissolve the lac in a clean pipkin ; when nearly fluid add the resin, stir well together, then add the turpentine, and, lastly, the colouring matter. Stir all the time, and pour it out into a mould of oiled tin, or upon an oiled marble slab, to cool.

The fine polish usually seen on the best sealing-wax is obtained by exposing the sticks, when removed from the moulds, to the heat of a fire, just sufficient to liquefy the surface.

The composition of black sealing-wax is the same as the red, with the exception of the colouring-matter, which for black sealing-wax is usually lamp-black, or ivory-black; but we may use the composition without any colouring matter. For general use it will be well to procure sticks of the very best sealing-wax from the makers. Whatever colouring matter be employed for sealing-wax it should be ground down on a marble slab with a fine muller, as in the preparation of delicate oil colours.

(76.) *Mode of Application.*—In cementing brass caps, sockets, or other metallic surfaces to glass, the glass must be heated over a charcoal fire, or a hot iron, until the sealing-wax cement rubs off freely upon the part to be cemented. The brass cap, or other surface, must be treated in the same way, and applied whilst hot to the glass. The two surfaces are then rubbed together by a gentle movement of the brass and glass upon each other, applying external heat to the metal to further liquefy the cement if necessary. The exact position of the metallic socket, or other body, being determined, the whole is left to cool. Before it is fixed, all superfluous cement should be carefully removed. When cold the joint is cleaned off with a little rectified naphtha. In this way a perfectly air-tight joint may be obtained. The metallic parts of the Thermo-Electrometer (178) are secured in this way, as also wood to glass or metal. All joints to be cemented are well-secured by the same process.

Formula 12.—Yellow resin . . . . 16 ounces.
Bees'-wax . . . . 2 ,,
Linseed oil . . . . 2 ,, by measure.

Melt these together in an earthenware pipkin so as not to boil and become frothy. Stir the ingredients well, and add, by degrees, red ochre, 4 ounces; or colour with any other better colouring matter if desired. If required to be of a very fine colour we may add vermilion. The best way of melting these ingredients together is to place the pipkin containing them in a sand heat. This cement is useful for ordinary purposes, and in cases where sealing-wax is not required.

Formula 13.—Yellow resin . . . . 5 ounces.
Yellow bees'-wax . . . . . 1 ounce.
Venetian red . . . . 1 ,,
Plaster of Paris . . . . $\frac{1}{2}$,,

Prepared and applied as before.

Formula 14.—Common resin . . . . 5 ounces.
           Bees'-wax . . . . 1 ounce.
           Plaster of Paris . . . . 2 tablespoonfuls.

This last is used for coarse purposes, and when applied the surfaces must be gently heated.

In preparing these cements, the colouring matter and the plaster
of Paris must be well-dried and gradually mixed with the resin
and bees'-wax whilst in a state of liquefaction. The heat should
be continued just above the boiling-point, until the mixture
becomes perfectly tranquil; it is then allowed to cool, but before
it is solid it is cast into cakes or rolls on an oiled marble slab,
and preserved for use.

For roughly cementing glass to wood a compound of plaster of
Paris, white lead, and spirits of turpentine may be employed.

Formula 15.—White lead . , ... 2 ounces.
           Spirits of turpentine . . . . . 1½ ounce.
           Plaster of Paris . . . . . . 2¼ ounces.

(77.) *Preparation of Electrical Plates and Cylinders.*—Cylinders
and plates of brimstone, lac, resin, and other resinous bodies are
often called for in experimental electricity, more especially in the
construction of the Electrophorus (92). Brimstone is particularly
available, as it easily detaches itself, by contraction, from the
moulds in or on which it is cast. If the moulds be very smooth
it comes away with a beautifully polished surface. Glass tubes
of various sizes furnish very efficient moulds for cylinders of
brimstone. If a little taper, so much the better.

(78.) *Preparation of Brimstone Cylinders, Cones, &c.*—Put
some common brimstone, broken into pieces (37), into a clean earthenware pipkin; place the pipkin in a strong sand-heat, raising the
temperature gradually until the brimstone liquefies. The heat
must not be pressed too far, but must be just sufficient to render
the brimstone freely fluid. It is then ready for casting. The moulds
being duly prepared, pour the fluid brimstone into them and let
it cool. After a short time the mass contracts and may easily be
removed from the mould, when it will have a beautifully polished
surface. If cast in conical glass moulds, such as wine-glasses,
we obtain very good insulating supports. Brimstone cylinders
are excellent for exciting negative electricity. When cast in a
short cylindrical lamp-glass, and a glass rod put vertically into

the fluid mass before cooling, to serve us an insulating handle, we obtain a convenient form of electric for negative excitation.

(79.) *Brimstone Plates.*—Prepare the brimstone in the way just described (78), and place a wooden ring of any given dimensions upon a plate of flat sheet glass, placed pretty level on a table. A few leaden weights may be put upon the ring, so as to press it firmly upon the glass. The wooden ring may be about half an inch or more in thickness, and should be well polished within; the brimstone being perfectly fluid, pour it upon the glass within the ring until level with its surface. When completely cool, the casting may be separated from the glass, and on being turned over will present a circular plate with a remarkably fine, smooth, polished surface. This casting should be supported on a plate of wood, having a raised edge. Brimstone castings produce negative electricity very freely, and the only objection to them is their liability to crack.

(80.) Cylinders and plates of resin, wax, and other substances of that kind, are produced in a similar manner ; the moulds, however, in which they are cast must be oiled to prevent adhesion.

Resinous plates, compounded of various resinous bodies, are of great value to the practical electrician; as in the case of sulphur (79), they may be cast upon a polished marble surface within a ring of wood or metal a little more than half an inch deep, or upon a glass or polished metallic surface. Block tin may be employed for this purpose. The surface on which the composition is to be poured, together with the sides of the ring, should be rubbed over with a piece of clean cloth slightly smeared with oil, to prevent the casting from sticking. If the ring be of wood, its interior surface should be covered with some very thin tinfoil : this admits of the plate when cast being readily disengaged from the wood ring.

The following composition for a resinous plate for ordinary purposes will be found sufficient :—

Formula 16.—Resin . . . . 8 parts.
    Gum lac . . . . 1 part.
    Venice turpentine . . . . . . 1 ,,
    As much ivory black as will give the mass a good appearance.

Break up the resin and lac into small pieces, or reduce to a coarse powder (37) ; put the powdered resin and lac into an iron kettle, or into an earthenware pipkin, and expose them to a sand heat, or to the heat of an oven, of a sufficiently high temperature to liquefy the resin and lac ; as soon as the resin and lac begin to melt, add the Venice turpentine, and continue stirring the mixture with a clean glass rod. When the whole is perfectly fluid, add colouring matter by degrees until a sufficiently deep colour is obtained. The fluid should be retained in a state of quiet fusion for a short time, so as to expel all the air from it. It is now poured out into the mould, and allowed to cool. When perfectly cold, the casting may be detached from the surface or mould, and turned over on a circular plate of wood.

A resinous plate of this kind, about 15 inches in diameter, is well adapted to experimental purposes; amongst others, to the separation of certain mixed powders, as brimstone, red lead, &c. &c., by positive and negative electricity.

The following is well adapted to an electrophorous plate :—

Formula 17.—Shell-lac, Venice turpentine, and resin, in equal parts.

Melt these ingredients in a clean, covered iron crock or pipkin placed in a sand heat, and run the liquid out upon a polished surface within a metallic or other ring of full half an inch in depth, taking care to prepare the surfaces by wiping them with an oiled cloth.

A plate of this kind, of from 12 to 15 inches in diameter, constitutes a very powerful electrophorus plate. The plate may be coloured with ivory black, prussian blue, vermilion, or red lead, as may be desired.

The following is another formula for an electrophorus plate :—

Formula 18.—Gum lac . . . . 10 parts
Resin . . . . 3 ,,
Venice turpentine . . . . . . 2 ,,
White wax . . . . 2 ,,
Pitch . . . . ¼ of a part.

Liquefy these ingredients as before, and cast them within a ring on a polished surface. This is an efficient combination. The liquid may be coloured with a little vermilion if desired.

Either of the formulæ for the coarser kinds of sealing-wax (81) answer very well for electrophorus plates. The ordinary ingredients for an electrophorus plate are resin and bees'-wax.

The following, by Pfaff, is a simple formula :—

Formula 19.—Gum lac . . . . 1 part.
Resin . . . . 8 parts.
Venice turpentine . . . . . . 1 part.

There is some little disadvantage in casting the plate upon an oiled surface, which if possible should be avoided. It is, however, necessary, to prevent adhesion. We may often succeed by casting the plate upon a highly-polished marble surface, especially if the liquid be cast at a low temperature. Compounds containing a large proportion of shell-lac congeal rapidly. In all

cases the surface should be freed as much as possible from
the oil upon them, which is best done by washing them over
with a little soft soap and warm water. The common method
of casting electrophorus plates is to pour the liquid upon a tin
plate with a raised edge, or upon a wooden plate covered with tinfoil, allowing the liquid to cool. If this operation be carefully
performed, we may obtain a very level, brilliant surface. The
following method of preparing an electrophorus plate will be
found efficient and available :—Melt the ingredients in a clean,
covered, iron crock, which is best done by exposing them to the
heat of a close oven, until they become perfectly fluid and free from
air-bubbles. Let a polished plate of marble be now closely
covered with a thin sheet of tinfoil, and carefully levelled. Place
upon the wooden ring—also covered with tinfoil, and which is to
serve as a mould—some leaden weights, so as to press the ring
closely upon the surface of the marble below. Let the melted
ingredients, now in a perfectly fluid state and free from air-
bubbles, be poured gently within the ring until the whole interior
be filled up to its edge. A perfectly level and brilliant surface
will in this case be the result. Sufficient time must now be given
for cooling, and nothing must be brought near the recently liquefied surface until it is perfectly hard. When cold, the resinous
plate is easily disengaged from the tinfoil surface and from the
ring surrounding it. The ring may be divided in one point so
as to admit of slight extension, it being temporarily held together
by a few turns of fine twine.

To produce an electrophorus plate by the last method, select a
composition such as Formula 16, which when cold will congeal
into a very hard substance. When the cake is removed from the
mould, the surface may be rendered perfectly level and fine in a
lathe. Brimstone, cast in the way above described, presents the
finest surface ; and is perhaps, of all substances, the most efficient
for an electrophorus plate. It ' is, however, very liable to crack
from change of temperature. This may be in some degree avoided
by repeated liquefaction of the brimstone before casting, and by
gentle annealing. With the view of annealing, the cake, on
removal from the mould, should be exposed to a moderately
low temperature in an oven. In this way we obtain a plate having
a very excellent exciting surface derived immediately from con-
gealation. The opposite surface is not usually so perfect. The cake,
however, if sufficiently hard, may be fitted into a wooden platter,
and polished in a turning lathe, when it will be easily adapted
to the purpose of an electrophorus, or to any other use for which

a resinous plate is required. We should be aware, however, of the liability of plates of this kind to yield or buckle under their own weight. Hence it is necessary, if we wish to preserve their form, to keep them supported upon a flat surface, which is best done by placing them on a circular plate of wood with a raised edge.

Whatever colouring matter be employed should, previously to mixing it with the other ingredients, be ground down with a muller upon a marble slab, as in the case of sealing-wax (75). No more colouring matter should be employed than is just sufficient for the purpose, more especially where black is used : a very little will suffice to colour the liquid mass. Light blue is perhaps the best colour for an electrophorus plate, although black has a very brilliant appearance, and for many experiments is to be preferred.

(81.) *Junction of Plates of Glass by Cementing.*—This is an important process in experimental electricity. Many kinds of thin glass, such as window-glass, being very excitable, two thin plates of window glass, when firmly united by means of good sealing-wax or electrical cement, may be substituted with advantage for plate-glass in the plate electrical machine, and may also be employed as a glass electrophorus. The best cement for the purpose is the common kind of sealing-wax, black or red. Formula No. 11 may be employed. The following formulæ for an inferior class of sealing-wax may also be used :—

Formula 20.—Shell-lac . . . . 8 parts.
       Venice turpentine . . . . . . 4 „
       Colophone, or black resin . . . . 3 „
       Ivory black, sufficient to colour the mass.

Formula 21.—Shell-lac . . . . 4 parts.
       Black resin . . . . 1 „
       Venice turpentine . . . . . . 2 „
       Vermilion . . . . 1 „
    Add the vermilion gradually, when the other ingredients have become fluid.

Formula 22.—Shell-lac . . . . 2 parts.
       Venice turpentine . . . . 8 „
       Black resin . . . . 4 „
       Vermilion . . . . 1 „
    Remove from the fire and add ½ ounce of rectified spirit.

There are a great variety of formulæ for sealing-wax, but the ingredients in nearly all of them are shell-lac, resin, Venice turpentine, and colouring matter. These are united in various proportions. The best sealing-wax is made without resin, the proportions being :—

Formula 23.—Shell-lac . . . . 4 parts.
                 Venice turpentine . . . . 1 to 1½ „
                 Vermilion . . . . 3 „

Melt the lac in a clean copper pan suspended over a clear charcoal
fire; then add the turpentine slowly, and afterwards the vermilion, stirring briskly all the time.

This forms a beautiful electrical cement for refined purposes.
For inferior kinds of wax we have less lac. The colour may be
chosen according to taste.

The colouring matters commonly employed are vermilion, cin-
nabar, red lead, red ochre, carmine, lamp black, ivory black,
Prussian blue, artificial ultramarine, and English umber. King's
yellow and Prussian blue produce a green colour.

The colouring matter must be rubbed down with a little linseed
oil and turpentine to a fine impalpable powder, by means of a
muller and marble slab. As already mentioned (76), the colouring
matter must be gradually added to the melted ingredients, and
carefully stirred in. The liquid wax thus prepared is to be run
into oiled moulds, which may be hollow cylinders of block tin from
half an inch to 1 inch in diameter, rather conical; or it may be
run out upon an oiled marble slab, and formed into rolls before
being quite cold. If the congealed wax does not readily escape
from the tin moulds, it may be easily detached by a gentle
heat. In cementing together glass plates of 18 inches to 2 feet
in diameter, the surfaces of the plates should be clean and dry;
the sealing-wax by which their union is to be effected should be
reduced to a moderately fine powder by the method already
described (37). The glass plates should be supported upon a
wooden platform, constructed of pieces of deal board, three-
quarters of an inch thick, screwed upon two stout cross-bars of
wood, at least two inches wide and one inch thick, placed edge-
ways. This enables us to get the hand under the platform, and
to move the whole in and out of a heating oven or drying-closet.
A few small circular or square pieces of deal, about 1½ inches in
diameter and half an inch thick, being now distributed upon
various points of the platform, one of the circular glass plates
to be cemented, covered with pulverised sealing-wax, uniformly
and carefully spread over its surface to a depth of about the
eighth of an inch, is laid upon the pieces of support, and the
whole placed in an oven, sufficiently heated to liquefy the wax.
When the wax is just fused and covers the glass, it is to be
removed, together with the platform upon which it rests. The
other circular plate to be united is subjected to the same process.

As soon as the wax is fluid this plate is also removed and the first plate inverted upon it, so as to bring the two wax surfaces together. A circular plate of glass of 3 or 4 inches in diameter, if the compound plate be intended for the plate of an electrical machine, should be now placed centrally between the two plates, so as to strengthen the centre. Four small circular plates of glass, of about one inch in diameter, should be also placed between the two plates near the circumference at cross diameters, in order to preserve an equable thickness of the compound plate; a few leaden weights should be now laid upon thin pieces of deal, and placed at various points upon the upper glass plate, so as to press the plates gently together; the whole should be now restored to the heating oven in order to completely liquefy the intermediate cement and run the plates together, as it were, under the gentle pressure of the leaden weights. The whole is then to be again removed from the oven, and allowed to cool gradually, taking care that the circumferences are fair one with the other. We may, during the cooling, easily move one of the plates upon the other, and adjust the position, giving gentle pressure with the finger if there should be any lack of contact with the glass. The union of large plates in this way requires much care. The temperature must not be pushed too far, only sufficient to melt the wax, and very little more. The joined plates, after being removed from the oven, should be meddled with as little as possible. When cold the superfluous edges of the wax are to be cut away with a warm knife, and any small interstices between the edges of the plates filled in with a little melted wax run out with a lip-ladle. Finally, we may smooth over the wax at the circumferences of the two plates with a heated iron.

The excitability of a plate of glass thus compounded is most remarkable, and is very superior to that of ordinary plate-glass. The vitrefied surface of flatted blown glass appears to be very favourable to electrical excitation. The plate may be drilled with care, and mounted on an axis in the usual way. The best colours for such a plate are either red or black. Black is perhaps the best for an electrophorus plate. A plate of this kind answers well. Smaller plates, from six inches to a foot or more in diameter, are easily joined by the same process. As also small square plates, from a few inches to a foot square. The best kind of glass for electrical excitation appears to be *crown-glass*, the pieces being selected as flat as possible.

(82.) *Isinglass, glue,* and *size* form a useful cement. It is easily obtained by dissolving the isinglass of commerce in clean water.

The solution should be gently boiled, and for nice purposes filtered; for coarse purposes, strips of parchment may be substituted for the isinglass. The proportion of isinglass will depend on the required strength of the size. As a preparatory size for wood or other bad conductors, previously to the attachment of metallic leaf by paste, one part of isinglass dissolved in twenty parts boiling water by weight will be found sufficient; for the more adhesive purposes of glue, one part of isinglass may be dissolved in ten parts water.

(83.) *Rice Glue, termed Japanese Cement.*—Mix together intimately rice in fine powder, and cold water. The rice flour and water should be well rubbed up in a clean porcelain mortar, and be then gently boiled. This forms a highly adhesive cement, beautifully white, and when dry nearly transparent.

(84.) *Paste.*—This is also a useful cement for practical electrical purposes, such as for attaching metallic leaf to glass, as in the preparation of the Leyden-jar; as also in covering wooden and other substances with tinfoil. Good flour paste may be made as follows:—

Formula 24.—Rub up some good flour in a porcelain mortar with as much cold water as will produce a liquid fluid.

Care must be taken not to leave any lumps of flour, but the whole must be perfectly smooth. Add about a drachm of finely powdered alum. Put into a clean tinned iron saucepan nearly as much clean water as there is of flour and water, and set it over the fire to boil; when boiling pour the mixture previously prepared into it, and keep stirring the whole with a large wooden spoon until it becomes an adhesive mass, which it will speedily do. When it shows symptoms of boiling remove it from the fire, continuing to stir the mass. Transfer it by the aid of a clean wooden spoon into a clean white open-mouthed jar, and set it by to cool. The paste used by curriers and shoemakers, which is especially good, is prepared somewhat in this way. It is adapted to the coating of the electrical jar with tin-leaf:—

Formula 25.—Good flour . . . . 2½ ounces.
           Coarse brown sugar . . . . 20 grains.
           Water . . . . 5 ounces.
           Isinglass size . . . . 5 ,,
           Essential oil of lavender . . . . 20 drops.

Rub up the flour, water, and sugar in a clean Wedgwood ware mortar, adding the water gradually, until the whole is in a perfectly liquid state; add now the isinglass size—which should be hot and perfectly fluid—together with the essential oil; rub the whole together so as to unite the ingredients more completely; pour the mixture into a clean saucepan containing a little boiling water, and boil the whole gently over a slow fire, stirring it until it thickens into a paste. Finally remove the paste into a covered jar for use.

Paste is usually a very perishable article, but made in this way it will keep in a covered vessel for a long time. The sugar renders it pliable, and the essential oil secures it from the minute vegetable growth which turns it mouldy. This kind of paste is very adhesive, and is especially adapted for covering wood with tin-leaf, the wood being previously coated with isinglass size. Even if this paste dry up into a hard mass, it may be recovered by diluting it with a little hot isinglass size.

(85.) *To cover Wood with Tin-leaf.*—It is desirable to prepare the wood by first giving it a coating of fine isinglass size, and allowing the size to dry. The tin-leaf with which the wood is to be covered is then cut to the required dimensions and form best adapted to different parts of the surface. The portion to be laid on the wood must be brushed lightly and equably over with the paste. The pasted foil is now laid on the wood surface, and pressed down upon it by means of a damp cloth or sponge, and then further rubbed down upon the wood by gentle friction with the handle of a bone paper-knife, wetted with water. When the whole surface has been covered in this way, clean it off with a wet cloth or sponge, and then wipe it dry with a soft muslin cloth. All projecting edges must be rubbed until they disappear. The surface will then be perfectly even and metallic, as if entirely of metal.

(86.) *Electrical Suspensions.*—For the perfect development of minute attractive force, great sensibility of movement is requisite. This is obtained by balancing light needles on fine agate centres, or suspending light bodies by filaments of unspun silk or other fibres; or otherwise, by the suspension of strips of leaf gold. The suspension upon hardened points and centres, after the manner of a delicate and fine compass-needle, admits of great freedom of motion. The centres may be of hardened brass or agate. The points or centres of motion may be of hardened steel, or hardened brass, which is often preferable. The points of suspension may either descend from the needle into the glass or agate centres, or the centres may be fixed in the needle, and the points ascend into them, so as, in either case, to have the centre of gravity of the needle a little above the point of suspension. The Chinese have an ingenious method of suspending a needle, so as to obtain

188   ELECTRICAL MANIPULATION.

remarkable delicacy of movement in their small compasses. It is represented in Fig. 108. In this construction, the centre of gravity of the needle is above the point of suspension; $d\,e$ is a small cap of brass, very thin and light; the needle $c\,s\,n$ passes through a small ring e of support, formed of a light slip of brass, and fixed on the top of the cap. The whole is sustained on a pin. Although the needle is in this case above the point of suspension, yet the centre of gravity of the whole system, namely, the needle, cap, and ring of support is somewhat below it. The Chinese needle is very small and light; it is not above 1 inch in length, and the one-fortieth of an inch in diameter; it is, however, very sensitive.

Fig. 108.

(87.) *Suspensory Filaments.*—The most sensitive and delicate of these is the spider's thread, which admits of being twisted some thousand times without exhibiting torsion. It is, however, difficult to handle and apply. The following method will be found successful. Construct two movable radii of light wood united upon a centre pin, after the manner of a pair of common compasses. Each leg should be about 12 inches or more in length, and tapered towards the extremities, and so closely united as to admit of being turned upon each other with some degree of friction. We may thus adapt the opening to any required length within the limits of the legs. When we are fortunate enough to discover a long and strong spider's thread hanging from the ceiling, or one of the long stretching lines of the geometrical web of the garden spider, we open the legs of the compasses to the length of the thread we require and can command. Then touching the extremities of the legs with a little weak gum-water, we intersect the thread between the legs of the compasses which immediately adhere to it; then by means of a little weak gum attach a fine pith ball to each extremity of the intersected thread, (which now becomes easily detached from its fixed points), holding one of the balls at the extremity of a fine needle, and allowing the other to remain pendent. The other extremity of the needle is now inserted in any fixed arm of support; and thus we obtain the most delicate suspension. A light pith ball may be successfully applied to the thread as it hangs from the ceiling, constituting a very delicate means of exhibiting attractive force.

A filament of unspun silk, as unwound from the cocoon of the silkworm, is another very delicate means of suspending light bodies, and may be managed in pretty much the same way as the former, especially where the thread is fine. It generally admits of being applied through the medium of a very fine sewing needle which may be passed through a fine reed, or may be otherwise attached by a little cement or gum, after being wound round the centre of the needle. A light needle of gum-lac is best constructed by inserting a fine thread of gum-lac in the extremities of a short, delicate reed, about half an inch in length, and then passing the suspension silk through the centre of the reed by means of a fine sewing needle, or otherwise attaching it by a central loop of stouter silk filament supporting the needle on each side of its centre. Filaments of glass thread drawn out in the lamp constitute very delicate suspensions for light bodies. They have an extremely small degree of torsion force, which is sometimes of advantage.

Very fine metallic wire may also be employed. The suspension wires employed by Coulombe were so fine as to admit of being twisted through eight circles without interfering with its elastic reaction. It was so fine, that 1 foot of the wire weighed only one-sixteenth part of a grain.

(88.) *Management and Handling of Gold-leaf.*—The suspension of strips of gold-leaf requires much attention and practice. The leaf is handled through the medium of the leather cushion and spatula knives already described (36), and in the following way:—The book of gold is laid flat on the cushion towards one of its extremities. Then lifting up the paper of the book by means of the thumb and fingers, so as to expose a leaf of gold, we insert the edge of the spatula for a short distance beneath the edge of the leaf, which must be carefully and gently done; then, giving the spatula half a turn with the hand so as to raise the edge of the leaf off the paper upon the spatula for a short distance, we drag the gold gently forward until it rests on the leather cushion. We may now blow very gently upon the leaf, which will press it equally upon the leather. This, however, requires a great deal of care and precision, for unless the breath fall perpendicularly upon it, the leaf is liable to be blown off the surface of the cushion. The leaf of gold being thus secured, and the width of the required slip determined on, we adjust the edge of the spatula to the given width or distance from the edge of the leaf, taking care to keep it parallel to the edge, which must be done by an accurate eye; after which the edge of the spatula is pressed down on the leaf, and drawn along its surface so as to divide it completely through; then taking a piece of gummed paper, such as is used for letter stamps, and cutting it to the required width and length, which may be about half an inch, and having very gently moistened it at one extremity with the lip, we lay the paper carefully upon the end of the strip of gold, and

thus it can be easily detached from the square of gold-leaf to which
it was previously united, and raised off the cushion. It may then be
placed in the holder of a Gold-leaf Electrometer (Chap. I., Note G),
or any other required point of suspension. Where two slips of gold
are required, as in Bennet's Gold-leaf Electrometer (Chap. I.,
33), a second equal slip of gold is prepared in exactly the same way, and
whilst one of the slips is resting on the cushion the other slip is laid
carefully on it, the two slips with the paper-holders being nicely
adapted one to the other so as to lie evenly and parallel, taking care
to place the surfaces of the paper to which the gold slip is attached
outward, so that the gold attachments may not touch. The double
slip of gold with the paper-holders are now transferred to any
given point of support, as in the holder of the Gold-leaf Electrometer above mentioned. If the strips are longer than necessary,
they may may be shortened by the same means as before. Gold
slips are attached to any other medium of support, such as small
wooden cylinders or other surfaces in much the same way, but in
all cases the gold-leaf must be handled upon the leather cushion
and with the gummed paper. The gold-beaters in handling their
leaves commonly rub a little dry whitening over the surface of the
leather cushion, and that in such way as to whiten the surface
equally over and then removing all superfluous whitening with
a dry leather. They also prefer to cut the gold with the sharp
edge of a piece of split cane.

The management of gold-leaf in the way described, although
requiring some practice and care, is easily attained. The squares
of leaf-gold as prepared by the gold-beaters are about 3¾ inches
square. Squares of silver-leaf are 4·5 inches square. A square of
leaf-gold, therefore, contains about 10·8 or nearly 11 square inches.
A square of silver-leaf contains 20·25 square inches.

(89.) *Selection of Glass for Electrical Tubes, and Electrical
Machines.*—All kinds of glass are not well adapted for electrical
excitation, indeed some kinds are not excitable at all. The best
kind of glass for electrical excitation appears to be white flint glass
of good quality ; this should have a fine vitreous polish, and be
clear and hard, without blemish. The common white bottle glass,
of which apothecaries' phials are made, will be generally found
to be very freely excitable—some apothecaries' phials powerfully
so. Common window or crown glass is also very excitable.

The constituents of glass generally are silica, lime, oxide of iron,
a little alkali, and smaller quantities of other matters. Flint glass
contains a large proportion of oxide of lead, which renders it
heavier and more fusible than crown-glass. Silica and borate of

lead are also employed as being the best ingredients. Glass for electrical purposes should contain as little alkali as possible, since it renders the glass soft and very liable to attract moisture from the atmosphere, and is consequently but little excitable. Glass plates containing an excess of alkali have been found after a short time to crumble in pieces. In olden times the glass of Cherbourg, the crystal glass of England, and the glass of Bohemia were much preferred for electrical excitation, and were observed to be more excitable when they had been long exposed to a high temperature. It was, for example, found by Bose, of Wittemburg, that glass exposed to heat in chemical distillations is incomparably more excitable than glass not so exposed. Hence it was recommended by Priestley to blow cylinders and globes for electrical machines, when the metal had been for some days in a state of fusion.

(90.) *Precautions to be observed in the Construction of the Cylindrical and Plate Machines.*—Almost every kind of glass is more or less applicable in the construction of electrical machines, but since there is much difference in the excitability of different kinds of glass, it is important to select a kind of glass found by experience to be the most excitable. Glass for electrical machines should be well annealed, that is to say, exposed to heat for a long time after manufacture. The crystal or white flint glass of England has great excitable power. Some differences of opinion have arisen amongst the older electricians relative to the glass of different countries, some giving a preference to the clear crystal glass of England, others preferring, more especially for cylindrical machines, the blown glass of Cherbourg, many recommend the clear yellow glass of Bohemia, others the white glass of Brittany. The excitation of green bottle-glass is difficult and weak. The vitreous surface of blown glass has certainly an advantage over the polished surface of glass plate, hence it is that the cylindrical machine has in some instances been found more powerful than the ordinary plate glass, the surface rubbed being of the same extent.

(91.) *Cylindrical Machine.*—Having been careful in the selection of the glass of which the cylinder consists, we have next to determine the relative proportions of the diameter to the length between the shoulders, that is of the surface to be rubbed. If the diameter be to the length between the shoulders as 8 to 10, that proportion will do extremely well. A cylinder of 10 inches in diameter, by about $12\frac{1}{2}$ inches in length, would constitute a powerful machine of a moderate size; but 12 inches in diameter by 15 inches in length between the shoulders, would form a very powerful machine. The cylinder should be carefully blown, so as

to turn as equably as possible upon its two necks, when a temporary
axis of wood is passed through it.   It should be rather flattened at
the shoulders.   The neck should be straight and open, but not more
open than is necessary to admit of nicely cleaning it on the inside.
In preparing the cylinder for mounting, it must be made clean
and dry, especially upon its inner surface.   There should not
be a particle of dirt or dust within the glass.   Two soft
silky corks should then be fitted so closely within the open
necks as to completely close them air-tight.   Before closing
the necks, the cylinder should be preserved in a very dry atmosphere, and a day should be selected for shutting up the cylinder
when the air is peculiarly dry, and the wind in a northerly or
easterly quarter.   When the cylinder is closed, the ends of
the corks should be covered with melted sealing-wax, which
should flow freely over them by the aid of heat, so as to exclude
the possibility of any damp air getting through their pores, which
it cannot do if the corks be firmly pressed within the necks after
the manner of bottle corks.   The closed necks should be covered
with caps of brass, which are to support the axis of the cylinder,
firmly cemented on them.   This should be done with the best red
sealing-wax (81).   Closing up the cylinder, having a perfectly dry
atmosphere within, is a most important part of this manipulation ;
the success of the whole business depends on it; for if the slightest
moisture be condensed upon the inner surface of the glass, or
eventually condense on it from any cause, the action of the machine
will be weak, and its efficiency all but ruined.   The inner surface
becomes more or less conducting, and hence carries off the electricity excited upon the outer surface.   This was probably the great
source of failure recorded by Dr. Priestley and the old electricians,
in the construction of their apparatus.   The most promising and
perfect of their glass globes would frequently fail to give a
single spark.   All possible means were tried with a view to promote the excitation.*   The author remembers a case in which a
cylindrical machine, very perfect in form, of very large dimensions, and blown of the best possible glass, totally failed in its
action, and could hardly be caused to emit a spark an inch in
length.   After numerous trials it was determined to remove the
caps covering the open necks, after which the cylinder was completely cleaned out with rectified spirit of wine, and wiped very
dry, at the same time very dry warm air was blown into it with
the bellows.   The necks were then again replaced, and cemented.
The consequence was, an extremely powerful development of

* Priestley's "History of Electricity," pp. 553, 554, &c.

electricity, and sparks were obtained from 12 to 15 inches in length, or more ; thus showing the vast importance of preserving the inner surface of the cylinder in a perfectly dry non-conducting state. It is most probably owing to the ruinous effects of moisture, or the humidity of the inner surface of the glass, that the old electricians observed a vast improvement in their globes on coating them internally with resinous cement, which was made to flow over the surface by means of heat. It is worthy of consideration even at the present day, how far it might be desirable, previously to closing the cylinder, to varnish over its inner surface with shell-lac varnish, and dry it off by internal heat. Sigaud de la Fond, a French philosopher who wrote about the year 1781, says that globes for electrical machines are greatly improved by returning them for a few days to the heating furnace, supposing they do not act at first very well; a remark which applies to glass cylinders.

(92.) *The Construction of an Electrical Rubber or Cushion.*—The best facing for a rubber is either red basil leather, the rough side of stout black oiled silk, or fine Morocco leather. The surfaces should be fair and even, and have some little degree of elasticity. The cushion is best constructed in the following way:—An oblong piece of mahogany about a quarter of an inch in thickness is first provided ; then a second similar, but somewhat thicker, piece, of the exact dimensions required for the cushion, is placed immediately upon it. One edge of the leather or silk to form the face of the rubber is then secured by very fine tacks along the side of the thinner piece of mahogany which is to form the back of the rubber, and then turned and strained firmly over the thicker piece of mahogany resting upon it as a mould. It is now secured in a similar way upon its opposite edge. The mould may now be drawn out from beneath the facing, leaving a fair and regular hollow within. Two or three layers of thick woollen stuff, making up the thickness of the hollow case, being now loosely and evenly sewn together and temporarily secured at one end to a little stout cord, are forcibly drawn by it into the hollow casing. The extremities of the rubbing surface are now neatly folded in and secured to the ends of the back piece. We have in this way a very regular and somewhat elastic surface, which, when in place, applies equally and fairly to the glass. Although thick woollen cloth may be employed with advantage, yet a still better substance is that known as spongeo-piline, prepared for medical purposes, and to be had at any druggist's, consisting principally of sponge and wool, to which are added various fibrous materials, backed by an

o

impermeable surface of gutta-percha or india-rubber ; the whole
constituting a light, porous, elastic substance, admirably adapted,
when cut into slips, for filling the cushions of electrical machines
in the way just described. The oiled silk flap (68), invented by Dr.
Nooth, should be carefully applied to the rubber about a quarter
of an inch from its edge, should be first secured by very strong
paste (84), and the edge of the silk should lie flatly and evenly
upon the cushion from one end of it to the other. It should be
finally fixed to the rubber by two rows of long stitching of waxed
sewing silk. To cover the heads of the tacks securing the facing
of the rubber, the edges of the cushion may be bound round with
narrow silk bind, fixed by paste, which gives the rubber a neat
appearance. The silk flap should be applied with its rough side
next the glass, and neatly rounded at its termination opposite the
row of points projecting from the conductor. The best way of
constructing the row of points is by the insertion of a thin edge
of brass along the line of the conductor, projecting from it about
one quarter of an inch, upon which are soldered short dumpy
points, about three-tenths of an inch apart.

(93.) *Plate Machine.*—Plate machines are generally constructed
of polished glass plate; the plate being first cast, and then
polished. In selecting glass plate for an electrical machine, the
polish of the surface should be the most perfect possible, and the
glass of the most clear and flinty kind.

Although well-polished plate is very efficient and generally
available in the construction of the plate machine, yet it must
be allowed that a polished surface is not, upon the whole, so
excitable as a vitreous surface, the vitreous surface of well-
blown glass, such, for example, as the best blown glass of Cherbourg, or the best blown flint-glass of England, being peculiarly
susceptible of excitation. Hence it becomes a question how far a
compound glass plate for an electrical machine, constructed in the
manner already described (81), is not preferable to a single plate of
polished plate-glass as usually employed, more especially if the process of cementing be improved by experience and practice, and the
blown glass be manufactured and selected expressly for the purpose.
The plates of several electrical machines, from 18 inches to 2 feet
in diameter, constructed in this way by the author, were found to
be wonderfully powerful, and certainly exceeded ordinary plate
machines of the same diameters. The process for obtaining an
electrical plate by the junction of two thin plates, in the way just
described (81), is as yet open to much further improvement. If
the best methods of uniting thin plates of blown glass by inter-

mediate cement were well understood, investigated, and practised by skilful mechanics conversant with such kind of manipulation, we should no doubt obtain plates for electrical machines of extraordinary power. A compound plate being thus obtained, is to be finished off, nicely rounded, drilled, and mounted on an axis in the usual way.

(94.) *Amalgam for exciting the Electrical Machine.*—The employment of an amalgam of tin and mercury by Mr. Canton, so long since as the year 1768 (a notice of which will be found in the 52nd vol. of the Phil. Trans.), was a most important step in practical electricity. Mercury amalgamated with tin, or zinc, or both, increases the excitation of glass in a most surprising manner. Canton's amalgam was of a very simple form. It consisted of an alloy of tin and mercury, which he directs to be well rubbed into the cushion with a little chalk. This will excite the glass very powerfully, and with little friction, especially if the rubber be somewhat damp.

(95.) Higgins's amalgam, as proposed by him, and noticed in the Phil. Trans. for 1778, consisted of mercury and zinc, in the proportion of 4 or 5 parts of mercury to 1 of zinc. This, he says, is better than Canton's amalgam of tin and mercury. The zinc and mercury is reduced by fusion or trituration to the consistence of butter, and converted into a powder by the addition of very dry chalk, well rubbed up in a mortar. An amalgam made of tin and mercury was treated by Canton in the same way.

(96.) The Baron Kienmayer, who compounded a powerful electrical amalgam, consisting of tin, zinc, and mercury, says, that in every kind of amalgam he had hitherto tried he found three inconveniences: 1st. The mercury became separated from the metal with which it was alloyed, and deposited small globules upon the electrical machine. 2nd. The excitation often decreased under the operation of the amalgam. 3rd. The friction was inconveniently great. This led him to employ a combination of the three metals in the proportion of zinc and tin each 1 part, mercury 2 parts, the mercury being equal to the tin and zinc together. This he converted into powder without the aid of chalk, and found it to produce a very powerful excitation. This method of preparation is as follows, and is continued to a greater or less extent up to the present day :—The tin and zinc having been purified according to the best chemical process, are melted together in a clean iron ladle or an earthenware crucible, and well stirred together with an iron rod. The mercury, equal to the weight of the two metals taken together, must be gently heated, and be poured into a cylindrical wooden box having a round hole in the cover.

The alloy of zinc and tin are now removed from the fire and poured upon the mercury through the hole in the box, which must be instantly closed with a cork, and the whole kept in a state of agitation by shaking the box until the amalgam is cold. It is then poured out upon a marble slab or table. The amalgam will be hard and of a silvery colour. It is to be reduced to a very fine powder in a glass or stone mortar. The mass must not be left long untriturated, otherwise it becomes so hard that there is a difficulty in reducing it to powder. If we triturate the amalgam for a long time, and at several repetitions, it becomes grey, and may, by frequent trituration, become black. But there is no rule for this; all we have to do is to obtain a fine powder, free from lumps. It should run through the fingers like dust. This powder may be preserved many years in a bottle. The mercury never separates, even by new trituration, which shows that it is intimately combined with the other metals. It is not desirable to make a large quantity at once; about half a pound is quite sufficient—that is to say, 2 ounces of zinc, 2 ounces of tin, and 4 ounces of mercury.

Kienmayer's amalgam has been more or less accepted by modern electricians; the different metals being amalgamated in different proportions. Mr. Singer observes that amalgam in which the proportion of mercury is considerable is more transient in its effect than when the proportion of mercury is less, at least within certain limits. It is therefore necessary to ascertain the best proportion, so as to obtain a steady and long-continued excitation without a frequent renewal of the amalgamated surface.

The following are a few formulæ which have been resorted to with success, by which it will be seen that the metals, tin, zinc, and mercury, form the basis of electrical amalgams, and that they have been combined in all sorts of proportions.

Formula 26.—Tin, 1; mercury, 2; fine chalk sufficient for
        pulverising . . . . Canton.
   „   27.—Tin, 1; mercury, 1 . . . .   „
   „   28.—Zinc, 1; mercury, 4 or 5 . . . . Higgins.
   „   29.—Zinc, 2; mercury, 5 . . . . Kienmayer.
   „   30.—Tin, 1; zinc, 1; mercury, 2 . . . . . .   „
   „   31.—Tin, 1; zinc, 2; mercury, 5 . . . . . .   „
   „   32.—Tin, 1; zinc, 2; mercury, 6 . . . . . . Singer.
   „   33.—Tin, 1; zinc, 2; mercury, $3\frac{1}{2}$ . . . . . .   „
   „   34.—Tin, 1; zinc, 2; mercury, 3 . . . . . .   „
   „   35.—Tin, 3; zinc, 5; mercury, 9; with the addition
        occasionally of a little sulphur . . . . . . Hearder.
   „   36.—Tin, 3; zinc, 5; mercury, 7 . . · . . . .   „
   „   37.—Tin, 2; zinc, 1; mercury, 5 . . . . . .   „
   „   38.—Tin, 1; zinc, 1; mercury, 1 . . . . . .   „

(97.) Much care is required in compounding electrical amalgam. The manipulation is not by any means easy. In many instances the zinc oxidizes rapidly in the course of fusion, and some part may fly off, hence a change in the proportion of zinc. Kienmayer's method, as already noticed, is no doubt efficient, but it is still open to failure unless it be carefully carried out.

(98.) The following process for an electrical amalgam may be relied on. Pure tin and zinc are to be carefully weighed out and put into a covered earthenware crucible, or into a clean, covered iron ladle. The zinc should be first heated, and when about to fuse the tin added. In the meantime the mercury is placed in a separate crucible or small iron ladle, and heated to something above the boiling-point of water. When the zinc and tin have become fluid, they may be removed from the fire, and the mercury poured gradually on the combined metals, stirring the whole at the same time with a clean iron or steel rod. We must, however, allow the melted metals to cool a little, for if the mercury be poured upon the alloyed tin and zinc at anything like a red heat, it may be driven into vapour before amalgamation takes place, and this vapour not only occasions a loss of mercury, but is very dangerous to inhale. The exact time when to add the mercury is a nice point in the process. The amalgamated mass is now allowed to cool further, and is then poured into a wooden box, as recommended by Kienmayer (96). The box should be kept in a constant state of agitation by shaking until cold. If the amalgam does not contain a large quantity of mercury, it will be commonly found as a fine black powder, which is a proof of the success of the operation. This powder is to be now rubbed down in an iron mortar, and passed through a fine muslin sieve, and then set by in a closed wide-mouthed bottle, or a clean covered box, for future use. If the amalgam contains a larger portion of mercury, this result does not ensue, but the mass is more or less of a pasty cha-racter, and of a bright colour. This is the case with Formula 32. It becomes more compact after some hours' cooling. It is desirable in this case to knock it to pieces in an iron mortar before it has become greatly consolidated. If not containing a very large portion of mercury, it will at length become so pulverised as to admit of being passed through a fine muslin or gauze sieve. All the different forms of amalgam specified (96) should be treated in this way as far as possible. It is not, however, always possible, where the portion of mercury is very large. In this case the amalgamated mass is very soft, and for some time after its formation may be spread almost like butter upon the rubber. The amalgam, Formula 32, is, as just stated, of this character. The actions of these amalgams being not always very certain, it is desirable to keep a number of them prepared in separate bottles, and occasionally combine them. If, for example, an amalgam containing a very small portion of mercury be found too dull, it may be stimulated by adding to it a portion of one of the amalgams containing a large portion of mercury. In this way we obtain an amalgam well adapted to our purpose. It is not always easy to prepare the amalgam to a certainty, whereas by this method we may ensure success. In every case of combination of different amalgams we should rub them well together in an iron mortar, and sift them immediately before using.

The author, after a series of careful experiments, found the following amalgam very powerful and effective, it being prepared in the way above-mentioned :—

Formula 39.—Tin, 1 ; zinc, 2 ; mercury, 4. The ingredients of tin,
zinc, and mercury being double of each other.

(99.) *Mode of Application of Electrical Amalgam.*—There are two modes of applying amalgam to the cushion of the electrical machine.

First, by smearing the rubber over with a thin film of pure lard, so as to give it a somewhat greasy surface, and then by means of a fine sieve sprinkling the surface freely with the powdered amalgam. We then pass a spatula, or thin, wide, stiff knife, with pressure over the surface, so as to smooth it down. The knife may be very slightly smeared with fine lard. The surface of the rubber will very often assume a bright metallic lustre, and will then be in a fit state for exciting the machine. Care must be taken not to allow the amalgam to spread over the edge of the silk flap attached to the cushion.

Second. Rub the powder up in an iron mortar with as small a portion of purified lard as will enable it to spread over and adhere to the surface of the cushion. It should be lightly and equably spread, and not passed beyond the edge of the silk flap.

This method of application is often a difficult one. The amalgam will frequently break off the surface and not adhere with sufficient freedom, so that the surface becomes irregular and lumpy. In order to spread the amalgam equably and thinly, a large proportion of grease or lard is necessary. This is by no means favourable to excitation. To correct this defect we should finish by sifting some dry powder upon the surface, and smooth it down with a spatula. When the amalgam has been applied for some considerable time, the excitation often decreases. In this case we find on examining the rubber that the coating of amalgam has become a hard, solid surface, and requires to be renewed. We may frequently restore the power of the machine at the instant by passing a piece of coarse brown paper or fine glass paper between the glass and the cushion, drawing it a few times forwards and backwards so as to rub up the hard surface and restore it to a more compressible state by a renewal of the original surface. When fresh amalgam is required, the old amalgam must be completely removed, and the surface of the rubber made clean. Mr. Wilson, an electrician of about the middle of the last century, was in the habit of gilding the rubber of his machine, or coating it with silver or some other metal, which he says answers very well.

The Professor of Natural Philosophy in the Royal College of Chartres, France, so long since as the year 1748, moistened his rubbers with water to promote excitation. Dr. Watson also, about the same time, moistened his rubber with a view to promote conduction.

(100.) *Aurum Musivum*, commonly termed *Mosaic gold*, has been occasionally employed with success in promoting electrical excitation. It is a bisulphide of tin; and is obtained by preparing an amalgam of 12 parts tin and 6 of mercury. This is reduced to a powder, and mixed with 7 parts of sublimed sulphur and 6 of sal-ammoniac. This mixture is exposed to a gentle heat in a flask with a long neck, and sulphurated hydrogen is driven off. When the smell of the sulphurated hydrogen is completely gone, the temperature is raised to a low red heat; eventually a scaly mass remains. This is a bisulphide of tin, and is of a beautiful yellow colour. If the chemical process has been properly conducted, this yellow mass acts powerfully as an electrical amalgam, and is a very clean preparation. On the contrary, if not perfectly prepared, the compound is a total failure in this respect. The quantity of sal-ammoniac which it may contain is very objectionable from its liability to attract moisture. To avoid this defect the mass may be reduced to a powder, and washed on a filter until all the sal-ammoniac is got rid of. It is then to be carefully dried upon *heated* paper, and applied as a powder without the aid of grease. It has been found advantageous to smear over the surface of the rubbers with aurum musivum. If the preparation be very pure it may be applied at once to the rubber from the lump, which soon yields up a golden surface. As yet, however, no other process has been devised for producing this substance by direct union

of the ingredients. In this process part of the ingredients employed
are driven off by heat, and seem to act partly mechanically in
giving the required texture to the product, and partly chemi-
cally by absorbing a portion of the heat which accompanies the
sulphuration of the tin and mercury, and which would otherwise
rise high enough to partially decompose the product.

(101.) *Excitation of the Cylindrical Electrical Machine.*—First
remove the cushion and conductors, and then proceed to clean
the glass, which must be effectually done. If the machine has
been long in use, black spots and lines will be found frequently
adhering to its surface. If they be very numerous and rigid,
they require to be removed by mechanical means. The fine edge
of a bone paper-knife may be employed. Mr. Higgins (Phil.
Trans., vol. 68) recommends the skin of a dog-fish, which he says
will not scratch the glass as many powders will. The glass may
be wiped over with finely-prepared chalk as used for medical
purposes, wetted with rectified or methylated spirits of wine, and
to which a little benzine, or a small quantity of solution of soda
or potash, has been added, in order to remove any grease which
may adhere to the glass. The surface of the cylinder should now
be rubbed off with a clean linen or muslin rag until all the chalk
has disappeared. The preparation called *Powdered Blue*, which is
very finely-powdered starch coloured with cobalt, has been also
used for this purpose, and was employed by Cuthbertson. The
cylinder should be finally wiped over with a fine muslin rag wetted
with rectified spirits, and then rubbed dry with a clean silk handkerchief, or some other soft substance.

Having effectually cleaned the glass cylinder, the cushion and
flap next demand attention. All the old amalgam which is
adhering to the rubber should be carefully scraped off, and its
surface wiped very clean. The dirt adhering to the silk flap
must likewise be removed. This should be further cleaned, if
necessary, by a muslin rag dipped in rectified spirits, with which
a few drops of benzine have been mixed, the object being to
remove any grease and dirt which may be adhering to the surface
of the flap; the silk should be finally wiped dry with a clean
muslin rag. We may now proceed to cover the surface of the
rubber with the amalgam already described; taking care not
to plaster it, as it were, upon the rubber in a thick and irregular
coat, but rather incorporate it with the surface of the rubber by
gentle pressure and friction, so that we may obtain a smooth and
equable exciting surface. Proceed now to fit the rubber in its place,

the negative conductor being wiped clean and dry, as also the insulating rod upon which it is supported. Finally, interpose a wide slip of foolscap paper between the rubber and the cylinder, and turn the glass round against it about a dozen times. The paper should have been previously made hot against the fire, so as to deprive it of all moisture. The glass will be now in a very excitable state, and will emit faint sparks, attended by a low rustling sound. If the paper be now drawn out from between the rubber and the cylinder, brilliant luminous sparks will soon fly round the glass, especially on connecting the negative conductor with the ground; and if the knuckle be presented to the surface of the cylinder at the termination of the silk flap, the evolved electricity comes like a wind from the glass upon the knuckle, and is often attended by a crackling, rushing noise. This is good evidence of the success of our manipulation.

We may now apply the prime conductor, taking care to wipe it, and its insulator, very clean. Brilliant sparks will then fly from the prime conductor to the knuckle, or to a large brass ball or other metallic body brought near to it, if the presented body be in connection with the earth or the negative conductor. The machine is now in a fit state for experiment.

(102.) *Excitation of the Plate Electrical Machine.*—In the construction of the Plate Electrical Machine, care must be taken in the selection of the glass, avoiding that which contains a large proportion of alkali. The more metallic the glass is, the better is it adapted for electrical purposes. Alkaline salts give glass a strong disposition to absorb moisture. This defect is said by the old writers to be remedied, in great measure, by returning the glass to the furnace for a short time, or by exposing it in some way to heat. Old looking-glasses have been found well adapted for glass-plate machines, from the circumstance of having been long exposed to a seasoning temperature, by which the alkaline salt in the glass has been dried out. It is to be further remembered that the surface of the glass of a Plate Electrical Machine is not what may be termed a natural vitrefied surface, but is a polished vitreous surface, produced after the glass has been manufactured. This is a most important difference in the plate and cylindrical machines, the natural vitrefied surface being superior to a polished surface in its adaptation to the purposes of electrical excitation. In the construction of the plate machine, therefore, we must be careful to select glass of a high and perfect polish. Common crown or window glass is admirably adapted to the purpose of electrical excitation, and consequently

to the Plate Electrical Machine.   It is, however, too thin and
fragile for the purpose in its ordinary state; but when two
plates of crown-glass are joined together by means of a liquefied
sealing-wax, we obtain a very efficient plate for the Plate Machine.
A plate thus prepared, with black sealing-wax, has a fine polished
appearance.   The author constructed a plate machine of 20
inches diameter in this way so long since as the year 1820.
The power was remarkable, although the glass was by no means
flat, having been cut out of the common tables then employed for
window glass.   It would be difficult to construct a plate machine
in this way, of a very large diameter, without very considerable
experience.   It is, however, to be accomplished, and would be no
doubt superior to ordinary plate-glass.   In mounting the plate
upon its axis, the compressing flanges must be set remarkably true
and perpendicular upon the axis.   One of the flanges, as observed
(77), is a fixture; the other is a nut which, by a screw upon the
axle, can be turned up against the fixed plate, and confine the
plate of the machine with pressure between them, and enable the
axle to hold it fast whilst revolving.   The screw of the flange to be
turned up against the glass, together with the screw upon the
axle itself, should be extremely fine, so that the pressure may
increase very gradually.

Glass-plate machines have been found very liable to crack by
pressure somewhere upon the central hole, which has been drilled
through them for the reception of the axis.   Too much care cannot be taken to avoid this, both in the drilling of the hole (which
should be smooth, without the sign of a splinter) and checking
the pressure of the flange beyond a certain point.   The pressure,
it is clear, will go on constantly increasing in turning the machine,
since we turn in one direction against the resistance of the rubbers
in the opposite direction.   The most effectual way of securing the
plate against fracture from the pressure, is to drill a hole through
the rim of the flange perpendicularly into the axle, tap it, and
turn up a fine screw, which will prevent any further movement
of the screw flange (a sufficient pressure for turning the plate
round having been obtained), as also by the judicious interposition
of a circular piece of fine, elastic leather between the flange and
the glass.   It is also very desirable to thicken the centre of the
plate by small plates of glass about 4 inches diameter, and from
one-eighth to one-fourth of an inch in thickness, having holes in
them to take the axle.   These plates are cemented to the glass
plate, one on each side, by adhesive varnish.

Fig. 47 (page 71 *ante*) represents the brass axle upon which the plate is mounted. It is about 20 inches in length, and from eight-tenths to one inch in diameter. *a b* are the brass flanges; *c d*, the glass plates cemented to the plate of the machine. The axle has a little play in its bearings, which are fixed to the transverse bars of support, Fig. 46. That part of the axle carrying the handle is encircled by a fixed brass socket, screwed fast to the frame of the machine, about 1 inch in diameter and 2 inches in length. A solid ring of brass is adapted to this cylindrical socket so as to turn upon it with easy friction, and gives support to the vertical conductor *a* R *b*, Fig. 51, and by which the electricity of the glass is carried off when the prime conductor is transformed into a negative conductor. [See Chap. III. (80)]. The axle projects on the side of the handle about 4 inches clear of the frame. The glass handle is full an inch in diameter at its thickest part, is a little taper, and about a foot in length between its brass caps.

In exciting the plate machine we have first to get the plate perfectly clean by the same means as recommended for the cylindrical machine. The skin of the dogfish may be employed with advantage for removing any black lines or spots upon the glass. A half-sheet of foolscap paper made hot before the fire, and interposed between the cushions and the glass, and the plate revolved between them, is a very beneficial process in the final preparation of the plate for excitation; indeed, papers of this kind should be always kept between the rubbers and the glass, or the rubbers themselves when not in use.

The preparation of the cushions or rubbers with the silk flap in the electrical plate machine demands considerable attention. In the machine now under consideration, the rubbers may be in length about one-third of the diameter of the plate, and in width about one-fifth their length. Fig. 48 represents the rubbers of the 3-foot plate as constructed by the author. They are 1 foot long, by 2½ inches wide, and consist of layers of thick cloth or felt, or better of spongeo-piline (92), laid upon a flat piece of mahogany, and faced up with fine morocco leather, or thick black oiled-silk, the rough sides of each being next the glass. The leather or silk is secured to the edges of the mahogany back by very fine copper tacks, as already described for the cylinder machine (92). The rubbers are loosely held in a central position in a sort of spring frame by projecting pins, screwed into the mahogany back. The frames supporting the rubbers are represented in Fig. 48, and

consist of two oblong side pieces, z z', of mahogany, slightly
hollowed to give them elasticity. These spring pieces fit loosely
by a dove-tail joint in a mahogany block, B; and are held
together with pressure upon the rubbers by a short cylindrical
bolt, a b, passing freely through them immediately in front of
the block, and two milled balls screwed one upon each end of it,
so as to press the mahogany holders together upon the rubbers
and the intermediate glass plate with any required degree of force.
All these points should be easy and movable—not by any means
rigid—so that the whole may easily play with the movement of
the plate. In the old plate machines it was customary, instead of
allowing the compressing bolt to go freely through both the holders,
to fix it to one of them, and by a nut in the other they were compressed. The rubbers cannot accommodate themselves too easily to
the movement of the plate. The flaps should be applied to the cushion
in the same way as described for the cylindrical machine (92),
that is to say, first securing its edge upon the face of the rubber at
a distance of about one-third of an inch, by a little strong adhesive
paste, and finally, by a double row of running stitches of very fine
waxed silk, so that the edge of the flap may be well secured to
the rubber, and lie fairly and evenly upon its surface. The flaps of
each pair of rubbers are eventually united over the edge of the
plate. In applying the amalgam to the cushions, care must be
taken to lay it on thinly, smoothly, and evenly. The amalgam
should not pass beyond the edge of the flap where it is secured
to the rubber.

OCCASIONAL MEMORANDA AND EXPLANATORY NOTES.

(V) Lac is a gum resin, the product of an insect termed the "Coccus
lacca," which deposits it on the branches of certain trees in India, especially
in Assam and Thibet, where it is found in a regular cellular structure, containing the eggs of the insect. It is one of the best electrical insulators
we have, and plays a most important part in electrical investigations. Its
constituent parts are resin, gluten, wax, and a peculiar red colouring
matter.

It is imported into Europe under three forms, *Stick-lac*, *Seed-lac*, and
Shell-lac.

Stick-lac is the first or rude state, as found encrusting the twigs and branches
of the trees on which it is deposited. For purification it is broken into
small pieces, put into a long narrow canvas bag, and exposed to a sufficient heat to liquefy the gum. The liquefied gum is forced out by twisting
the bag, and is allowed to flow over a plane smooth surface, to which it

does not adhere. The inhabitants of India allow the liquid to flow out and consolidate upon the convex surface of a plantain-tree expressly prepared for the purpose. The mucilaginous and smooth surface of this tree prevents the gum from adhering. In 100 parts of stick-lac in its crude state we find about 68 resin, colouring matter 10, wax 6, gluten 5·5, extraneous matter 10·5. The colouring matter is a valuable product, and forms the basis of the lac-lakes and lac-dyes. Stick-lac thus purified and consolidated, being pounded in a mortar, reduced into small grains, and a further portion of the colouring matter extracted by the process of boiling, constitutes the substance called *Seed-lac*.

Shell-lac is the lac in its natural state after the process just described of simple purification by heat, and is produced by liquefying, straining, and forming the liquid lac into thin plates, from whence its name *Shell-lac*.

PART II.

ON THE LAWS OF ELECTRICAL FORCE.

CHAPTER I.

BRIEF ENUMERATION OF FACTS AND PHENOMENA OF ELECTRICAL
ACTION, AS DEDUCIBLE FROM AN INVESTIGATION INTO THE THEO-
RETICAL AND PRACTICAL NATURE, OPERATION, AND LAWS OF
ELECTRICITY.

1. The nature of electricity has always been a source of much
intricate discussion and theoretical investigation. No solid conclusion, however, has ever been arrived at, calculated to determine
the precise nature of this wonderful invisible agency. Many
philosophers, both ancient and modern, have considered electrical
force as depending on a highly elastic fluid. Gilbert thought
that electrical attraction was the same as the attraction of cohe-
sion, and was occasioned by effluviæ proceeding from bodies
excited by friction.

An hypothesis by Digby, as set forth by Brown, supposes that
a kind of electrical effluvium extends by friction from an electrified
substance which, after extension, again retracts, and brings with
it light particles of matter. Gassendi adopts a notion of this
kind. Similar views have been countenanced by Boyle and other
philosophers, both ancient and modern.

The Abbé Nollet supposes that in electrical excitation there is
an efflux of electrical effluviæ from, and an afflux toward, the
electrified body. By the afflux toward the excited body, attraction is the result; by the efflux from, repulsion ensues. Nollet
further supposes that in all bodies there are two kinds of pores,
one for receiving, the other for giving out the effluviæ.

2. Notions of this kind, however, can only be considered as
mere wanderings of the imagination, and we are left to base our
views of electrical action on substantial experiment.

3. We are indebted, as we have seen (Part I., 17), to M. du Fay, Intendant of the Gardens of His Majesty the King of France, in 1730, for the first great step in theoretical and practical electricity. This ingenious philosopher discovered that electrically-excited bodies develop two kinds of force, one produced by excited vitreous, the other by excited resinous substances. Hence he terms the one *vitreous*, the other *resinous* electricity. On further investigation these different electricities were found to be attractive of each other, but repulsive of themselves (Part I., 27).

4. The wonderful discoveries of Newton, and his determination that the laws of gravity were universal, naturally led philosophers to entertain the notion that every other force associated with ordinary matter was subject to the same law; that the law of gravity was, in fact, the dominant law of all attractive forces, of whatever kind or species, of electricity and magnetism more especially; and a great variety of experimental inquiries were instituted with the express view of verifying this conjecture. Cavendish, after Æpinus, was certainly the first philosopher who investigated experimentally, and threw light on this question, as appears by his celebrated paper in the sixty-first volume of the Philosophical Transactions for the year 1772, and likewise by his MS. but as yet unpublished papers, handed down to us by the Earl of Burlington, a descendant of the Cavendish family. Cavendish directly proves that on the supposition of electricity being an elastic fluid, the mutual repulsion of its particles is in the inverse duplicate ratio of the distances. That in the case of a sphere being charged with electricity, all the charge should be found on its surface, whilst the action upon any interior point would be nothing (Part I., 55). He further describes a great variety of original and very beautiful experiments in support of this hypothesis. It is, therefore, a misapprehension on the part of some writers when they say that " Cavendish had assumed no particular law of electrical force, but merely that the action decreases as the distance increases." Dr. Robison, in 1769, read a paper professing to show that the law of electrical attraction was identical with that of gravity : the same result had been arrived at by Mayer, and had been eagerly sought for by other philosophers of that period. The beautiful and profound physico-mathematical researches of Coulombe, in the Memoirs of the French Academy for 1785, all went to confirm this law of the inverse duplicate ratio of the distance as stated by Cavendish. The views of electricity which Coulombe has handed down to us have since received the sanction of the most eminent philosophers of Europe.

The experimental inquiries of Coulombe have been made the basis of a comprehensive analysis by Poisson, as also by Le Place, Biot, and other eminent mathematicians.   Coulombe's theory now stands pre-eminent as the Coulombian theory of electricity. Such a galaxy of eminent names, countenanced by so wide and general a reception, would necessarily go far to discourage doubts of the accuracy of Coulombe's views; still the progress of phy- sical discovery must not be limited ; since from the moment we establish a limit, and say that limit has been attained, from that same moment we close the door upon all further research into principles, and we remain immersed in a sort of philosophical orthodoxy, very unfavourable to a more complete knowledge of those unseen yet astounding powers of nature, the effects of which we daily experience.   (Note A.)

That the law of the inverse square of the distance, as deduced by Coulombe and other philosophers, is a law of electrical force, and a very general and important law in its consequences, we cannot for a moment doubt ; although the experiments by which it was sought to establish this law were few, and not always satisfactory, being derived principally from the operation of what is termed electrical repulsion, an element liable, in many instances, to much uncertainty.   Admitting, however, this law of the inverse dupli- cate ratio of the distance to be a general law, it is still important to examine under what circumstances the law obtains ; and the peculiar conditions requisite to its full development.   Such an inquiry is calculated to bring us further acquainted with the nature and operation of electrical action, and to develop numerous laws of this force, of which we might otherwise remain in ignor- ance.

5. It may not be hence altogether unprofitable to define clearly what we really understand when we say that a force increases or decreases in the inverse duplicate ratio of the distance of its action, since upon this a large amount of sound reasoning upon the laws of physical forces mainly depends.   It is highly probable, if not morally certain, that every physical effect is in a simple proportion to its cause.   If the cause increase in any proportion, the effect will increase in the same proportion. The effect cannot, for example, be as the square or the cube of the cause ; for that would be, as remarked by the learned Dr. Samuel Clarke, to refer the effect partly to the cause, and partly to nothing. The amount of illumination, for example, of a given point, will depend on the quantity of light concentrated in that point.   If we double the quantity of the agency in operation, we should

P

double the illuminating effect; we should certainly not quadruple it. In referring, therefore, the decrease of a force to an increase of the distance, or more especially an inverse square of the distance, we are fairly led to inquire into the cause of so wonderful, and apparently so inconsequent, a phenomenon. Suppose, for example, we endow distance, that is to say, a mere imaginary line or space, with all the properties of entity, and consider its extension as the immediate cause of the observed decrease in the force; still we have no ground for supposing that a double extension in length would cause more than a proportionate reciprocal decrease of the force. We could have no ground for assuming that the force would become four times, nine times, &c., less, as the distance became twice, three times, &c., greater; that is, upon the hypothesis that mere distance was the cause of the decrease of the forces. Reasoning much in this way, many philosophers—eminent for their intellectual powers—looking upon gravity as a central force, or an emanation from a centre, have considered this emanation to expand in all directions, and to become more feeble in proportion to the increased spherical surfaces over which the emanation may be supposed to expand; and since the concave surfaces of the imaginary spheres, intercepting the emanation at given distances, would increase directly as the squares of those distances—that is, as the squares of their radii—therefore the quantity of the emanation or force in any point of space would be in the same ratio; so that at double the distance from the centre we should only have one-fourth the force in any one point, and, therefore, only one-fourth the effect, and thus the force would be said *to vary in an inverse ratio of the squares of the distances.*

Dr. Halley, for example, reasoned in this way;* and the reasoning, it must be admitted, is perfectly clear and intelligible; the effect here is in a simple ratio of its cause, not as the square of its cause. The force, for example, decreases with the cause of the force; and we see so far how it is that distance enters as an element into the general expression of the law of the force of gravity.

6. The question then is, whether electrical or magnetic force is such a force as this?

Some philosophers object to this kind of reasoning, and contend for the absolute existence of force, varying in an inverse duplicate ratio of the distance between indefinitely minute points of vanishing magnitude, in which they say that no attenuation of

* Whewell's "Hist. of Inductive Sciences," vol. ii. p. 158. 2nd edition.

the agency upon which the force depends can be supposed to obtain; they take the points as points, without magnitude or parts. How far this involves a metaphysical subtlety is left to be determined. We certainly know nothing of forces, or the laws of their action, except through the medium of observation and physical experiment; but as yet no physical experiment in either electricity or magnetism, on the attractive or repulsive force of points of vanishing magnitude, has been ever realised. Then, again, we must not leave out the peculiar condition of power on such points. If we suppose, for example, all the electricity upon a given sphere condensed into a point, and all the electricity of another sphere condensed into a point, still we may conceive the physical powers of these two points to be very different.

Reasoning in this way, all the matter of the sun condensed into a point would necessarily constitute a very different point of power from that of all the matter of the earth condensed into a point; and, therefore, at last we have a somewhat complicated question to deal with, when we apply this sort of analysis to such forces as those of electricity and magnetism, exerted at different distances from each other.

7. Our present object, however, is to investigate the physical conditions under which the forces of attraction and repulsion apparent in electricity obtain; what are the general laws of the operation of such forces; how far we may consider them as central forces, such as gravity; or whether they are to be considered more in the light of parallel forces destructive in their character, in all their relations to common matter, and in the elementary conditions of their operation.

8. If we examine the physical data upon which our general views of the laws of these forces rest, we do not find the experimental investigations to be extensive, nor are they always satisfactory; certainly they are not at all comparable with the precedents of other branches of science. The data nearly, if not all, depend on a few selected and especial cases of repulsive force, and are very dependent on the perfect operation of a small insulated disc termed a tangent plane (Part I., 143), which, being brought into contact with various portions of a charged body, is assumed to be an element of the surface of that body, so that when withdrawn from it, the tangent plane is considered as coming away charged with a quantity of electricity proportionate to the quantity absolutely existing in the point to which it has been applied. From such experiments a certain distribution of electricity, considered as a highly elastic fluid, is inferred; all of which is, to a greater or less extent, still

within the limit of conjecture, and it must be considered as being a somewhat remarkable fact, that up to a late period there existed scarcely any electrical instrument of research, based upon attractive force. A very large range of electrical phenomena remains, therefore, to be investigated by varied experiment, and which does not appear as yet to have been brought under the dominion of experimental analysis.

9. Now it is to be here observed, that if we imagine an insulated conductor to exist in space perfectly freed from the influence of common matter—supposing, for example, it were the only body in space—then it would really be impossible to say—supposing it charged with what we call electricity—how the charge exists upon it. For anything we should know to the contrary, the electrical agency might be uniformly distributed throughout its substance or over its surface. We could really know nothing, nor could we predicate anything of its electrical condition, except through the instrumentality of other bodies brought to share in the charge; we might, indeed, upon a variety of hypotheses and refined methods of analysis, arrive at what should be its especial electrical condition—supposing the hypothetical basis of such calculations to be correct; but we should still have to verify the results upon the broad basis of observation and experiment.

10. It may here be as well to offer a few observations on the operation of what has been called the *Tangent Plane*, as employed by Coulombe and the French philosophers, with a view of determining the distribution of electricity on charged bodies, on the assumption that electricity is an agency depending on two fluids of high elasticity, repulsive of their own particles, but attractive of each other.

One of the most striking and important features of electrical attractive force is not the phenomenon of the apparent attraction itself observed to arise on presenting a charged to a neutral surface, or to another body in an opposite electrical state, but the previous or sympathetic, or what is termed the *induced* actions which take place between the bodies, and upon which the subsequent attractive force altogether depends. In fact, both in electricity and magnetism, bodies are first rendered attractable or repellent, before attraction or repulsion can ensue; and if such changes do not obtain, no attractive or repulsive force is apparent. If, for example, we suspend an extremely thin conducting disc from an arm of a balance, by insulating threads of shell-lac, we shall in vain attempt to attract it by opposing to it a similar disc charged with electricity; or at least the force will be extremely small; nor will such an insulated neutral disc be in a perfect condition to receive or abstract electricity from a charged disc (Part I., 44).

11. Now we find on increasing the thickness of the suspended disc the attractive force becomes more fully developed; and if we give it a conducting communication with the earth, we then complete its susceptibility of being rendered attractable; and the force, with a given charge at a given distance, is then the greatest possible; but if we take the suspended disc insulated and indefinitely thin, we have really no development of attractive force at all.

Take the tangent disc in any way you please, it really cannot be considered in any other light than that of a neutral conductor, brought to share in the electricity of a charged conductor.

12. Now I observe, in touching various points of a long cylindrical and charged conductor with a small insulated conducting body, that the proportion of the charge it receives by contact with different points of the charged body is very dependent on its thickness; so that the touching body takes up less and less electricity as we diminish the thickness. When the thickness of the disc is inconsiderable, the charge which it receives at the centre vanishes, or nearly so. Now I should wish to have it considered whether at a certain limit of thickness this touching body might not come away altogether neutral, or nearly so, from any or every point of the charged body. That is a point which I think deserves great consideration before we are entitled to draw any legitimate conclusion as to the distribution of the electricity on the charged body. Now it is notorious that if you take a small insulated conductor of great extension in thickness, such a conductor comes away equally charged with whatever point of the charged body it be brought into contact. If we coat a small disc of talc with tinfoil—so as to complete the conditions of the Leyden jar or coated pane—and apply it to various points of a long conductor charged with electricity, it will be charged equally high from every point of such a conductor; or if we bring an electrometer in contact with such a conductor, it will evince the same degree of power at all points of application; so that the vital source of the power of one body to receive electricity from another body, is the susceptibility of induction in the recipient body; and I cannot conceive the possibility of any body taking electricity from another without undergoing previous inductive change. This consideration leads me to ask attention to another series of facts which I conceive to be intimately associated with the operation of the tangent disc. It is well known that if a small insulated carrier ball be plunged

within a charged hollow sphere, it comes away perfectly free from all charge; its contact with the interior surface of the sphere has been productive of no electrical change in the ball whatever; it has been hence inferred, as we all know, that no electricity exists on the interior surface of the sphere. That, I admit, may possibly be the case; but certainly this result is not conclusive of the fact. The failure of the small ball to become electrified, may as well be supposed to arise from the total incapability of all inductive change in the small ball under such circumstances as from the absence of electricity on the interior surface of the sphere. In evidence of which I made the following experiment, and which will be found fully described in the Royal Society's Transactions. A hollow globe of glass, with a long neck, was arranged so as to have the coatings of mercury. The glass was then charged, and the mercury drawn off. The outside of the jar was bristling with electricity, yet no electricity was to be obtained on the inside. (B.)

13. It is I think evident from this, that the forces here so operate upon the carrier ball in opposite directions when placed within the charged sphere, as to fetter completely its inductive susceptibility; hence it fails to share in the electrical charge. That there is really no essential difference between this and the case of the hollow conducting sphere, as regards electrical charge, and that the absence of inductive change is really the cause of the neutrality of the ball, is shown by a repetition of the same experiment, with the exception of substituting a plane of glass for the hollow sphere. If we communicate a charge to a coated pane, and remove the coatings, we may take up electricity freely from the charged side, but in this case the carrier ball can, by external exposure, undergo inductive change.

That the induction upon the ball should be zero within the sphere, supposing the force to be in an inverse duplicate ratio of the distance, is admitted, and that is so far an evidence of the truth of such a law, which I beg leave again to state I am not disputing; but it is still consistent with the phenomena of the hollow charged conducting sphere, in which also the inductive change on the carrier ball is zero, and hence it cannot receive electricity from the interior surface, even supposing the interior surface to be capable of communicating it.

14. I think we may conclude from these facts that a tangent plane will receive more or less electricity, in proportion as it is more or less freed from the opposing forces of surrounding electrical particles. According to Volta it would be the greater or less emersion of the body in the electrical atmosphere of the charge.

In illustration of this, let the square or circle A, Fig. 109, and the rectangle B, represent two plane conductors of equal area, charged with the same quantity of electricity, and let us imagine for the moment that the electricity is equally distributed, as designated by the crossed straight lines on those figures. Now it is quite evident that if we apply a small insulated plane to the centre of the square, the small plane would be more immersed in the surround-

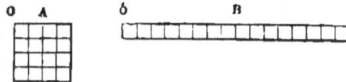

Fig. 109.

ing electrical particles than if applied at the middle of the rectangle B, and certainly more embarrassed in its inductive efficiency than if applied at the points o ó of either of these conductors. In fact, at the points o ó it would have as it were one exterior portion free from electrical particles; hence, on this account only, it could take up more electricity at the extremities than at the centre of the conductor, independent of any consideration of distribution.

15. That the grouping together of electrical particles to a greater or less extent influences their activity, or what may be termed their state of quietude, is evident from the fact that a given quantity of electricity will repose with more quietude on a given surface, in proportion as the surface becomes extended in length. Thus, although the areas of the circle or square A, and of the rectangle B are the same, yet B has a greater linear extension than A, and an electrometer will evince a very much greater effect when connected with the circle than when connected with the rectangle. As I have already observed, in calling attention to these facts I would by no means be considered as professing to determine authoritatively the question of the accuracy of the proof plane as an instrument of research, and as employed by Coulombe and other philosophers. It would be unwarrantably presumptuous in me to pronounce dogmatically upon the absolute merits of such investigations as those to which I have already adverted. If I venture, therefore, on submitting the question of the proof plane for further consideration, I trust it will be believed that I do so with all becoming diffidence, and a due appreciation of the many beautiful researches in which the method of the proof plane has been so largely employed.

16. Having now adverted to the question of the proof plane as important in the investigation of electrical force, I will briefly consider the way in which the forces of attraction and repulsion in electricity and magnetism manifest themselves, together

with the laws to which they are subjected. The dependence of the electrical and magnetic forces of attraction and repulsion on the elementary changes by induction, which the attracting or repelling bodies evince, at once gives a distinctive and separate character to these forces, as compared with the force of gravity. In the force of gravity we know nothing of what may be termed a dual force; nor do we perceive that the attractive power exerted between one mass of matter and another is at all dependent on any previous or elementary change in the attracting masses, or in the existing condition of the agency on which the attraction depends, that is, so far as we at present know; and, therefore, so far the forces of electricity and magnetism are essentially different in their very nature from the force of gravity. That these forces become less in some inverse ratio of the distance of the bodies through or upon which they are exerted, is certain, but according to what law they thus increase or decrease was for a long time a problem of some difficulty.

17. *The early Experimentalists.*—Sir Isaac Newton conceived magnetic force to vary as the cubes of the distances, inversely. Hawksbee, Whiston, Brook Taylor, Muschenbröck, Martin, and several other eminent philosophers, all arrived at different experimental results. Hawksbee found the force to vary in an inverse sesquiduplicate ratio of the distances, that is, as the $\frac{5}{2}$ power of the distance inversely. Martin deduced the law of an inverse sesquiplicate, or $\frac{3}{2}$ power of the distance. Brook Taylor conceived that the force did not change according to any regular law, but varied at different distances from the magnet. Whiston, however, and subsequently Brook Taylor and Hawksbee, again found it as the $\frac{5}{2}$ power of the distance. Æpinus, and after him the Dutch experimentalist, Muschenbröck, found the force to be in the simple inverse ratio of the distance. In some cases, however, Muschenbröck obtained all sorts of laws. Finally, Cavendish, Lambert, Mayer, and Coulombe arrived at the laws of the inverse square of the distance, as in the case of gravity. Now, it has been, as I think, too prevalent a practice to discard the experiments of the earlier experimentalists, in which other laws appear to have been deduced. It is not likely that such men as Hawksbee, Brook Taylor, and Whiston would have instituted inaccurate experimental investigations. It remains, therefore, to conciliate these differences, and show the source of the differences in the respective results.

18. Now, in examining the operation of these forces as exerted between given masses, we have first to examine the laws of those

changes we have termed inductive, upon which the forces depend. Now I think I have shown in my former papers, contributed to the Transactions of the Royal Society, that when an electrified conductor is placed in the presence of a neutral conductor in a free state, a change takes place in the neutral conductor, such as to induce in it an opposite force, which force increases in a simple inverse ratio of the distance between the opposed surfaces. Then this force, so called into operation, re-acts upon the charged body, by what I have termed a species of *reflection*, or a *reflected induction*, which reflected induction also varies in an inverse simple ratio of the distance between the opposed surfaces, that is, if the charged surface have likewise great capacity. When these two elementary forces increase or decrease together, the total force varies with the two conjointly, and thus we obtain a law of force in an inverse duplicate ratio of the distance; but if from any cause either of the two elementary inductive forces becomes impeded, as in the case of the force between a charged conductor and an insulated neutral conductor of small extension, then we have no longer the law of force just stated; the law may be any of the laws arrived at by the early experimentalists, depending entirely on the law of variation of the previous inductive changes.

These direct and reflected inductions in magnetism I am enabled to illustrate, to a certain extent, by the apparatus before me,* and this is equally true for electrified bodies.

19. This understood, it is easy to see that the limit of the law of electrical or magnetic force is a simple inverse ratio of the distance —that is to say, if we take two masses affected electrically or magnetically to saturation (that is to say, so affected that neither undergoes any further inductive change by approximation), we have then a force varying as the simple distance inversely. In approaching the limit of saturation, therefore, we may readily conceive the induced force to change gradually; still the convergence may be so slow that for a long series of terms the induction may be taken as constant. It is only, then, between certain limits, and under certain free states or conditions of magnetic or electrical change, that we obtain the law arrived at by Coulombe. In the case of electrical force between a charged and neutral conductor, the force will be always in the inverse ratio of the squares of the distances between the attracting surfaces, so long as the neutral, or what is considered the attracted body, has free communication with the earth, and the charged body has also an unlimited extent of charge.

* See " Rudimentary Magnetism," part iii., chap. i.

Thus, for example, if a plane surface be in communication with a charged Leyden jar, and another plane and equal surface in connection with the earth be opposed to it, the force will vary in an inverse duplicate ratio of the distances with great precision, and it will be as the number of attracting points directly. If the opposed surfaces be spherical, as in the case of two spheres, we have then to determine a point within each hemisphere in which we may conceive the whole force to be collected, and to be the same as if operating from every part of the opposed hemisphere. These points, designated by $p\ p'$ in the annexed figure, may be easily determined on the demonstrable hypothesis that the force is as the areas of attraction directly, and as the squares of the distances inversely. When the spheres touch, these points will be in the surface; when at an infinite distance they will be at the centres of the spheres. I have exhibited, by means of the Electrical Balance, the precision with which the attractive force of spheres varies in an inverse duplicate ratio of the distances between these points, so as to leave no doubt on the mind of the truth of this law.

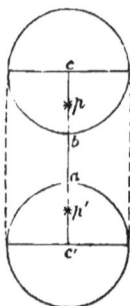

Fig. 110.

20. Now it is a remarkable fact, that in order to obtain this law of force it is essential that both the attracting surfaces have unlimited electrical capacity. We cannot, for example, obtain this law with two thin conducting planes, nor can we obtain it by opposing to each other mere positive and negative electricity accumulated on a non-conducting surface. I have lately contrived an experiment which seems to verify this perfectly. If two light discs of talc be furnished with temporary coatings, and be charged as in the case of the Leyden pane, we may, on removing the coatings, obtain a purely positive and negative surface. If these two surfaces be opposed to each other, and the forces of attraction at given distances measured by some adequate instrument, the force will be in an inverse simple ratio of the distance.

21. I have already stated that the measurement of electrical and magnetic force, through the instrumentality of repulsion, is by no means satisfactory; and that will, I think, be apparent on a very little consideration. When two bodies similarly charged, either electrically or magnetically, are brought within each other's influence, the elementary inductive forces, already adverted to, are still to be considered, and the immediate tendency of each body is to reverse the existing electrical or magnetic state of the other—that is to say, the inductive and repulsive

forces are in opposite directions. From the induction, attraction ensues ; and hence it is that two bodies similarly but unequally charged will attract each other at some distances, and repel at others ; and it is only, therefore, in cases in which the bodies are so permanently charged as to maintain their existing state in opposition to the inductive actions, that we can expect to arrive at consistent results. The inductive forces in opposition to the existing state of the bodies is in fact the immediate source of the apparent repulsion. The effort really is *attractive force* after all ; but in the effort to bring about this attractive force, the bodies, if movable, recede from each other. Hence it follows that the amount of the observed repulsion is only the difference between the repulsion maintained and the attractive force induced ; and this is really a very serious consideration in operating with the Torsion Balance.

OCCASIONAL MEMORANDA AND EXPLANATORY NOTES.

(Note A. p. 209.) The following note by the author appeared in the *Philosophical Magazine* for August, 1857 :—

It is not without regret that I am led to offer a few remarks on the several notices of my papers on electricity by Professor Thomson and other writers, which have from time to time appeared in the *Philosophical Magazine*. In the first place, I observe (Phil. Mag., vol. viii., p. 42) Mr. Thomson states, that Dr. Faraday and myself have undertaken researches with a view of invalidating the theory of Coulombe,—certainly a gratuitous and unproved assumption to begin with. Then the experimental result I obtained, $F \propto Q^2$, is represented as having been adduced by me in opposition to the theory ; and the Report made to the British Association in 1837, by the Rev. Dr. Whewell, is referred to in support of Mr. Thomson's views. In this report I am represented as having been surprised at finding the force to be as the square of the quantity ; and in another place (" Hist. of Induct. Sciences," vol iii., p. 28, new edition), it is said that I considered the result as " inexplicable." Now if Mr. Thomson or the Rev. Dr. Whewell, for whom I must ever entertain the greatest regard and respect, will be so good as to point out where, in any of my works, even the shadow of all this is to be found, I should feel myself greatly obliged. So far from having been surprised at the result, or considering it inexplicable, I distinctly say in my paper in the Philosophical Transactions for 1834, referred to by Mr. Thomson, p. 266, that " I do not advert to these experiments as containing any unexpected results, but rather in explanation of particular methods of research," &c. ; and at page 236 I refer the law in question to electrical induction upon the attracted disc, &c., being just what Mr. Thomson has been so obliging to repeat, virtually in as many words, and with a view of exposing the delusion he assumes I labour under in regard to the Coulombian theory. In Mr. Thomson's paper (Phil. Mag. for 1854, vol. vii., p. 193), I observe the following remark:—"The amount of heat is proportional to the square of the quantity discharged, as was first demonstrated by Joule, although it had been announced by Sir Snow Harris, as an experimental result, to be simply proportional to quantity." Riess, Joule, and Clausius appear to agree as to my having made this announcement, and confidently allude to the inaccuracy and to the refutation of my deductions. Now I do here most emphatically and most positively deny ever having stated anything of the kind; but since these gentlemen say I *have* done so, and pretend to have corrected my error, I call upon them, as a point of honour, to say when I made such an announcement, or where, in any of my published works, it is to be found. The fact is, that so far from having stated the law of electrical heat to be simply as the quantity of electricity discharged, I was really the first to discover with precision, and demonstrate with exactitude, the law in question, and that, too, long before Mr. Thomson and the other gentlemen above mentioned were at all known in the world of science. If Mr. Thomson will turn to pages 67 and 68 of the Transactions of the Plymouth Institution, published in 1830, and quoted in the Journal of the Royal Institution, 1830-31, p. 380,* he will find in the former work a series of original experiments on the heating effects of the electrical discharge, and at page 68 the following announcement:—" It may be hence inferred that the effects of an electrical discharge on a metallic wire, all other things remaining the same, is *directly as the square of the quantity.*" See also Journal of Royal Institution, p. 381 (vi.). Moreover, the date of my paper is so far back as November, 1825; and I may say, in the words of Dr. Riess (Phil. Mag., 1854, vol. vii., p. 348), that " an assertion of such a general character as that ventured by Mr. Thomson ought to be the consequence of a careful examination," especially of what has been done at home.

As this subject is of importance, I may further remark, that the experiments of Cuthbertson and others, referred to by Becquerel, and quoted by Joule (Phil. Mag., October, 1841), can scarcely be said to have established the law in question, or even its probability. In the first place, they had no accurate measure of the quantity of electricity accumulated; they were unaware that twice the quantity accumulated on a given coated surface would counterpoise four times the weight, regulating Cuthbertson's own steel-yard electrometer, in which a charge of 30 grains was taken as twice that of 15 grains. Their methods of research by the fusion of wires appears to have been anything but exact, and all sorts of tricks were played with the battery, as by breathing into the jars, &c. Cuthbertson says (" Practical Electricity," pp. 180 to 186), " If 18 inches of wire be taken, and a given charge just causes it to run into balls, much shorter lengths will still be only converted into balls; if only 7 inches be taken, nothing but balls will appear; the only difference will be that the balls will be smaller, and will be dispersed to a greater distance, which may be easily overlooked." Now what confidence can be placed in all this as experimental research? Van Marum found his batteries produce a heating effect proportional to the coated surface. Cuthbertson, by his most exceptionable method of breathing into his jars, when he wanted a greater effect, did, it is true, obtain a higher ratio, yet no direct satisfactory comparison between the quantity discharged and the heat produced was ever arrived at, and various results ensued. This we see admitted at page 182 of Mr. Cuthbertson's work, Exp. 149: here the law in question evidently failed. At page 185 we observe that when the quantities of electricity were said to be as 2 : 3, another result ensued; for experiments 150 and 151 show that the lengths of wire fused, instead of

* See also a copy of my paper (printed in 1825), in the Library of the Royal Society.

being as 4 : 9, were as 2 : 6; that is, as 1 : 3. It is quite impossible to repose
any confidence in such a state of things. Indeed, we have only to examine
Mr. Cuthbertson's experiments attentively, as given in his work, and in
Nicholson's Journal (4to., vol. ii., p. 218), to be assured of the inexactitude
of the experimental processes. In the latter we find the lengths of wire
melted to be as the quantity of electricity (see p. 218). However true, therefore, it may be that Cuthbertson obtained results which led him to imagine
that twice the quantity of electricity would melt four times the length of
wire, he cannot be said to have demonstrated and established that law; and I
may, therefore, without any philosophical injustice, claim to have been the
first to have clearly developed that law by exact electrical measurements,
and by new methods of research, as my paper, dated 1825, and quoted in the
Journal of the Royal Institution, 1830-31, fully shows. There, also, will
be found the hypothesis advanced by Mr. Joule, that increased velocity is probably the source of the quadruple heat. I endeavour to show that the heat
is as the velocity with which the unit of charge traverses the wire, that a
double quantity passes with a double velocity, and the effect is "as the
momentum," or quantity into velocity. I again, therefore, ask Mr. Thomson
to state when or where I ever announced, "as an experimental result,"
that the heating effect of an electrical discharge was "simply proportional
to the quantity of electricity." Mr. Joule, in referring to my paper in the
"Transactions" for 1834, appears to have confounded this question with
that of the *same* quantity accumulated under different electrometer intensities,—a question discussed in my late differences with Dr. Riess. What I
said in my paper in the "Philosophical Transactions," and which I still
insist on, was simply this, viz., that under whatever electrometer intensity
you accumulate a given quantity of electricity, provided the battery surface
be undivided, that quantity, when dischared through a metal wire, will still
excite in it the same degree of heat. Thus, if a quantity of electricity = A,
for example, be collected on a Leyden jar, exposing 2·5 square feet of coating,
and then be collected on a jar exposing 5 square feet of coating, I say that,
notwithstanding the electrical intensity in these two cases may be as 4 : 1
nearly, yet that the discharge of the quantity A in each case will excite the
same heat in a metallic wire; that is, if the same charging rod and circuit
be employed. And I ask the gentlemen to whom I have just alluded to
make that experiment. Until they do so, they have no right to talk of "the
memoir in which Riess refuted my statement," &c. (Phil. Mag., 1854, vol.
vii., p. 297).

(B. p. 214.) *A hollow globe of glass about five inches in diameter, with a
short neck about an inch in diameter, is filled up to the neck with dry mercury,
and placed in an outer glass vessel also filled with mercury. We have here all
the conditions requisite for imparting a charge to the interior of the glass.
For this purpose let a light insulated charging rod, surmounted by an electrometer of repulsion, be introduced within the globe, and the mercury in
the exterior vessel be connected with the ground by a metallic communication. Let this system be charged in the usual way: when charged to any
degree of intensity, as shown by the electrometer, remove the communication with the earth, and also the charging rod, by means of its insulating

---

* The substance of this note is from a paper by the author, entitled "Researches
in Statical Electricity," contained in the *Philosophical Magazine* for 1857. The
paper is illustrated by an engraved Plate containing 22 Figures. Figs. 5, 6, and 7
refer to the experiment alluded to at p. 264, and described in this note.—ED.

support; run off the mercury first from the outer vessel, and then from the interior of the globe by means of a glass siphon, and place the now empty globe on an insulating stand. We may be now assured, on the faith of Franklin's celebrated experiment of the electrical jar with movable coatings, that all the interior surface of the globe is covered with electricity. Introduce now a small insulated carrier-ball into this charged globe, so as to touch the interior electrified glass, and again withdraw it. The carrier-ball comes away quite neutral, as in the case of the hollow metal globe, notwithstanding that it has been actually brought into contact with a dense stratum of electricity, the presence of which may be made evident by simply attaching the carrier-ball to the lower point of the insulated charging rod, and introducing it as before. The electrometer will, if delicately hung, be immediately affected; or otherwise the charge may be shown by the medium of an ordinary gold leaf.

That the failure of the carrier-ball to take up electricity is in no degree dependent on the circumstance of what may be considered in the light of electrical accumulation on an insulating surface as distinguishable from the case of the hollow metal sphere, may be clearly shown by charging a plane glass surface having movable coatings, and treating the plane charged glass in a similar way. In this case electricity is freely taken up by the carrier-ball from the charged side: it is hence evident that the globular form of the surface is the immediate cause of the failure of the carrier-ball to take up electricity from the glass. The carrier-ball, in fact, cannot assume that induced electrical state requisite for its reception of free electricity; the forces operating on it being in contrary directions, its natural electricity cannot recede from any point of its surface, hence all induced change in the distribution of its own proper electricity is impossible.

This experiment with charged glass may be effectively managed by employing water instead of mercury; or we may envelop the outer surface of the globe in tinfoil and electrify the glass internally by means of a point connected with the electrical machine, and projecting within the globe.

That induction would go on within the globe were it free to do so, may be exemplified by one or two striking illustrations.

Let a charged hollow globe of metal or glass be placed on an insulating support as before; introduce within it the small insulated carrier-ball, and whilst within the globe touch the carrier-ball with a light insulated wire, projecting freely into the air; remove this wire, and then withdraw the carrier-ball; the carrier-ball will be found charged with electricity opposite to that of the globe; if the globe be plus, the carrier-ball will be minus, as might be expected. In this state introduce the carrier-ball again within the globe, and so as to touch the interior surface; it comes away now quite neutral,—that is to say, it has taken up positive electricity from the interior surface, either by immediate contact with the electrical particles in the case of a charged globe of glass, or through the medium of the metal surface in the case of a charged metal globe.

The same result ensues if we touch the carrier-ball with the free wire when in contact with the interior surface. The insulated free wire comes away positively charged with the electricity which had retired from the carrier-ball. But if both the touching wire and the carrier-ball be raised together and removed without the globe, then the whole evinces positive electricity; for the carrier-ball, whilst in contact with the interior of the sphere, having first become negative, immediately takes positive electricity from the charged sphere and becomes neutral, and probably remains so whilst in contact with the sphere; whilst the exterior ball of the touching wire evinces positive electricity, being necessarily charged with the electricity superinduced upon it by the first induction. On raising the whole system out of the sphere, however, this superinduced electricity expands over the whole; for the original conditions are restored, whilst the new electricity taken up remains; hence the carrier-ball will now evince positive electricity. And this is really what happens when an insulated wire and ball are introduced within the globe of sufficient length to project into the air.

Although the two cases of charged globes to which I have thus called attention may at first appear different, the one being a case of a hollow globe of metal, the other of glass, yet a very little reflection will show that both cases are virtually the same thing; the difference is a difference of degree, not of kind; they are, in fact, both reducible to the elements of the electrical or Leyden jar. Indeed, every case of what is commonly called a charged conductor, resolves itself into the form of a coated electric, and is the result of a peculiar disposition and combination of electrics and conductors; it is, in fact, the accumulation of electricity upon the terminating strata of a dialectric medium, bounded by, and in direct contact with, conducting matter, either near or distant. Thus, in the case of what has been termed a charged hollow metallic globe, in which all the charge is conceived to be impelled, as it were, by the repulsive force of its particles from a centre of force, and so finds its way to the surface of the metal, we find on an attentive examination the following arrangement of conductors and electrics, and into which every case of electrical charge may be finally resolved. We have, first, a metallic surface; secondly, exterior to this an insulating medium, viz., atmospheric air, in a stratum of which, immediately surrounding the globe, there is a dense electrical accumulation ; thirdly, beyond this stratum we have in continuation other air not so immediately electrified, and susceptible of further inductive change : the external air is in its turn bounded by other conducting matter. When, therefore, we impart free electricity to the hollow sphere, we do nothing more than cause an electrical accumulation to ensue upon the dense stratum, according to the well-established principles of the Leyden experiment. We do not, in fact, charge the sphere at all, any more than we charge the coating of an electrical jar; indeed, it is doubtful if in any case we could charge a metallic conductor taken apart from, or in the absence of, a dialectric boundary. The globe itself can be regarded in no other light than that of the inner coating of a given dialectric bounded by distant conducting matter, and which we may consider as the opposed coating; the metal of the globe is merely the conductor to the charge. The inductive action upon which the charge depends, may be shown to extend to great distances. Cavendish traced it from the centre to the walls of a room 16 feet in diameter.* Faraday traced it from a ball suspended in the middle of a room to the walls, 26 feet distant. †

If we examine the experiment of the glass globe with mercury coatings, we find the elementary conditions precisely the same. Here we have an interior coating of mercury, then an external dialectric, which in this case is glass; finally, an outer coating of mercury. Here, as in the preceding case, we do not charge the interior mercury, as is well known, although it may possibly, on being removed, be slightly electrified: the result of the operation is to cause an electrical accumulation on the interior surface of the glass. The great difference in the two cases simply consists in the more or less perfect application of the coatings to the dialectric medium upon which the charge depends. If we could suppose our hollow metal globe surrounded by a second external globe of metal, the two globes being near each other, but not anywhere touching, then we should have the two cases

---

* Cavendish, MS.   † "Experimental Researches," 1303.

identical, as seen in the very ingenious apparatus employed by Faraday.*
If, further, we imagine that, subsequently to the charging of this system,
both the globes were removed, as in the case of the mercury coatings, and
the intermediate air to remain, as it were, fixed and immovable, then
would be developed upon the boundaries of this stratum all the phenomena
of the hollow glass globe above described; the charge would remain with
the air. If, in the case of the charged glass globe, we allowed the interior
metallic coating to remain under the form of thin metallic leaf attached to
the glass, then the final experimental conditions would be identical. As it
is, we operate upon the charge in the case of the hollow metal sphere,
through the medium of the coating; in the case of the glass globe, we come
into contact more immediately with the charge itself.

The theoretical view, therefore, of the celebrated experiment of what has
been termed a charged hollow sphere, and which appears the best adapted to
explain the phenomena, is that of charged electrics generally. The free
electricity first communicated to the inner coating, viz., the metallic sphere,
operates by induction upon the nearest matter susceptible of electrical
change, and thereby develops and calls into operation the opposite electricity. The opposite forces thus brought into life tend to combine and
exhibit attractive force, and consequently come as near together as want of
conducting power in the intervening restraining dialectric will permit: the
imparted electricity must therefore necessarily find its way upon the exterior
surface of the hollow metal globe without the aid of any kind of repulsive
force to which the phenomenon has been hypothetically attributed. It is
in virtue of this kind of action that we are said to charge simple insulated
conductors generally. The amount of charge, however, or quantity of electricity which can be sustained by them under a given electrometer indication,
can never be so great as in the case of systematically coated electrics of comparatively small thickness. The case of simple conductors is much the
same thing as the case of extremely thick glass, or the limiting of the free
action of one of the coatings of an electrical jar; in either case the quantity
of electricity which can be accumulated under a given degree of the electrometer is greatly diminished. The experiment with an electrified hollow
globe therefore appears to have been expressed in very inexact terms, and
the phenomenon of the exclusive appearance of the charge upon the exterior surface somewhat misapprehended.

It follows from these demonstrable conditions of electrical charge, that a
stratum of what may be termed electrical particles must always necessarily
exist upon the surfaces of a charged conductor, as is clearly demonstrable
by experiment; and the electrical agency, whatever it be, penetrates to a
greater or less degree the substance of the air itself, or other dialectric
medium in contact with the conducting surface, as is well shown in Faraday's most comprehensive researches (1245). This is really the acceptation
of the term *electrical atmosphere;* a term correctly applied by the celebrated
Volta, who most thoroughly comprehended the practical nature of electrical
force, notwithstanding that his power of rigorous thought has been questioned, and his theoretical views of electricity rather severely remarked on
by an eminent writer not altogether unbiassed by theoretical opinion, and
evidently not a little impatient of dissent.† That Volta was most perfectly
correct in attributing the phenomena of charged electrical conductors to the
presence of electrical atmospheres surrounding them, taken in the sense in
which I have just explained the term, is absolutely demonstrable by the
most conclusive experiments: we remove the metal, and there remains the

* "Experimental Researches," 1195.
† "Bibliothèque Universelle," article "Volta."

charged stratum; in other words, the atmosphere of electrical particles, as it may be termed, exists entirely without and independent of it.

The electrical stratum thus found to exist on electrified conductors appears firmly held to the surface by attractive force, and is inseparable from it by any movement of the body. Franklin whirled a charged ball attached to a silk cord many times round in the air, and with great velocity, still the ball retained its charge.* An electrified conductor, therefore, when transferred from one place to another, may be supposed to carry the electrical stratum along with it, just as the metallic coating of charged glass would do. It is true that the term electrical atmosphere has been occasionally used in a vague and unsatisfactory sense, and has hence been justly discountenanced by many eminent physicists and mathematicians. When taken, however, in the sense in which I have applied it as expressive of a demonstrable fact, the question assumes quite another form. Volta, therefore, in referring the phenomena of electrical attraction and repulsion, and the operation of electrified bodies generally on each other, to the immediate action of the electrical particles themselves held firmly on the surfaces of bodies, did in no way violate sound deduction from rigorous thought, or evince in any degree a vague and imperfect apprehension of the probable nature of electrical force. When we remove the coatings of a charged electric, something is evidently left behind—some agency or source of power. What is that something? In what does it consist? It is evidently external to the metallic surface with which the dialectric surface was previously in contact, although inseparable from it so long as the two remain combined; and it is really from this something, which we express by the term electrical stratum or atmosphere, that the phenomena we observe arise, and not from a hypothetically charged conductor.†

* If we charge a plate of glass through plane movable coatings of gilt wood, the coatings will adhere to the glass.

† The experiments of Beccaria and Franklin with the smoke of resin and colophonia, and which they observed to collect about electrified conductors, so as to envelop them, although perhaps no very satisfactory evidence of the existence of a similar atmosphere of electricity, are still not without very considerable interest. It is not easy to explain the adherence of these atmospheres of smoke to the charged surface, admitting the theory of electrical repulsive force.

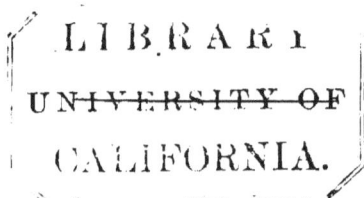

CHAPTER II.

THE most approved view of electricity as a physical agency, not only in remote but in modern times, assumes the existence of an extremely subtle fluid form of matter of high elasticity, termed the *electrical fluid.* So long ago as the year 1748, M. le Morin, Professor of the Royal College of Chartres, in France, had, after Lucretius's view of magnetic agency, and in common with many other philosophers, the idea that this electrical fluid displayed itself in the shape of exhalations from the surfaces of bodies; and he endeavoured to show that all bodies exhale and inspire this electrical fluid.* The Hon. Robert Boyle, in 1760, seems to have had a similar notion. He imagined that electrical bodies emit a glutinous effluvium which seizes upon small and light particles of matter, and then carries them back to the electrically excited substance. Wilson, in 1756, Muschenbröck, in 1769, M. Sigaud de la Fond, in 1781, together with many other eminent philosophers, all agreed in the assumption of an existing material agency which they termed an *electrical fluid.* M. Avogardo, an acute and distinguished physicist, who wrote in more recent times, viz., in 1806, thought the assumption of the existence of an electrical fluid diffused throughout common matter, not only a necesssity, but a great theoretical fact—since without it we can form no adequate conception of electrical phenomena. Coming down to the present day, philosophers adhere for the most part to the same notion ; they are, however, divided in opinion as to the more immediate constitution of this fluid. They all appear satisfied of its extreme, I may say, vanishing rarity, and of its almost immeasurable elasticity and expansive power ; but they differ as to its character—whether it be of a simple elementary kind, or compounded of other elementary electrical fluids. On the hypothesis of its being an elementary fluid, repulsive of itself, but attractive of common matter, whilst the particles of common matter, deprived of this fluid, are also

* Nouvelle dissertation sur L'Electricité des Corps.

repulsive of each other, a large amount of electrical action is explicable. So, on the other hand, by considering the electric fluid as a compound of two elementary fluids, repulsive of themselves, but attractive of each other, and without any affinity for common matter, we are enabled also to elucidate an equally large amount of electrical phenomena. The objection to the first hypothesis, adopted by Æpinus, appears to be the necessity for attributing a repulsive force to the particles of common matter, in order to obtain a new force requisite to complete the condition of what has been termed electrical equilibrium, but which inconvenience would be obviated by conceiving the electrical fluid as made up of two other elementary fluids—each repulsive of its own particles and attractive of the other. But since those who advocate this latter doctrine do not pretend to have proved the actual existence of an electrical fluid at all, but place it in theory with its two elementary fluids, merely as aiding us to predict results and helping us to better comprehend electrical phenomena,* it seems of very minor importance, taken in this point of view, whether we make up the number of required forces in the one way or the other. Hence, any rigid criticism of the theory of a single elementary fluid, as interpreted by Æpinus, is certainly uncalled for, and superfluous. The time, as it appears to me, is fast approaching, if not actually arrived, when it may be found desirable to abandon the theoretical notion of an existing electric fluid or fluids, and base our researches on a much less hypothetical and more solid ground.

Newton, in his masterly researches, examined and deduced the laws of gravity, but he did not stop to inquire whether gravity were a fluid or fluids endowed with given properties; he was content to treat gravity simply as a physical force, and deal with it accordingly, upon principles of a sound practical philosophy, and a profound geometry. We gain but little, so far as I see, by resorting to any hypothetical view of the nature of electricity as an occult power. Take the case of force, as exemplified in the action of an elongated or compressed spring, which either expands or contracts. No one would, I believe, build much on the assumption of a *fluid elasticity* as the source of the observed forces. Far better would it be to deal with electrical phenomena as evidence of some, at present, inexplicable physical power of nature, the laws of whose action we seek to determine; and so build upon the substantial ground of demonstrable facts, involving no assumption whatever. Moreover, this pretended electrical

* Haüy, vol. i., p. 103.

fluid or fluids, repulsive of their own particles, involve very
anomalous and contradictory elements, inconsistent with the nature
of things. We assume, for example, the existence of an extremely
rare species of matter, devoid of gravity, and yet exerting forces
of repulsion and attraction, through a mechanical agency, greater
than any similar force observed in ponderable matter. We give
to the particles of this imponderable electrical fluid a species of
repulsive force, essentially different from any kind of repulsive
force of which we have the slightest cognizance from experience. In
all elastic fluids, such as air, steam, and the like, the repulsive force
of the particles only extends to small distances, and to particles
in the immediate vicinity of each other ; whilst their attractive
power, by which they obey the law of a universal gravitation,
extends to all distances. Now electrical repulsion seems the very
reverse of this. Here the force of repulsion extends to great
distances, and we observe a detached mass of electrical particles in
one place, exerting repulsive force upon another mass of electrical
particles in another, without any apparent proximity ; and whilst
thus exerting an apparent force at very sensible distances, the
force between the molecules themselves at insensible distances is
so feeble as to be incapable of expansion into space, and often
inoperative between similarly charged bodies when brought into
ordinary contact with each other.

It is to be regretted that so few experiments have been instituted
of a simple and direct kind, with a view of elucidating the laws of
electricity, considered simply as a physical power, and without
reference to any theoretical assumption as to its occult qualities.
The experimental investigations upon which our fine mathematical
superstructures are based, are really very few, and certainly not
very satisfactory. They consist principally in determining the
law of distribution of a hypothetical electrical fluid on or between
electrified bodies of different forms, and also what has been
termed the *density* or thickness of the stratum in various
points of the charged surface. This is supposed to be effected
by touching the charged body in different parts with a small
insulated circular metallic disc of little thickness, and then
by transferring the disc to a balance of torsion, so determine the
quantity of electricity with which it has become charged, upon
the assumption that it is virtually an element of the surface it has
been caused to touch, and hence informs us of the proportionate
quantity of electricity at the point of contact. Now, although it
cannot be questioned but that the repulsive force indicated in the
torsion or bifilar balances is a true measure of what may be

called the quantity of force actually in operation, yet it is extremely doubtful whether the tangent disc can be considered as an element of the surface of the charged body in the point of contact, and still more doubtful whether it comes away charged with a quantity of electricity proportionate to the quantity actually existing in that point, as I shall presently endeavour to prove.

The basis, therefore, of our present knowledge of electricity being so far limited, it appears to me desirable to treat the question of electrical force in a more general and less hypothetical manner, and to deal with it by simple and more direct experiments, much in the way that we are accustomed to examine other physical forces of whose peculiar nature we have no adequate conception. It may be, perhaps, convenient and useful to employ analogical expressions in interpreting the phenomena observed, and to facilitate description, as when we talk of the quantity of electric matter, of its tension, intensity, thickness of stratum, density, the addition or abstraction of electricity considered as a fluid, and so forth : and it is, nevertheless, most important to remember, that it is with force simply, and the laws of force, that we are dealing, and not with electrical fluids and their supposed distribution on the surfaces of bodies.

It has been well observed by an eminent writer* that *number*, *weight*, and *measure*, form the foundation of all exact science ; that no branch of human knowledge can be held as being out of its infancy which does not in some way or other frame its theory or correct its practice with reference to these elements. Impressed with this great truth, I have for a long time endeavoured to contrive and perfect such quantitative measures and processes in electricity as shall give us, in a speedy, direct, and simple way, the information we require relative to the laws and operations of electricity, taken simply in the light of a physical force, without any care about its precise nature or its occult quality. The instruments I employ, with the exception of the bifilar balance, are now before us in their most recent and perfect state. I am aware that I have before called attention to some of these instruments, and to some few experiments connected with them ; but their more perfect and refined state, with a great number of new results, I have still to notice.

These instruments have already been described in the previous pages, and they consist—1st. Of the Scale-Beam Electrometer, with new and recent improvements for estimating force, distance, and other elements. 2nd. The Hydrostatic Electrometer. 3rd.

* M. Quetelet on "Probability."

The Unit Measure, with its mode of operation. 4th. The Quantity Jar. 5th. The Thermo-Electrometer.

Having explained the nature of these instruments for bringing the observed phenomena of electricity under the dominion of number, weight, and measure, I proceed to some of the leading features of electrical action.

First. It is to be observed that electricity, considered as a simple physical power, assumes two forms or modes of action. These have been commonly termed *vitreous* and *resinous* electricities, and have been further designated by the mathematical signs *plus* and *minus.* These two forms or modes of action are always present in every electrical development. They are really not distinct and separate elements, except when taken in a certain sense, but are in themselves identical as to active power. Thus, for example, when we shoot an arrow from a bent bow, considered as a spring, the particles of matter of the bow upon the concave surface are compressed; those upon the convex surface are strained and extended. In this case there are two opposite forces. We cannot have one without the other. The expression for the kind of tension of the one may be different from the expression for the tension of the other; but still the force which propels the arrow is resolvable into one thing, viz., the force of what we call *elasticity.* As I have already remarked, we should gain little or nothing by assuming the existence of two elastic forces in order to investigate and explain the nature or mode of action of this exertion of mechanical force.

Secondly. We have to observe that electrical force does not operate in several directions at once with the same degree of power as if exerted in one direction only. In the planetary system, the attractive force of the sun on the earth is, so far as I understand it, quite independent of its force on the other planets. The sun would, for example, attract the earth at a given distance with the same force as at present were the other planets all annihilated, and the sun and earth the only bodies in space. Now, this is not the case with electrical force. In the exertion of electrical force between two bodies, the introduction of a third body causes the force to decline, as has been already proved by experiment.

Let us now see on what this depends. With this view we observe—

Thirdly. That in electrical attraction the bodies are first rendered attractable by a peculiar operation of force termed *induction,* and then it is that attraction follows, but not before; and if any disturbing force or impediment to the full and free exertion of this influence arise, the amount of force is either greatly limited or does not appear at all. I wish the reader to observe that I am now taking the phenomena as they present themselves without any relation whatever to the possible occult qualities of electricity, and that he should dismiss all impression with reference to the actual nature of electricity, of electrical fluids and such like.

This species of inductive action admits of being illustrated by experiment, or possibly in a more direct way by an approximation of what may be termed the charged to the neutral body, instead of the neutral to the charged body.

Now, it is this kind of influence, apparently a sympathetic action, exerted between the bodies at very sensible distances, that constitutes the great characteristic of electricity as a physical power; and, although most probably resolvable into a species of propagation of action through the particles of the medium interposed between the attracting bodies, is still the great and first phenomenon to be examined.

According to what law, then, does this action between the bodies progress? That is the first point to which I call attention.

Now, we are enabled to determine this by the instruments before us, and by the torsion or bifilar balances : observing, as stated in my Bakerian Lecture in the Philosophical Transactions for 1839, that not only is the inductive influence exerted by the electrified on the neutral body, but that the previously neutral body so influenced reflects back a similar action upon the electrified body; and that it is actually upon the simultaneous operation of these two elementary forces combined that the whole force of attraction between the two bodies depends. If either be interfered with, if both be not freely and fully exerted, we have no longer an invariable amount of force. Murphy, in his fine mathematical tract on the Laws of Heat and Electricity, embraces a view not very dissimilar to this, alluded to in my paper in the Royal Society's Transactions, under an hypothesis of a principle of *successive influences*. He obtains numerical approximations to the representation of inductive forces between two bodies, by calculating the effects of four or five successive acts of influence.

Professor Thomson has, subsequently to all this, placed the question under the form of a principle of reflected images, and has obtained a series easily applicable, in some instances, to the exertion of electrical attractive force between spheres and other bodies.

I have elsewhere described my method of measuring these

inductions, and have shown that they are inversely as the simple
distance between the bodies. Calling H the force of induction,
we have

$$H \propto \frac{1}{D}$$

These previous inquiries lead us more especially to a consideration of the law of force as observed between two bodies—one
electrified, the other neutral, and both susceptible of what may be
termed *electrical change by influence.*

Now some considerable anomalies appear to have arisen in the
experimental results obtained by philosophers at different periods
relative to this question, some finding the force to vary in
the inverse ratio of the simple distance, others in the inverse
sesquiplicate ratio of the distance, that is, the square root of the
cube of the distance; others in the inverse sesquiduplicate ratio of
the distance, that is, the square root of the 5th power of the distance; others inversely as the square root of the distance. Cavendish,
Coulombe, Lambert, and some other philosophers, asserted that
electrical force was in the inverse duplicate ratio of the distance
in common with gravity; others, such as Brook Taylor and
Muschenbröck, were led at one time to the conclusion that the
operation of these forces did not follow any particular law. The
question now is, how are we to reconcile such anomalies? It is
next to impossible to imagine that men so distinguished in
physico-mathematical science should have made clumsy and ill-
contrived experiments, or that their results should not have been
faithfully recorded by them. It has, I believe, been too much the
practice to call the experimental skill of these philosophers in
question ; more especially in every case in which they failed to
establish the favourite law of force,—viz., the inverse square of
the distance. I believe, however, I can conciliate all these apparent
anomalies, and show that both in electrictity and magnetism, any
law of force is possible so long as the bodies are in any way so
circumstanced as to limit the inductive change of which they are
susceptible. For pure electrical attraction we require both bodies
to be fully and freely susceptible of inductive change. In this
case the actual force will always vary in the inverse duplicate ratio
of the distances—not as a simple elementary force—but as the
result of a combination of the elementary inductions.

When a magnetic bar, A, Fig. 111, is opposed to a similar, but
smaller, bar of iron, B, then a new polarity, n, is induced in the
near parts n of the iron, opposite in kind to that of the opposed
polarity A, whilst another polarity, q, arises in its more distant

parts, similar in polarity to that of the polarity A, but opposite to that of the induced polarity n. This, however, is not all. On further examination, we find that the temporary polarity n, thus induced in the near surface of the iron, operates in its turn on the near surface p of the magnet, producing there, by a species of reflection or reverberation, what may be considered as a new polarity, p, opposite in kind to that of the induced polarity n, but similar to that of the permanent polarity A ; that is to say, a portion of the force, which under the ordinary conditions of magnetized steel is directed towards the centre of the magnet, becomes now determined toward the iron in the direction $p\,n$.

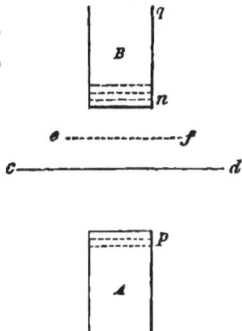

Fig. 111.

The sublime discoveries of Newton, and his verification of the general law of universal gravitation, naturally enough led philosophers to entertain the notion that all forces in nature were subject to the same law, and that this law was a dominant law of every species of attractive force. Hence a strong bias existed in favour of such a conclusion. Philosophers have not unfrequently gone out of their way, and twisted experiments, solely with a view of verifying this law in the exertion of electrical and magnetic force; and have, as I think, betrayed some want of respect for the investigations of those philosophers who failed to establish it. But let us see how this case stands.

Since the total force between the bodies—suppose between the two bodies P N—will be dependent on the direct and reflected actions we have just described; then supposing they both go on *sine limite;* that P is a charged body and N a neutral body, the first operation is a direct action of influence of P upon N. Now let the number of particles of excited force in the proximate surface n of N = a, and the number of particles of reflected force in the opposite surface p of P = b, the total force at the distance, which we may call $p\,n$ = unity or 1, will be represented by $a \times b = a\,b$. For the attraction of one particle in the proximate surface n of the body N, to all the particles on the surface p of P, will be as b. The attraction of two particles will be

Fig. 112.

$2\,b$, of three particles $3\,b$, &c., and so on until we have the attraction of the whole number = $a\,b$, for the direct force. Well now,

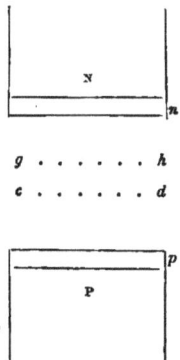

suppose we diminish the distance between these bodies, say to the line $c\,d$, and reduce it by one half in this case; the induced force is doubled, a becomes $2\,a$, the reflected force b becomes $2\,b$, and the total force is $2\,a \times 2\,b = 4\,a\,b = 2^2\,a\,b$. If we again reduce the distance $p\,n$, say to $\frac{1}{3}\,n$, the direct force a becomes $3\,a$, the reflected force $= b$ becomes $3\,b$, and the total force is $3\,a \times 3\,b = 9\,a\,b = 3^2\,a\,b$. Thus, whilst distances decrease in the series $1, \frac{1}{2}, \frac{1}{3}, \frac{1}{4}$, &c., forces increase in the proportion $1, 2^2, 3^2, 4^2$, that is, inversely, as the squares of those distances.

So far, the results are plain and intelligible. But now let us imagine that by an imperfection of insulation, or a limited surface, we curb these elementary forces, and that the law of inductive force between them changes, which, by experiments similar to those we have described, it may be shown to do, we may, for example, with a limited induction obtain a force in the inverse ratio of the square root of the distance. If instead, therefore, of a free state of the neutral body, we insulate it, and its limited capacity is so reduced as to cause the induced and reflected forces to vary in the simple inverse ratio of the square root of the distance; then, reasoning as before, and taking a unit of force $= a\,b$, at a unit of distance $= p\,n$, we have no longer at the distance $\frac{1}{2} = c\,d$, the induced force $a = 2\,a$, but equal $1\cdot4\,a$, so that the reflected force instead of being $2\,b$, is $1\cdot4\,b$, that is, inversely as the square root of $\frac{1}{2}$. In this case the total force is $1\cdot4\,a \times 1\cdot4\,b = 2\,a\,b$, and the total forces are as $1 : 2$, that is to say, the force varies as the simple distance inversely—a result arrived at by Muschenbröck and Æpinus; so that we may actually obtain almost any law of total force, depending on the limiting elements of the inductive actions upon which the force depends.

Suppose, by way of further example, the induced force should approach the inverse ratio of the $\frac{2}{3}$ power of the distance; then at the distance $\frac{1}{2}$, the forces of induction are $1\cdot68\,a$, instead of $2\,a$; and $1\cdot68\,b$, instead of $2\,b$. Then the total force being $1\cdot68\,a \times 1\cdot68\,b$, will be $2\cdot8\,a\,b$ nearly. So whilst the distances are 1 and 2, the forces are as 1 and $2\cdot8$; that is, in the inverse sesquiplicate ratio of the distances, as found by Martin in three very unexceptionable experiments on magnetic force.

I cannot but here express my conviction that, taken as a simple elementary or fundamental law, there is no such law in nature as that of an inverse duplicate ratio of the distance. That we can have an effect in any other proportion than that of its cause, is, as observed by the learned Dr. Samuel Clarke, to suppose the effect to depend partly on its cause and partly on nothing. Take the case of gravity or of light, which is but a force emanating in all directions from a centre. We have the light in a given point directly proportional to the quantity of light in that point; that is to say, the effect is directly proportionate to the cause, but still represented by the expression $\dfrac{1}{D^2}$

Now one hitherto difficult physical problem in electricity is the law of force as exerted between spheres and other curvilinear bodies, and many theoretical views of this case have been advanced by M. Biot and others. Nevertheless, taken in the way just explained, the question becomes a very simple one, and we arrive at very beautiful experimental results.

We may demonstrate with perfect success, that the quantity of force being constant, the total force will vary directly as the areas of the attractive surfaces, and in the inverse duplicate ratio of the distance between them. (C.)

OCCASIONAL MEMORANDA AND EXPLANATORY NOTES.

(C. p. 235.) The distribution of force in the charge and discharge of the electrical jar admits of geometrical representation in a way calculated to elucidate in a satisfactory manner the precise condition of this question.

Let the line A C, Fig. 113, move forward parallel to itself upon the line C D, and generate the equal rectangular spaces A m, e n, fo, g D; then these spaces may stand for and represent equal and successive quantities of electricity communicated to the electrical jar; and the total rectangular space A B C D may stand for, and represent, the total charge. Now,

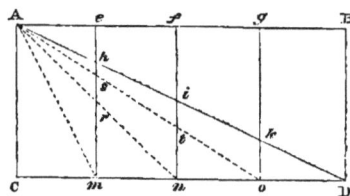

Fig. 113.

since the force between the coatings continually decreases as the force through the external circuit increases, therefore these forces may be represented by any magnitude supposed to flow, and at the same time continually increase or decrease. Let therefore the line A C move forward parallel to itself as before, and continually decrease by the linear magnitudes e h, f i, g k, B D (which are proportional to the lines A e, A f, A g, A B, that is, to the respective quantities of electricity, C e, C f, C g, C B), and so generate the triangular space A D C, which may stand for and represent the total force in the direction of the coatings up to the point D, where it vanishes in explosive discharge. Similarly, let the extremity A of the line C A flow and gradually increase by the same proportional lines e h, f i, g k, B D, generating the triangular space A B D, which may stand for and represent the total force in the direction of the circuit up to the same point of explosion, D. This being understood, we may observe that the first quantity accumulated being represented by the space $C\,e$, the force in the direction of the coatings will be represented by the space $A\,C\,m\,h$, and the force in the direction of the external circuit (that is, the electrometer indication) by the triangular space $A\,e\,h$. Similarly a double quantity of charge will be represented by the double space $C\,f =$ twice $C\,e$; the force in the direction of the coatings by the space $A\,C\,n\,i$; and the force in the direction of the external circuit by the triangular space $A\,f\,i$, and so on. But the triangles $A\,e\,h$, $A\,f\,i$, are to each other as the square of $A\,e$ to the square of $A\,f$; and since $A\,f$ is double of $A\,e$, these triangular spaces are as 4 : 1. In the same way it may be shown that the triangles $A\,g\,k$, $A\,e\,h$, are as 9 : 1, and so on. Now, as these triangles stand for and represent the force through the external circuit, that is to say, the respective electrometer indications, whilst the sides $A\,e$, $A\,f$, $A\,g$, &c., are proportional to the respective accumulations $C\,e$, $C\,f$, &c., we see that the force in the direction of the circuit is (all other things remaining the same) directly proportional to the square of the quantity accumulating. It is easy to see by the diagram that the decrements of force upon each added quantity will be as the rectilinear spaces $A\,C\,m\,h$, $h\,m\,n\,i$, $i\,n\,o\,k$, $o\,k$ D, whilst the increasing force in the direction of the circuit will be represented by the triangular spaces $A\,e\,h$, $A\,f\,i$, $A\,g\,k$, $A\,B\,D$.

In like manner let the lines $A\,e$, $A\,f$, $A\,g$, &c., stand for the intervals, or explosive distances between the balls of the Lane's discharger, which the force in the direction of the external circuit can break through, the increasing force being represented by the triangles $A\,e\,h$, $A\,f\,i$, &c.: then, since the attractive forces between the exploding points of the balls of the discharger, with a given accumulation, are in the inverse duplicate ratio of the distances $A\,e$, $A\,f$, $A\,g$, &c., these attractive forces will be inversely proportional to the same triangular spaces, $A\,e\,h$, $A\,f\,i$, $A\,g\,k$, &c. If, therefore, when force in the direction of the circuit is $A\,e\,h$, discharge takes place, quantity being $C\,e$, distance $A\,e$, and attractive force between discharging balls $= 1$, then supposing $A\,e$ to become $A\,f =$ twice $A\,e$, the force between the balls at distance $A\,f$ would be only one-fourth as great; that is to say, it would be inversely as triangle $A\,e\,h$ to triangle $A\,f\,i$. Hence with the same quantity accumulated $= C\,e$, discharge could not occur at distance $A\,f$. Let now the first quantity accumulated $= C\,e$ become twice as great; that is to say, let it be represented by rectangular space $C\,f = 2\,C\,e$: in this case the force in the direction of the external circuit would be represented by triangle $A\,f\,i = 4$ times $A\,e\,h$; and since the attractive force between the exploding points of the balls of the discharger is as the squares of the quantity accumulated, the attractive force through the external circuit with the double accumulation $C\,f$ becomes four times as great; and is the same at the distance $A\,f$ with a double accumulation as at the distance $A\,e$ with a single accumulation; in this case explosive discharge again ensues.

In a similar way it may be shown that when the distance $A\,e$ is extended to $A\,g$, the attractive force between the balls with the single accumulation $C\,e$ is reduced to one-ninth, in which case no explosive discharge could occur at the distance $A\,g$. Let the quantity accumulated, however, become three times as great, that is to say, let the rectangular space $C\,e$ become $C\,g$; in this case the force through the external circuit is represented by the triangular space $A\,g\,k = 9$ times the triangular space $A\,e\,h$; but since these spaces are inversely proportional to the attractive forces at distances $A\,e$, $A\,g$, the attractive force at the distance $A\,g$ is the same with a treble accumulation, as at the distance $A\,e$ with a single accumulation; explosive discharge will therefore again occur, and so on. Hence the interval at which discharge occurs, as measured by a Lane's discharger, L, Fig. 7, will be directly as the quantity accumulated; whilst the electrometer indication or force through the external circuit will be as the square of the quantity, being as the triangular spaces A e h, A f i, A g k, &c. M. De la Rive, in his comprehensive treatise on Electricity, considers this result as somewhat remarkable.* It is evident, however, from the geometrical diagram above referred to, that it could not be otherwise, and is a necessary result of the forces in operation. We may infer from the equality of the triangles A D C, A D B, that at the instant of explosive discharge the force A B D through the external circuit has superseded the force A D C in the direction of the coatings. If, therefore, we suppose discharge to occur with the successive accumulated quantities C e, C f, C g, &c., then discharge with first quantity $=$ C e taking place at distance A e, the force through the external circuit will be represented by the triangle A m e, and the force in the direction of the coatings by the triangle A m C. When discharge occurs at the distance A f $=$ 2 A e with a double accumulation C f $=$ 2 C e, the force through the circuit will be represented by triangle A n f, and force in the direction of the coatings by triangle A n C, and so on. We have here to observe, however, that, in estimating the forces through the circuit, the force for the quantity C e, with the distance of discharging balls A f $=$ 2 A e, will not be represented by the triangle A m e (when the quantity becomes C f $=$ 2 C e), but by the triangle A r e, or one-fourth of the triangle A n f $=$ force through the circuit at the instant of discharge with a double quantity. When, therefore, the distance A e is increased to A f with the unit of charge represented by C e, then force acting through the external circuit is represented by triangle A r e, and not by triangle A m e.

In order, therefore, to measure a double accumulation by means of a movable electrometer, we must oppose to the force through the circuit four times the resistance: thus, in the operation of Cuthbertson's ingenious steelyard discharging electrometer, we must, in order to obtain a double accumulation, set the slider of the balance-arm to four times the number of grains; in order to obtain a treble accumulation, we must set the slider of the balance-arm to nine times the number of grains, and so on. When, in the application of Lane's discharger, therefore, discharge occurs at a double distance, the quantity of electricity accumulated is twice as great, and the respective forces through the circuit as 1 : 4, as represented by the triangles A e r, A f n; when discharge occurs at a treble distance, the quantity of electricity accumulated is three times as great, and the force through the circuit nine times as great, as represented by the triangles A e s, A g o, and so on,—the force through the circuit being, as already observed, as the square of the quantity.

It may perhaps be as well to further remark that, although, according to the diagram, the forces A B D, A D C are at the instant of discharge considered as equal, we cannot however infer their *precise* equality, or suppose the decreasing force A D C in the direction of the coatings to absolutely vanish at the point D : hence some little residuum, commonly called *residual discharge*, may remain in the jar in consequence of the vanishing attractive force which the coatings exert upon each other, as is found by experiment.

* " Ce qu'il y a d'assez remarquable, c'est que la distance à laquelle une décharge entre deux balles chargées d'électricités contraires peut avoir lieu, est simplement proportionnelle aux quantités d'électricité, tandis que les forces attractives sont proportionnelles aux carrés de ces forces."—*Traité d'Électricité*, tome i. p. 66.

FURTHER INQUIRIES CONCERNING THE LAWS AND OPERATIONS OF ELECTRICAL FORCE.*

1. By quantity of electricity we are to understand the amount of the unknown agency, whatever that may be, constituting electrical force, and which, being accumulated, may be represented by some arbitrary quantitative electrical measure.

The actual quantity, as thus represented, which any insulated conductor can sustain, under a given electrometer indication, was not unaptly termed by Cavendish the charge of that body, or electrical charge; this, in fact, is the only true measure of what may be called electrical capacity.

Electrical charge, therefore, is the quantity of electricity which can be placed upon a given insulated conducting surface under a given electrometer indication.

Electrical intensity, on the contrary, is the electrometer indication answering to a given quantity upon a given surface.

2. The term electrical intensity has often been employed and received in a very indefinite and vague sense, as I have endeavoured to show elsewhere,† and has been made to signify certain hypothetical qualities assumed to be peculiar to the occult nature of electrical force. It has been used to express, for example, different degrees of elasticity or tension, different degrees of density, and such like; thereby taking it for granted that the electrical agency is capable of altering its state or condition under certain circumstances. We have, however, no substantial evidence of this.

I have in vain sought for experiments which could at all lead us to infer the possibility of effecting a change in the constitution of electrical force without, at the same time, varying the quantity of the agency in operation, or the surface on which it is distributed,

* This forms the subject of a paper read before the Royal Society, June 16th, 1864. The reference numbers within brackets refer to this chapter only.—ED.

† Phil. Mag. for December, 1863, Supplement.

and have been hence led to conclude that the term intensity, when rigidly interpreted, is, after all, only another form of expression for quantity, signifying nothing more or less than the quantity of electrical force operating at a given point of the charged surface. It is in this sense I employ the term intensity in the following pages.

3. Electricians have observed, from a very early period of the progress of modern electrical discovery, that, although the quantity of electricity which insulated conductors can receive depends not upon their solidity or volume, but upon their surfaces; yet bodies will not take up electricity in proportion to their surfaces. This important fact was first noticed by Le Monnier, in "L'Histoire de l'Académie," so long since as the year 1746. Cavendish, in his manuscript papers (1770), notices the same fact. The most interesting notice, however, of past times relating to quantity, as regards surface, is to be found in Volta's memoir on the capacity of electrical conductors, which appeared in *Rozier's Journal* in 1779. Volta shows that if any plane surface be extended in length, it can sustain a greater quantity of electricity, although the area of the surface remains the same; Exp. 10 (25). Thus a conducting plate, A, Fig. 114, 1 foot square, of small thickness, would not sustain so great a charge as a rectangular conducting plate, B, 48 inches long by 3 inches wide (and of the same thickness), although the areas do not differ. Volta attributes this result to the circumstance that the electrical particles are, in the case of the elongated surface, "placed further without each other's influence," in consequence of which the electrical excitation or

Fig. 114.

action of the particles upon each other is diminished; it being, in fact, demonstrable, by familiar experiments, that when two similarly charged bodies, with attached electroscopes, are brought near each other, the angular divergence of the electroscopes increases, and the accumulated electricity does not rest upon them with the same quietude as it does when the two bodies are at a distance from each other. Let, for example, two equal spheres, equally charged (having attached electroscopes), be caused to approach each other, the angular divergence of the electroscopes will be greatly increased; conversely, when the same two spheres are widely separated, the angular divergence of the electroscopes will be greatly diminished. Supposing the equal spheres, therefore, to represent or stand for approximated or separated particles, Volta's view, as must be admitted, is very consistent with experiment. The greatest intensity (2), according to Volta, is when the electricity is accumulated on a given circular area, and the least when the same area is expanded into a right line of small width. In the former case the electrical particles are congested or packed, as it were, more closely together; in the latter this congestion is relieved, and the particles are at greater distances from each other's influence. Volta has, no doubt, shown much ingenuity in this explanation of the phenomena.

4. Some further elucidation of the question, however, is derived from more recent inquiries into the operation of what has been termed neutral electrical induction. We cannot possibly extend the charged surface A, Fig. 114, in length as B, without at the same time exposing it to a greater amount of neutral inductive action, arising from the influence of surrounding neutral matter, which, on the principles of the condenser, diminishes the electrometer indication or intensity, hence the intensity is less: and the given surface can receive an increased quantity, before the electroscope again rises to the same angle. We have to consider, therefore, not only how the particles are disposed or grouped in relation to each other, but how they are disposed or grouped in relation to external bodies, by which they are subjected to a greater or less amount of neutral inductive action. If, for example, we have an insulated charged sphere, with an attached diverging electroscope,* and an adjacent neutral surface, the divergent electroscope, under the influence of the neutral surface, approximated in a greater or less degree towards the charged surface, declines, and the given surface receives a higher charge before the electroscope would again diverge to the same angle; showing the influence of the adjacent neutral matter on a charged surface, and the consequent increase of charge, under the same electrometer indication. In accordance with this fact, Volta shows, as already stated (3), that a given conducting surface extended in length takes an increased charge. We may in fact perceive, by reference to Fig. 114, that a given surface under the form of a square plate, A, is exposed to a less amount of exterior inductive action than when elongated or drawn out into a rectangular plate, B, whose length equals four times the side of the square, although the surfaces are in each case actually the same; the intensity, as illustrated by the sphere,

* Figs. 27, 38, and 39, in Part I., show the kind of figures referred to in this paper.—ED.

is consequently diminished, and the quantity under the same electrometer indication increased. This latter condition involves the extent and disposition of the linear boundary, or perimeter, of the given surface, as we shall presently show.

5. No very satisfactory experiments appear to have been instituted, showing the relation of quantity to surface, at least none from which the general laws of this relation have been deduced : the quantity upon a given surface has been often vaguely estimated, without any regard to a constant electrometer indication or intensity. For example, we can scarcely infer, from the beautiful experiment of Coulombe, in which a charged sphere was brought into contact with an insulated neutral circular plate of twice its diameter, that because the charge was shared between them in the proportion of the two exposed surfaces of the plate to the one exposed surface of the sphere, that therefore the capacity of the circular plate is twice the capacity of the sphere. If the electrical capacity, or charge of the plate, were fully twice that of the sphere, we could, under the same electrometer indication, place twice the quantity on the plate that we could on the sphere ; but this is not possible, as we shall eventually show (26, Exp. 16).

In the meantime we may observe, that the sharing of electricity between two bodies is a different question from that of electrical charge, according to our previous definition of it (1).

6. In further investigating the laws of electrical charge and intensity, together with Volta's discovery of the increased capacity of conductors when the surface is extended in length, I was led to infer that the quantity of electricity which any plane rectangular surface could receive under a given intensity (2) depended, not only on its surface, but also on its linear boundary extension, or perimeter. Thus, the linear boundary of 100 square inches of surface under a rectangle of 37·5 inches long, by about 2·66 inches wide, is about 80 inches ; whilst the linear boundary of the same 100 square inches of surface under a plate 10 inches square is only 40 inches. The linear extensions of the two surfaces in this case may be taken as 2 : 1 ; so that the linear boundary of the rectangle is twice that of the square, in accordance with which fact we find, as already noticed (3), that the charge of the rectangle is much greater than that of the square, although the surfaces are equal, or nearly so.

7. A rigid experimental examination of this question (25, Exps. 7, 8, 9, and 10) led to the conclusion that electrical charge (1) in certain cases depended on the surface and linear extension conjointly.

R

It appeared from a series of careful experiments with improved instruments, to be hereafter described, in which rectangular conducting plates of small thickness and various forms and extensions were employed—

1st. That there exists in every plane surface what may be termed an electrical boundary, having a relation to the grouping or disposition of the electrical particles in regard to each other and surrounding matter. In circles and globes the perimeters, or boundaries, are represented by the circumferences.

2nd. That if this boundary be constant, the charge varies with the square root of the surface; if the surface be constant, the charge varies with the square root of the boundary; if the boundary and surface both vary, the charge varies with the square root of the surface multiplied into the square root of the boundary. Thus, calling C the charge, S the surface, B the boundary, and u some arbitrary constant, depending on the electrical amount of charge, we have $C = u\sqrt{S\,B}$ (25), which will be found, with some excep‧ tions, a general law of electrical charge.

It is evident from the above formula, $C = \sqrt{S\,B}$, if when we double the surface we also double the boundary, the charge will be twice as great; that is to say, the charge in this case may be said to vary with the surface, but not otherwise, as we shall presently see (25, Exp. 10). According to this formula, if l and b represent the length and breadth respectively of a plane rectangular surface, then the charge of such a surface is expressed by $u\sqrt{2\,l\,b\,(l+b)}$, which is the law agreeing with experiment.

We have, however, to remember in all these cases the difference between charge (1) and intensity (2).

8. On examining the electrical intensity of plane rectangular surfaces, that is to say, the electrometer indication of a given quantity in operation at any one point of the charged surface, it was found to vary inversely as the boundary multiplied into the surface. Thus, if the surface be constant, the intensity is as the boundary inversely; if the boundary be constant, the intensity is inversely as the surface; if both vary, it is inversely as the surface, multiplied into the boundaries. Thus, calling intensity, that is, electrometer indication $=$ E, we have $E = \dfrac{1}{B}$. the surface being constant; if the boundary be constant, we have $E = \dfrac{1}{S}$. If, therefore, both vary alike and together, E may be said to vary inversely as the square of either. If, for example,

when S is double, B is also double, E may be said to vary as $\frac{1}{S^2}$ (25, Exp. 10).

The electrical intensity of a plane rectangular surface being given, we may always deduce therefrom its electrical charge under a greater given intensity, since we only require to determine the increased quantity requisite to bring the electrometer indication up to the given required intensity. This we may readily effect, the intensity being by a well-established law of electrical force as the square of the quantity (21, Exp. 1).

Let, for example, the intensity of two surfaces, which we will call A and D, each charged with a given quantity, m, be as 1 : 4, the intensity of A with quantity, m being equal to 1, and the intensity of D with the same quantity m being equal to 4; and let it be required to determine what quantity must be accumulated on A to produce the same electrometer indication = 4 as that shown by D, with the given quantity m. Then if m' represent the total quantity to be accumulated on A, in order that the intensity may become four times as great, that is, equal to the intensity of D, intensity of A with quantity m is to its intensity with quantity m' as is $m^2 : m'^2$; or $m^2 : m'^2 : : 1 : 4$; we have consequently $m : m' : : \sqrt{1} : \sqrt{4}$; therefore $m' = 2\,m$, which is twice as great as that with which the surface D is charged; so that the charge of A is twice that of D; or in other words, A can receive twice the quantity under the same electrometer indication, as is demonstrable by direct experiment (25, Exp. 10); and we have the charge as the square root of the intensity inversely, or $C = \dfrac{1}{\sqrt{E}}$.

9. It is important to observe that these laws relating to charge, intensity, surface, &c., apply more especially to continuous surfaces taken as a whole. If a surface be divided into equal separated parts, the formulæ we have just quoted require correction. Let, for example, a plane rectangular surface, A, Fig. 115, of a given length and breadth, be divided into two equal portions, a, b, separated and placed distant from each other, as at B; further, suppose a given quantity of electricity = 1 accumulated on A to be disposed on the two equal half surfaces a and b, so that each of the separated half surfaces, a, b, may in this case have one-half the total quantity accumulated on the total continuous surface A; on comparing the two half surfaces, a, b, with each other, it is clear that the total quantity = 1 disposed upon them, occupies two equal surfaces, having separate and equal linear boundaries; so that the given

quantity equal 1, may be considered to be accumulated on a
double surface, $a + b$, as compared with one of the half surfaces
a, taken singly. Hence the quantity upon the two half surfaces
united as at A, would be double that of the quantity upon one

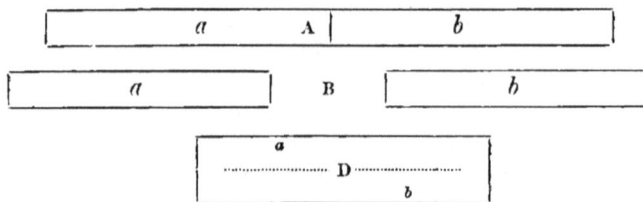

Fig. 115.

of the half surfaces, a, taken singly; and the intensity of the total
quantity accumulated upon the separate surfaces, a, b, would,
when taken together, be the same as the intensity of one-half the
quantity upon one of the half surfaces, a, only. We therefore
perceive, that if as we increase the quantity we also increase the
surface and boundary, the intensity does not change. If three
or more separated equal spheres, for example, be charged with
three or more equal quantities, and be each placed in separate
connection with the electrometer, the intensity of the whole is not
greater than the intensity of one of its parts. Similar results
ensue in charging any united number of equal and similar electrical jars. An electrical battery, for example, consisting of five
equal and similar jars, charged with a given quantity = 5, has
the same intensity as a battery of ten equal and similar jars
charged with a quantity = 10; the intensity of the ten jars
taken together is no greater than the intensity of any one of
the jars taken singly. Referring to B, Fig. 115, suppose the
whole quantity = 1 to be now accumulated on one of the half
surfaces, a, only, the other half surface, b, remaining charged
with quantity $\frac{1}{2}$ as before, then, since the intensity varies as
the square of the quantity (21, Exp. 1), the intensity of a with
the quantity 1 will be to the intensity b with the quantity $\frac{1}{2}$ as
$1^2 : (\frac{1}{2})^2$ or as $1 : \frac{1}{4}$; that is to say, the intensity of a with quantity = 1 will be four times the intensity of b with quantity = $\frac{1}{2}$, the
surface being in each case the same. When, therefore, the quantity
= 1, instead of being accumulated upon one of the half surfaces
a only, is accumulated on both the half surfaces, $a + b$, so that
each may contain quantity $\frac{1}{2}$, the intensity of the two separate half
surfaces, a, b, is only one-fourth the intensity of the total quantity
accumulated on one of them. It appears from this, therefore, that

in accumulating a given quantity of electricity upon a double surface, B, in separate and equal jars, a, b, the boundary being also double, the intensity varies as the square of the surface inversely, and we have $E \propto \dfrac{1}{S^2}$ (8); hence the surfaces, a, b, can receive, when taken together, under the same electrometer indication, twice the quantity which either can receive alone, and the charge varies with the surface, the boundaries being equal. In this case, as we increase the surface in separate equal parts, we also increase the linear boundary, and consequently we increase the charge.

Taking, therefore, a given surface in equal and divided separate parts, we may conclude that the intensity varies with the square of the quantity directly, and with the square of the surface inversely, from which we derive the formula $E \propto \dfrac{Q^2}{S^2}$, which fully represents the phenomenon of a constant intensity attendant upon the charging of equal separated surfaces, with quantities increasing as the surfaces in the case of charging a united series of equal electrical jars. Thus, if $Q = 1$, and $S = 1$, we have $E = \frac{1}{1} = 1$; if $Q = 2$, and $S = 2$, we have $E = \dfrac{2^2}{2^2} = 1$, and so on. In cases, however, in which, from any circumstance, E does not vary with $\dfrac{1}{S^2}$, this formula does not apply.

If, for example, E varied as $\dfrac{1}{S}$, which under certain circumstances is found to be the case (26, Exp. 15), the intensity is no longer constant, as the quantity increases with the surface. This leads to a consideration of the case in which separate surfaces, a, b, Fig. 115, are united into one continuous whole, as at A. Let, for example, the two half surfaces, a, b, be joined at their extremities so as to form a continuous surface, A, and suppose a quantity $= 1$ to be accumulated upon them thus combined, whilst either of the separate half surfaces, a, b, remains charged each with a quantity $\frac{1}{2}$; we have then, as in the preceding case, a double quantity accumulated on a double surface, A, the only difference being that the double surface in the former case, B, is made up of two equal separate parts, whilst in the present case, A, it is made up of one continuous whole. We might hence be led to conclude that the intensities of a and A will be the same, and that we could accumulate on A, under the sam

electrometer indication, twice the quantity we could accumulate
on a or b.   Such, however, is not the fact; the intensity of a
double quantity on A exceeds the intensity of half the quantity on
a, whilst the charge of A is considerably less than twice the
charge of a; or the charge of a and b when apart from each
other.   The cause of this appears to be the difference in the
relative linear boundaries of the figures.   We have already observed that, in order to accumulate a double charge, we must have
not only a double surface, but also a double linear boundary, as
shown (25, Exp. 10).   Now the linear boundary of A cannot
possibly be double the linear boundary of a.   Thus, calling L the
length, and U the breadth of the figures, we have for the linear
boundary of A, $2L + 2U$, and for the linear boundary of a,
$2(\frac{1}{2}L + U) = L + 2U$.   But it is evident that $2L + 4U$ is equal
to twice the boundary of a; the boundary of A, therefore, is less
than twice the boundary of a by $2U$; that is, by twice the breadth;
consequently, the intensity of A with twice the quantity, is greater
than the intensity of a with one-half the quantity.   The charge
(1) of A, therefore, is less than twice the charge of a, although the
surface of A is twice the surface of a.   If we suppose the two half
surfaces a and b, instead of being joined at their extremities, as A,
to be joined at their sides, as D, Fig. 115, the linear boundary of D
is still less than twice the linear boundary of a by L; that is, by
twice the length of a; and is less than the linear boundary of A
by $L - 2U$; that is, by the length minus twice the breadth.   The
intensity therefore of the two half surfaces a and b, when joined
at their sides, as at D, is much greater than when joined at their
extremities, as at A; the charge therefore of D is less than the
charge of A, and is in either case much less than the charge of
the two equal half surfaces $a + b$ apart from each other, as at B;
so that, as we have observed (3), (Fig. 114), a given conducting surface, D, will, as stated by Volta (3), receive a greater charge when
extended in length, as at A.   From these analytical inquiries also
it is further evident, as observed by the early electricians (3),
that conducting bodies do not take up electricity in proportion
to their surfaces, except under certain conditions of surface and
boundary (25, Exp. 10).   If the breadth U and the length L of a
given surface, A, be capable of unlimited variation, the surface
remaining constant, then, as observed by Volta (3), the least
quantity that can be accumulated on a given conducting service
under a given electrometer indication, is when the given surface is
under the form of a circular plate, that is to say, when the
boundary is a minimum; and the greatest when extended in

a right line of small width, that is, when the boundary is a maximum.

10. From these facts it is further evident that in the combination of two similar surfaces by contact, as for example, two circles, two spheres, two rectangular plates, two cylinders, &c., we fail in either case to obtain twice the charge of one of them taken separately. If we bring them into contact either at their extremities or sides, we fail to decrease the intensity, the quantity being constant, or increase the charge, the intensity being constant. It being evident that whatever tends to decrease the intensity or electrometer indication, must increase the charge, that is to say, the quantity that can be accumulated under that intensity; conversely, whatever tends to increase the electrometer indication must decrease the charge, that is to say, the quantity that can be accumulated under that electrometer indication. As a further illustration, let two equal separated spheres, for example (5), equally charged, be made to touch each other. In this case the electrical particles of the charge will be more congested and within each other's influence when touching each other, and differently grouped in relation to external bodies, than when separated and at a distance from each other (3) (4); consequently, the intensity of either sphere in a separated state will be much less than the intensity of either of the spheres in contact. The total intensity, therefore, of the two spheres in a separated state will be much less than the total intensity of the two spheres in contact; the sum, therefore, of the charges of the two spheres, when taken together in a separated state, will be greater than the sum of the charges of the two spheres when in contact. Hence, two equal spheres when in contact, similarly charged, have not the same intensity as one of the spheres taken alone; the charge of the two spheres in contact is consequently not double the charge of one of them taken alone—a result which applies to any two similar figures, either separated or in contact, as is found by experiment. Take, now, the case of two spheres in which the surface of the one is double that of the surface of the other, and suppose a given quantity of electricity = 1 to be accumulated on each; then, in any one point of the larger sphere, there will be only one-half the quantity there is in one point of the smaller sphere. We might infer from this that the intensity in any one point of the double surface is only one-fourth the intensity in any one point of the single surface, according to the formula $E = \dfrac{1}{S^2}$ (8); so that if we double the quantity upon the double surface, the intensity

would be the same as half the quantity upon the single surface. Such, however, is not the result of experiment, owing to the relative congested states or groupings of the electrical particles in regard to each other and external bodies in each of the spheres; as also from the circumstance of the total quantity upon the smaller sphere being grouped upon a double surface on the larger, taken as a whole, and not upon a double surface in two separate equal parts, as with spheres of equal size.

In the small sphere the electrical particles are more congested, or within each other's influence relative to its circumference, than they are in the large sphere relative to its circumference; the circumferences of the two spheres, in fact, are not double of each other, but are as $1 : \sqrt{2}$, that is to say, as their diameters; the surfaces of the spheres being double, whilst the quantity in any one point of the double surface affecting the electrometer is only one-half the quantity affecting the electrometer in any one point of the single surface: a sphere of twice the surface, therefore, does not take a double charge. The charges of spheres, in fact, are found to be as their diameters, that is, as the square roots of the surfaces (11); hence, the charge of a sphere of diameter equal 1 : the charge of a sphere whose diameter equals $2 : : 1 : 2$, as just observed, whilst the intensity varies as the surface inversely, that is, as $\dfrac{1}{S}$. If we could obtain spheres whose surfaces varied with their circumferences or boundaries, then the intensities with quantities increasing as the surfaces would be constant; but this is not possible. We cannot, for example, obtain a sphere whose surface and circumference are each double the surface and circumference of another sphere.

11. It follows from these and similar considerations that if in any case it should so happen that the grouping or disposition of electrical particles, in regard to surrounding matter, be such as not to materially influence the amount of external inductive action on the electrical boundaries of the surface, then this external induction would not materially affect the result; for example, in all similar figures, as squares, circles, &c., the electrical boundary of the figure is, in relation to surrounding matter, pretty much the same, whatever be the extent of the surfaces.

Thus, for example, if we suppose two circular plates to be surrounded by an outer circle of inductive rays, both circles will be similarly placed in regard to those rays, so that when charged with electricity, the particles would be grouped, or disposed upon their surfaces, in regard to this induction, pretty much in the same way, and little or no difference might ensue ; the difference depending merely upon a slight increase or decrease of distance. We should hence only have to consider the grouping or disposition of the particles upon the surfaces in regard to each other, and might, therefore, take into calculation the surfaces only, without regard to the boundaries or circumferences of the circles ; a consideration which materially influences charge, surface, and intensity, as regards similar figures ; thus, in all similar figures, such as square plates, circular plates, globes, &c., their relative charges were found to be as the square roots of the surfaces only, without regard to the boundaries. The charges of square plates were found to be as their sides ; the charges of circular plates as their diameters, as also the charges of globes (26, Exps. 11, 12, 13, &c.). In cases of rectangles also, in which the boundary extension is the same whilst the surfaces vary, the same result ensues ; the charges are as the square roots of the surfaces simply. In cases of square or circular plates and globes it is easy to see, as we have already observed, that the boundary extensions of each are similar and parallel. A circular plate, for example, of 18 inches diameter is, with reference to its surface, not under a much greater inductive influence in respect of its boundary than is a plate of half that diameter in respect of its surface and boundary, the surrounding influential bodies being in each case similarly circumstanced. The difference, as is evident, consists merely. in a slight change of distance from surrounding matter ; hence the induction arising from the boundary extension does not materially affect the result. In cases of hollow cylinders and globes, in which one of these surfaces is shut out from external influences, only half the surface may be considered as exposed to the inductive action of neutral bodies. If, therefore, we suppose a square plate of any given dimensions to be rolled up into an open hollow cylinder, by which we shut out one of the surfaces from the influences of external bodies, the charge of the cylinder will be to the charge of the plate, into which we may suppose it to be expanded, as $\sqrt{1} : \sqrt{2}$, that is, as the square roots of the exposed surfaces, being as 1 : 1·4 (28, Exp. 17). In like manner, if we take a hollow globe and a circular plate of twice its diameter, in which the surfaces are the same, surface for surface, the charge of the globe will be to the charge of the plate as the square root of the one exposed surface of the globe is to the square root of the two exposed surfaces of the plate ; that is, as in the former case, as 1 : 1·4, as already observed (5) relative to the ingenious researches of M. Coulombe. This is the general relation of the charge of closed to the charge of open surfaces. The charge of a circular plate is to the charge of a square plate, whose side is equal to the diameter of the circular plate, as 1 : 1·13. According to Cavendish, it is as 1 : 1·15, which is not far different. It is not unworthy of remark that the relations of a square plate to a circular plate, as determined by Cavendish nearly a century since, are in near accordance with the formula we have already given for circles, squares, and similar figures generally, $C \propto \sqrt{S}$ (7). Since, on examining the relation given by Cavendish, 1 : 1·15, we see that is very nearly the proportion of the square roots of the surfaces of a circular and a square plate whose diameter and side are the same; for example, the surface of a circular plate of 10 inches diameter is 78·54 square inches, whilst the surface of the square plate of 10 inches side is 100 square inches. Now, $\sqrt{78\cdot54} : \sqrt{100} : : 1 : 1\cdot5$; that is to say, we have 8·85 : 10 : : 1 : 1·15 very nearly. If we substitute the proportions given by my experiments, we have 8·85 : 10 : : 1 : 1·13, which is very nearly exact. Cavendish further states that the electrical relation of a circle whose diameter is equal to 1 is to that of a circle of the same area as a square whose side is equal to 1 as 1 : 1·02; not a very appreciable difference. I have, therefore, not been enabled to detect any very sensible difference by direct experiment. Cavendish further states that the relation of a circle to a globe of the same diameter is as 12 : 18·5; that is, as 1 : 1·54.

My inquiries give the relative charges as 1 : 1·4. In fact, the surface of a globe is four times one of its great circles; hence the two exposed surfaces of a circular disc of equal diameter are exactly one-half of the surface of the sphere; that is, the surfaces exposed to inductive action in the disc and sphere are as 1 : 2, and therefore the charges under a given intensity are respectively as $\sqrt{1} : \sqrt{2}$, as in other similar cases. The charge of a cube approaches the charge of a globe whose diameter equals the edge of the cube, but not exactly; the charge of the globe being to the charge of the cube as 1 : 1·2 nearly; which is not far from the relation of a circular plate to a square plate whose side is equal to the diameter of the circle, as just observed.

12. On examining the intensity of circular plates, globes, and closed and open surfaces generally, the charges being constant, I found the intensity (2) to vary inversely as the exposed surfaces (26). The intensity of a circular plate, for example, is only half the intensity of a globe of the same diameter; the exposed surfaces being as 1 : 2 (26, Exp. 15). The intensity of circular planes are in a simple inverse ratio of their surfaces; thus, a circular plane of 18 inches diameter has only one-fourth the intensity of a

circular plane of 9 inches, or half that diameter, the surfaces being here as 4 : 1. The intensities of globes are inversely as their surfaces. Thus, calling E the intensity and S the surface, we have

$$E = \frac{1}{S}.$$

13. The following enumeration of formulæ deducible from these inquiries embrace the general laws of quantity, intensity, surface, and boundary extension, and will be found practically useful in deducing the laws of statical electrical force.

Symbols.—Let C = electrical charge ; Q = quantity ; E = intensity, or electrometer indication ; S = surface ; B = boundary extension, or perimeter ; Δ = direct induction ; δ = reflected induction ; F = force ; D = distance.

FORMULÆ.

C $\overset{..}{\alpha}$ S when S and B vary equally together   .   .   (25) Exp. 10)

C α Q, E being constant or equal 1· .   .   .   .   (1)

C α $\sqrt{S}$, B being constant or equal 1·   .   .   .   (25) (26)

C α $\sqrt{B}$, S being constant or equal 1·   .   .   .   (25)

C α $\sqrt{SB}$ when S and B vary together   .   .   .   (25)

E α $\frac{1}{SB}$, Q being constant for all plane rectangular surface (8)

E α $\frac{1}{B}$, S being constant .   .   .   .   .   .   (8)

E $\overset{..}{\alpha}$ $\frac{1}{S}$, B being constant .   .   .   .   .   .   (8) (26)

E α $\frac{1}{S^3}$, when S and B vary together   .   ,   .   (8)

C α $\frac{1}{\sqrt{E}}$.   .   .   .   (8)

E α Q^2, S being constant, or equal 1·   .   .   .   (21)

E α $\frac{Q_2}{S_2}$.   .   .   .   (9)

In square plates C α with side of square   .   .   .   (11)

In circular plates C α with diameter .   .   .   .   (11)

In globes C α with diameter   .   .   .   .   (11)

Δ, or induction α with S, all other things remaining ⎫
the same ; the same for δ, or reflected induction . ⎭   (4)

In circular plates, globes, closed and open surfaces.

E α $\frac{1}{S}$ or as $\frac{1}{\Delta}$ (12) (4)

F or E α Q^2 .   .   .   .   (21)

F or δ α $\frac{1}{D^2}$, Q being constant .   .   .   .   (21)

Generally we have F α $\frac{Q}{D_2}$.   .   .   .   (22)

14. It is easy, from the laws of charge for circles and globes (11),

to calculate a series of circular or globular measures of definite values, taking the circular inch or globular inch as unity, and calling, after Cavendish, a circular plate of an inch diameter charged to saturation a circular inch of electricity; or otherwise charged to any degree short of saturation, a circular inch of electricity under a given intensity. In like manner a small globe of an inch diameter may be designated as a globular inch of electricity. In the following table are given the quantities of electricity in particles or units of charge, contained in circular plates and globes, together with their respective intensities for diameters varying from 0·25 to 2 inches.

In calculating this table, the circular inch—that is, a circular plate of an inch in diameter and one-fifth of an inch thick—is taken as unity, and supposed to contain 100 particles or units of charge.

Diameters or Units of Charge.	Circle.		Globe.	
	Particles.	I.tensity.	Particles.	I.tensity.
0·25	25	·062	35	0·124
0·50	50	0·250	70	0·500
0·75	75	0·560	105	1·120
1·00	100	1·000	140	2·000
1·25	125	1·560	175	3·120
1·40	140	1 960	196	3·920
1·50	150	2·250	210	4·500
1·60	160	2·560	224	5·120
1·75	175	3·060	245	6·120
2·00	200	4·000	280	8·000

15. These elementary data being premised, I shall now endeavour to submit them to the rigid philosophy of experiment. It will, however, be first requisite to explain the nature of the processes resorted to, together with the instruments employed.

I must here crave permission, as being essential to the progress of this inquiry, to enter upon a brief review of the Hydrostatic Electrometer, recently perfected and improved. This instrument, I am aware, has been before adverted to, in its primitive state, in my paper in the Philosophical Transactions for 1839, plate 3. I find it, however, requisite to particularly explain it under its recent and more perfect form.

> [The author here describes the instrument, already noticed at sufficient length in the previous pages, Figs. 86 to 89, para. 147. But the drawings in the paper in the Royal Society's Archives are even more minute and elaborate than those given in the text just referred to. Should any inquirer

at some future time wish to construct the hydrostatic electrometer, it may be interesting to him to consult the manuscript,
although, for all general purposes, the description given at
p. 122 is sufficiently ample. The instrument, as applied to
magnetic force, is described as the *hydrostatic magnetometer*
in "Rudimentary Magnetism," published in Weale's Series,
in 1850.—ED.]

16. The nature of the electrometer as an instrument of measure
being fully apprehended, it remains now to notice certain instruments of quantitative measure to be employed in connection with it.

First. A series of globular and plane circular transfer measures,
specified in the foregoing table (14), varying from 0·25 to 2 inches
in diameter, and being about one-fifth of an inch thick (14).
These globes and plates were carefully turned in a lathe, out of
very light wood, and smoothly gilded with extra thick gold
leaf. Each transfer measure has a loop of fine silk gut, firmly
secured to it, by which it may be easily taken up on a long
slender rod of varnished glass or vulcanite.* In circular plates it
is requisite to secure the suspension loop effectually in the following way:—A fine thread of silk gut being selected, a small open
loop is tied centrally in it; small holes are drilled obliquely upwards
from the lower edge of the plate at each extremity of a diameter,
so as to extend from the edge of the plate to a little within its
surface: the extremities of the suspension loop are passed through
these holes in the surface towards the circumference, so as to bring
the loop in the centre. Being drawn tight, the thread of silk gut,
with the loop of suspension, is secured by means of fine wooden
pegs. The extremities of the pegs and ends of the gut are cut
off by fine nippers, fair with the circumference, so as not to leave
the least projection. In globes, this loop of silk gut is easily
secured by drilling a hole in the globe, and inserting both extremities of the gut in it, after which the gut is firmly secured by
a small wooden peg. The vulcanite rod for suspension is slit open
by a thin saw-cut at one of its extremities, which is pointed, in
which slit the loop of suspension is held fast; or the extremity of
the glass rod is a little bent up, by applying the heat of a spirit
lamp, to prevent the suspending loop from slipping off it. Thus
prepared, the given measure is charged to saturation, or to a given
intensity, by contact with the ball of a charged electrical jar. This
jar I term a "quantity jar," the construction of which it is important to notice.

[The quantity jar is described at p. 119, Fig. 85, par. 141.

* See Fig. 85, page 119.—ED.

The author then goes on to describe the unit jar, which has already been figured and described at p. 106, Fig. 75, par. 120. These descriptions occupy paras. 17 and 18 in the original paper.—ED.]

The quantity jar being charged with a given number of measures from the unit jar, and the unit jar, or the quantity jar itself, withdrawn, a transfer measure, suppose a circular or globular inch, insulated on its vulcanite rod (16), is brought into contact with the ball of the jar, a communication being made at the same time with its outer coating, through the slip of tinfoil passing over the edge of the salver on which it rests;* the charged circular plate, or other measure, is now transferred to the subject of experiment; for example, an insulated conducting sphere or plate, in connection with the electrometer. This being effected, the transfer measure is returned to the quantity jar, contact being made again with its outer coating : a second measure is now deposited on the plate, and so on ; and thus, by a repetition of the operation, we may deposit on the plate any number of circular or globular inches of electricity of a given intensity we require. The error arising from the small quantity of charge left upon the transfer measure after each deposit may in most cases be neglected, or otherwise calculated and avoided altogether. In the example just given, suppose the relative diameters of the transfer measure and the plate to be as 1 : 12 ; in that case their respective charges would, by the preceding data (14, Table 1), be twelve times as great upon the large plate as upon the small one. To put in evidence how small a quantity is in this case left upon the transfer plate, after contact with the plate, we may observe, as just stated, that the charge of which the larger plate is susceptible is at least twelve times that of the smaller plate ; and therefore, after contact, the charge of the smaller disc will be shared between the two ; the larger one consequently taking twelve-thirteenths of the whole charge, leaving only a residue of one-thirteenth part of the transfer charge on the circular inch ; which, according to the preceding table (14), would be considerably less than could be sustained by a disc of one-tenth of an inch diameter. There will of course be a slight increase after each transfer, but for the few transfers generally employed, the error arising therefrom would be of no moment. It is easy to estimate the amount of the total deficient quantity, by placing a hollow sphere having an open mouth in connection with the electrometer, and after each contact of the transfer measure with the given insulated conducting plate, deposit the transfer measure within the hollow sphere.

* This is seen at N, Fig. 85.—ED.

We know in this case it will be deprived of all its remaining electricity, and may be withdrawn in a perfectly neutral state: the electrometer will consequently indicate the total sensible quantity of electricity which the small transfer measure has carried away after any given number of transfers to the plate, the distance between the electrometer discs remaining constant. This kind of experiment will be quite sufficient for estimating the greatest error which could possibly arise from the small quantity carried off by the transfer measure, after any given number of deposits.

19. These mechanical conditions being fully apprehended, the experimental manipulation may next be discussed.

[For this (which occupies paras. 20 and 21 in the original paper) we may refer to Part I., par. 120, pages 150-152. For the details of experiments made to illustrate the law of electrical attractive force, as regards quantity, to par. 152, Exp. 38; and as regards distance, to Exp. 39.—ED.]

22. Similar results ensue when, instead of connecting the electrometer with a simple conducting surface charged by transfer measures in the way just explained (18), we connect it with an electrical jar, charged through a unit measure from the electrical machine. The force between the attracting plates will be in this case also as the squares of the quantities directly, and as the squares of the distances inversely, according to the formula $\dfrac{Q^2}{D^2}$ (13), and these laws are rigidly exact, and are obtainable with the greatest ease and precision. Experiments in this way in connection with the electrical jar are rather more certain and manageable than with simple and open exposed surfaces; the electrical charge being more readily retained by the jar than by a simple surface. The index of the electrometer, under the operation of the jar, remains unmoved for a considerable time, evincing little or no dissipation of charge. This is very important in the employment of every instrument of quantitative measure, whether of attraction or repulsion, such as the instrument in question, or the "Torsion" or "Bifilar" balances ; and we must be especially cautious to perfect the insulations, and in the case of simple surfaces not to employ quantities of electricity which, under a given insulating state of the air, cannot be retained on them without dissipation, at least for a longer time than we require for the experiment. The best method of perfecting the insulations is to dry them off before making the experiment, with a curved iron heated to redness.*

23. Since in the ordinary application of the electrometer we

* See Fig. 103, page 158.—ED.

connect the attracting plate p with a given surface, P,[*] it is requisite to further ascertain what differences may possibly arise in consequence of the extension of the surface under examination, as well as the influence of the reflected induction from the suspended neutral plate N. We may conclude, however, that in cases in which the plates p n are small in proportion to the charged surface, no sensible error will arise in such cases. If, for example, we add to a circular plate, P, of 1 foot in diameter, a small plate, p, then the total surface will be the surface of the plate p plus the surface of the attracting plate P; but as the surface of p is very small in proportion to P, the addition of p, together with the reflected induction of the suspended neutral plate n, may be in most cases neglected. The reflected induction of the suspended disc n, however, must, as is evident, tend to diminish the intensity of the charged surface under examination, and enable it to sustain a somewhat greater charge than it could otherwise, but as this in the ordinary course of experiment may be considered as common to every case, it cannot introduce any considerable error in our calculation; moreover, the result of the induction must be necessarily small. Suppose, for example, the circular plate P in connection with the electrometer were under the influence of a similar neutral plane of equal diameter placed at a very short distance from it, with a resisting dielectric between, after the manner of a condenser. Even in this case we should fail to neutralise all the electrometer indication, and this influence of one plate upon the other would decrease rapidly in some inverse ratio of the distance between the plates, and in some direct ratio of the area of the condensing plate. Now, as the area of the suspended electrometer disc would in any case bear a small proportion to the surface under experiment, the intensity of the charged surface would be but little changed by the influence of the small neutral electrometer disc n; for example, a disc of only $1\frac{1}{4}$ inch in diameter, and whose area does not greatly exceed 1·76 square inch, operating by induction at a distance of half an inch, or perhaps at double that distance, on a circular plate, P, of eight times its diameter, exposing a surface of 113 square inches (at least sixty-four times the surface of the small plate), could not increase the charge of such a plate by any very appreciable quantity. If, for example, we communicate 5 circular inches of electricity to an insulated plate, which we may call P, of 9 inches diameter, apart from the electrometer, and then connect it with the attracting plate p of the electrometer, the intensity is the same, or very nearly so, as when

* See Fig. 90.—ED.

charged in connection with the attracting plate, p, under the influence of the suspended disc n. Nevertheless cases may possibly occur in which it might be desirable to estimate the increased charge liable to arise from the reflected induction of n. Suppose, for example, the augmented charge were less in proportion to a large surface than to a small one, then it might perhaps be necessary in certain cases to take this into calculation. Now the reflected induction of the suspended disc is constant, whatever be the extent of the charged surface, it being limited to the area of the suspended disc. Imagine, by way of illustration, the reflected induction of the suspended electrometer disc, n, on a circular plate, P, of 9 inches diameter, to be such as to enable the 9-inch plate to receive under the same intensity (2) one-half a measure more upon five measures than it could if no such auxiliary induction were present, and that a circular plate of 18 inches, or double that diameter, with a double charge or 10 circular inches, receives under the same influence only a quarter of a measure more ; cases, therefore, might arise in which the augmented charge of a large plate consequent upon the electrometer induction, may at last become extremely small in regard to the augmented charge of a small plate, which would in such a case interfere with an exact result ; for since the half measure bears a much greater proportion to the charge of the 9-inch plate, the charge of the 9-inch plate would not be actually as great in proportion to that of the 18-inch plate as it appears : still it is not difficult to estimate this correction. When, however, the approximations are so close as to involve such small differences, we can have but little doubt of the law we seek.

24. The following experiments may be adduced in evidence of the small influence of the operation of the instrument on the accuracy of its indication.

Exp. 3. Withdraw the slider M,* together with the fixed electrometer plate p, and let the suspended plate n hang immediately over an insulated circular plate, P, of 1 foot in diameter, insulated on one of the travelling stages, a, the two plates, p, n, being accurately adjusted to a given distance, suppose half an inch. Charge plate P with a given quantity, suppose 8 circular inches, so as to produce an attractive force of say 10 degrees, then by a movement of the electrometer on the rails, or by a movement of the travelling stage, a, supporting P, bring the suspended plate n over different points of P in order to observe if the electrometer is differently affected by different points of the charged surface.

* In the description of these experiments, the letters refer to such figures as Figs. 86 to 89, 90, and 91, in Part I.—ED.

Result of this Experiment.—The force appeared invariable; it is everywhere the same, being 10° at the distance of five-tenths.

Exp. 4. Discharge and withdraw plate *p*. Replace slider M and attracting electrometer plate *p*. Connect the plate P with the electrometer plate *p* in the usual way, and let the distance between *p*—*n* be set accurately to five-tenths, as in the former case. Charge the plate P with 8 circular inches as before. *Result.*—The indicated force is the same, that is, 10°, showing that the addition of *p* does not in this case sensibly interfere with the electrometer indication, and that it is of no consequence what part of the combined surfaces is opposed to the suspended plate *n* of the electrometer.

Exp. 5. Repeat the last experiments 3 and 4, substituting for the circular plate P a long rectangular plate, R, of copper, or any other conductor, which may be from 2 to 3 feet long, and from 2 to 3 inches wide. It is requisite for this experiment, to place the rectangle upon strong insulating supports, D, before alluded to, (15 Section K), on the platform at the side of the rail, and turn the electrometer a little aside upon its pillar in order to admit of the suspended plate *n* being immediately over the surface of the rectangle; the suspended disc *n* may be thus easily moved along over the surface of the rectangular plate by the movement of the electrometer upon the rails. The result in this case is the same as in the last experiment; the force is invariable throughout the whole length of the rectangle, the distance of the attracting surfaces being everywhere the same; and this is observed to be the case at any distance, however small, at which we can make the experiment, a result perhaps not to be anticipated, but nevertheless in strict accordance with ordinary electroscopic indications; thus, at whatever point of a charged surface any ordinary electroscope of repulsion be placed, its regular divergence is the same.

Exp. 6. Let the attracting plate *p* and the suspended disc *n* be now replaced in position as before, and the rectangle R connected directly with the electrometer as in ordinary cases. Adjust the distance of the plates *p*—*n* to five-tenths as in the last case; charge the rectangle R with the same quantity : here the result is again the same as when the electrometer disc *n* hung immediately over R, being still 10°, so far showing, as in the preceding case, Exp. 4, that the additional surface *p* has little or no influence on the result; and that it is of little or no consequence what part of the combined surfaces operate on the electrometer. It may, however, be well to remark, that in any case of the fixed plate

p of the electrometer bearing a large proportion of the surface under experiment, the suspended plate n of the electrometer should be immediately over the surface itself, that is, the surface under experiment should be substituted for the fixed plate p.

25. *Laws of surface and boundary as regards rectangular plane surfaces.* *Verification of the formulæ* $C = \sqrt{SB}$ (13) *and*

$$E = \frac{1}{SB} \ (13).$$

In order to investigate the charges of plane rectangular surfaces together with the verification of the formula $C = \sqrt{SB}$ (13), a series of smoothly polished plates of copper were employed, varying from 10 inches square to 40 inches in length, by from $2\frac{1}{2}$ to 6 inches wide, and about one-eighth of an inch thick, exposing from 100 to 200 square inches of surface. The charges of these plates were carefully determined, the attracting plates of the electrometer being at a given distance, suppose half an inch.

Exp. 7. Comparison of a copper plate, A, Fig. 116, 10 inches square, with a rectangular plate, B, 40 inches long by 2·5 wide, the thickness

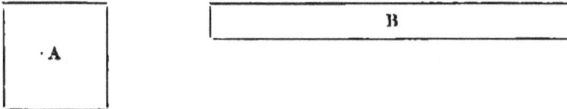

Fig. 116.

of the plates being the same. In these plates we observe that the surfaces are each 100 square inches, whilst the boundaries are 40 and 85 inches, so that the boundaries may be taken as 1 : 2, without any very sensible error, the difference not being of any practical moment to the experiment.

On connecting these plates with the electrometer successively, the charge of the square plate A was found to be 7 circular inches under an intensity of 8°. The charge of the rectangular plate B under the same intensity of 8° amounted to 10 circular inches nearly; by which we perceive, that the surface being the same, the charges were 7 and 10, whilst the boundaries are as 1 : 2. The charges, therefore, are as 1 : 1·4, or as $\sqrt{1} : \sqrt{2}$—that is, as the square roots of the boundaries.

Charge as determined by Intensity (2).—Seven measures deposited upon the square plate A had an intensity of 8°, as in the last experiment; the same seven measures deposited upon the rectangle B had an intensity of only 4°, or one-half the former intensity, being inversely as the boundaries, according to the formula $E = \frac{1}{B}$ (13).

When about three additional measures were added to the rectangle, the intensity was increased to 8°, or double the former intensity, the intensity being as the square of the quantity (21, Exp. 1); so that the relative charges as thus determined were 7 and 10 nearly, or as 1 : 1·4, as in the former case.

Exp. 8. Rectangular plate c, Fig. 117, 37·5 inches long by 2·7 inches wide, surface 101 inches, boundary 80·5 inches ; compared with a rectangular plate D, 34·25 inches long by 6 inches wide, surface 205 square inches, boundary 80·5 inches.—Here the boundaries are the same, whilst the surfaces may be taken as 1 : 2 nearly.

These plates being connected with the electrometer, the charge of the plate c, surface 101, was 8·5 circular inches under an inten-

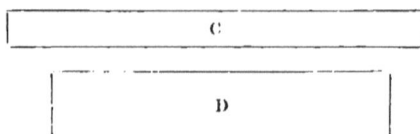
Fig. 117.

sity of 8°; the charge of the plate D, surface 205, was found to be 12 circular inches nearly, under the same intensity of 8°—that is to say, whilst the surfaces are as 1 : 2, the charges are as 8·5 : 12 ; that is as 1 : $\sqrt{2}$ nearly, or as the square roots of the surfaces.

Charge as determined by Intensity.—8·5 measures deposited on the rectangle c, surface 101, evinced an intensity of 8°, as already observed. The same 8·5 measure on the rectangle D, surface 205, evinced an intensity of 4° only, or one-half the former, being inversely as the surfaces; according to the formulæ $E \propto \frac{1}{S}$ (13) ; when 3·5 measures, or a little more were added, the intensity became twice as great, or 8°, the intensity increasing as the square of the quantity (21, Exp. 1). The relative quantities or charges, therefore, as thus determined, for the same intensity ($= 8°$), were 8·5 and 12 circular inches nearly, being as 1 : 1·4 ; that is, as 1 : $\sqrt{2}$, or as the square roots of the surfaces, the surfaces being in the ratio of 1 : 2.

Exp. 9. Comparison of a rectangular plate, E, Fig. 118, 26·25 inches long by 4 inches wide, surface 105, boundary 60·5, with a rectangular plate F, 40 inches long by 5 inches wide, surface 200, boundary 90.—Here the surfaces are as 1 : 2 very nearly, whilst the boundaries are as 2 : 3.

On examining the charges of these plates, the charge of the rectangular plate E, surface 105, boundary 60·5, was 7 circular inches under an intensity of 10°, and the charge of the rectangular plate F, surface 200 inches, boundary 90 inches, was 12 circular inches, nearly under the same intensity of 10°. The charges here

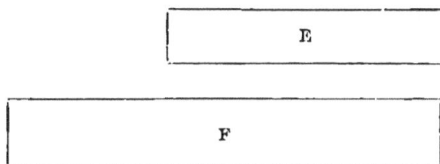

Fig. 118.

are as 7 : 12 nearly, or as 1 : 1·7, being as the square roots of the surfaces multiplied into the square roots of the boundaries of the two rectangles, very nearly.

Charge as determined by Intensity.—Seven circular inches on the rectangle E evinced an intensity of 10°, as just stated. The same 7 circular inches on the rectangle F gave an intensity of something more than 3·5°, being inversely as the surfaces multiplied into the boundaries according to the formula $E = \dfrac{1}{SB}$ (13). When 5 circular inches more were added, the intensity became 10° as before, being as the square of the quantity (20, Exp. 1). The relative charges, therefore, as thus determined for an intensity of 10°, were 7 and 12 nearly; that is to say, as the square roots of the surfaces multiplied into the square roots of the boundaries, as just seen.

Exp. 10. Plate A, Fig. 119, 10 inches square, surface 100 inches,

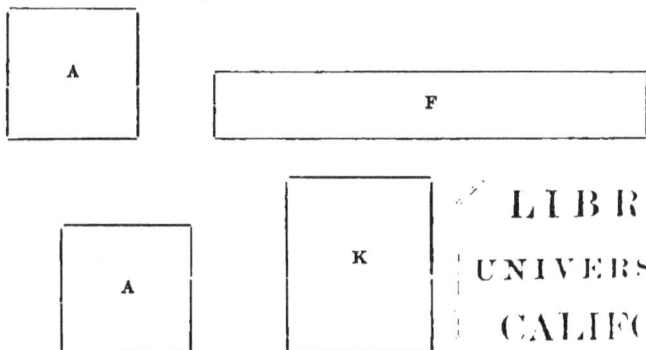

Fig. 119.

boundary 40, compared with a rectangular plate, F, 40 inches long by 5 wide, surface 200 inches, boundary 90.—Here the surfaces are double each other, and the boundaries nearly double each other, but not exactly; we may, however, take them as double; for since the charge varies as the square root of the boundary, Exp. 7, a difference of a few inches in the boundary would not introduce any very sensible difference in the experiment; so that in taking the boundaries as 40 and 80, instead of 40 and 90, we do so without any sensible error.

On examining the charges of these plates, the charge of the square plate A, surface 100, was found to be 6 circular inches under an intensity of $10°$; the charge of the rectangular plate F, surface 200, was 12 circular inches nearly under the same intensity of $10°$, or double the charge of the former; so that taking the boundaries as 40 and 80, the charges are as the square root of the surfaces and boundaries conjointly, according to the formula $C = \sqrt{S \cdot B}$ (13), as just determined, Exp. 9. A double surface, therefore, having a double boundary, takes a double charge, but not otherwise; neglecting all considerations of the boundary in each, the surfaces and boundaries varying equally together, the charge will vary as either directly, and therefore may be said to vary with the surface.

Charges as determined by Intensity.—Six circular inches on the square plate A had an intensity of $10°$, as already stated; the same 6 circular inches on the rectangle F of double surface and double boundary, gave an intensity of $2·5°$ only, or one-fourth the former intensity, being inversely as the boundaries multiplied into the surfaces; or the surface and boundary varying equally together, inversely as the square of the surface, according to the formula $E = \dfrac{1}{S^2}$ (13).

When another 6 circular inches were added, or twice the former quantity, the intensity $2·5°$ of the rectangle became $10°$, being as the square of the quantity (21, Exp. 1).

The relative charges, therefore, as thus determined, for an intensity of $10°$, were 6 and 12, being as 1 : 2 ; that is, as the square roots of the boundaries, multiplied into the square roots of the surfaces of the two rectangles ; or neglecting all considerations of the boundaries, directly as the surfaces, according to the formula $C \propto S$ (13).

A variety of similar experiments, which, to avoid tedious detail, I omit, confirmed these general laws of electrical force, in all cases of rectangular plates whose boundaries and surfaces involve considerable difference of extension and exposure to external inductive influence.

Charge and intensity of square plates, circular plates, spheres, and closed and open surfaces.   Verification of the formulæ $C = \sqrt{S}$, *and* $E = \dfrac{1}{S}$ (13).

26. *Exp.* 11. A copper plate, A (Fig. 119), 10 inches square, surface 100, boundary 40, compared with a copper plate, K, 14 inches square, surface 196, boundary 56.—Here the surfaces are as 100 : 196, that is, as 1 : 2 nearly, whilst the boundaries are as 40 and 56, that is, as 1 : $\sqrt{2}$ nearly.

On examining the charges of these plates, the charge of the plate A, surface 100, was 8 circular inches under an intensity of 10°. The charge of the plate K, surface 196, was 11 circular inches under the same intensity of 10°. Here the charges are as 8 : 11, whilst the surfaces may be taken as 1 : 2 ; that is to say, the charges are as the square roots of the surfaces, neglecting the difference of boundary (11), or as 1 : 1·4.

Charge as determined by Intensity.—Eight circular inches upon the plate A, 10 inches square, surface 100, evinced an intensity of 10°, as already stated. The same 8 circular inches upon the plate, 14 inches square, surface 196, evinced an intensity of 5° only, or one-half the former intensity, being as the surface inversely, according to the formula $E = \dfrac{1}{S}$ (13). When another 3 circular inches were added to the plate K, the intensity became 10°, as in the plate A, being as the square of the quantity (21, Exp. 1). The relative quantities, or charges, therefore, under an intensity of 10°, as thus determined, were 8 and 11, being as 1 : 1·4, as before ; that is, as the square roots of the surfaces nearly.

Exp. 12. A circular plate, which we will call M, of 9 inches in diameter, surface 63·6 square inches, compared with a circular plate N, of 18 inches, or double that diameter, surface 254.—Here the surfaces are as 1 : 4, being as the square of their diameters, whilst the circumferences or boundaries are as 1 : 2.

The charge of the 9-inch plate M was found to be 6 circular inches, under an intensity of 10°, the charge of the 18-inch plate N, 12 circular inches, under the same intensity of 10°. Here the charges are as 1 : 2, whilst the surfaces are as 1 : 4. The charges, therefore, neglecting the difference of circumference or boundary (11), are as the square roots of the surfaces, according to the formula (13) $C = \sqrt{S}$, B being neglected.

Charge as determined by intensity.—Six circular inches upon the 9-inch plate M evinced an intensity of 10°, as already stated. The

same quantity deposited upon the 18-inch plate N, evinced an intensity of only 2·5°, being inversely as the surfaces, according to the

formula $E = \dfrac{1}{S}$ (13). Another 6 circular inches being added to the
18-inch plate, the intensity became 10°, as at first, being as the
square of the quantity.—Here, as in the last experiment, the
quantities or charges are as 1 : 2, whilst the surfaces are as 1 : 4.
The charges, therefore, are as the square roots of the surfaces.

Exp. 13.   A circular plate, M, of 9 inches in diameter,
surface 63·6, compared with a circular plate O, of 12·72 inches in
diameter, surface 127·2 square inches.—Here the surfaces are as
1 : 2, whilst the boundaries are as 3 : 4, nearly.   On examining
the charges of these plates, the charge of the 9-inch plate, M, was
5 circular inches, under an intensity of 8°.   The charge of the
12.72-inch plate, O, was 7 circular inches under the same intensity
of 8°.   The charges here are as 5 : 7, whilst the surfaces are as
1 : 2; the charges, therefore, are, neglecting the difference of circumference or boundary (11), as the square roots of the surfaces,
being as 1 : 1·4, as before.

Charge as determined by Intensity.—Five circular inches on the
9-inch plate M evinced an intensity of 8°.   The same 5 circular inches deposited on the plate O, 12·72 diameter, evinced an
intensity of 4° only, or one-half the former, being inversely
as the surfaces; when 2 more circular inches were added, the
intensity of the plate O became 8°, being as the square of the
quantity (21, Exp. 1).   The relative charges or quantities,
therefore, under the same intensity (= 8°), were 5 and 7 ; the
surfaces being as 1 : 2.   The charges, therefore, are as the square
roots of the surfaces, as in the former case; that is, as 1 : 1·4.

Exp. 14.   Comparison of a sphere P, Fig. 31, of 4·5 inches
diameter, surface 63·5, with a sphere R of 9 inches, or double
that diameter, and surface 254 square inches.—Here the surfaces are as 1 : 4, whilst the circumferences or boundaries are as
1 : 2.

On examining the charges of these spheres, the charge of the
sphere P was found to be 4 circular inches under an intensity of
9°.   The charge of the sphere R, 8 circular inches under the
same intensity of 9°.   Here the charges, neglecting the difference of boundary (11), are as 4 : 8; that is, as 1 : 2, the surfaces
being as 1 : 4.   The charges, therefore, are as the square roots of
the surfaces, being as $1 : \sqrt{4}$.

Charge as determined by Intensity.—Four circular inches, deposited upon the sphere P, surface 63·5, evinced an intensity of

9°. The same quantity of 4 circular inches deposited upon the sphere R, surface 254, evinced an intensity of 2·5° only, being inversely as the surfaces, very nearly. Another 4 circular inches being added, the intensity of the sphere R became also 9°. The quantities, therefore, under the same intensity, were as 1 : 2, that is, as the square roots of the surfaces.

Exp. 15. A circular plate M, of 9 inches diameter, compared with a sphere, R, of the same diameter.—Here the surfaces are respectively, for the *plate and sphere*, 63·6 square inches, and 254 square inches, being as 1 : 4 ; whilst the boundaries are the same.

On examining the charges of these surfaces, the charge of the plate M was 8 circular inches under an intensity of 12°. The charge of the sphere R was 11 circular inches under the same intensity of 12°.

We have here to observe, that one surface of the sphere is closed or shut up; consequently, the two exposed surfaces of the circular disc are exactly one half the exposed surface of the sphere. The charges, therefore, being as 8 : 11, or as 1 : 1·4, are as the square roots of the exposed surfaces.

Charge as determined by Intensity.—Eight circular inches on the plate M had, as just noted, an intensity of 12°. The same 8 circular inches upon the sphere R evinced an intensity of 6°, being inversely as the surface, according to the formula $E = \dfrac{1}{S}$ (13).

When 3 circular inches or a little more were added, the intensity of the sphere became 12°, or double the former, being as the square of the quantity (21, Exp. 1). The charges, therefore, as thus determined, for an intensity of 12°, were as 8 : 11, or as 1 : 1·4, being as the square roots of the exposed surfaces, as before.

Exp. 16. Comparison of a sphere s of 7 inches diameter, with a circular plate, T, of 14 inches, or double its diameter. In this case it is to be observed that the surfaces of the sphere and plate are actually the same; that is to say, the inner and outer surfaces of the sphere will be equal to the two surfaces of the plate. It is, however, to be observed, as in the last experiment, one surface of the sphere is closed or shut up. The surfaces of the sphere and plate, therefore, electrically considered, are not equal; the plate having twice the exposed surface of the sphere. The surfaces open to induction, therefore, are as 1 : 2.

On examining the charges of these surfaces, the charge of the sphere s was 10 circular inches under an intensity of 20°. The charge of the plate T was 14 circular inches, under the same intensity of 20°. The charges here are as 10 : 14, or as 1 : 1·4. The charge of the sphere, therefore, as compared with the charge of the plate, is as 1 : $\sqrt{2}$; that is, as the square roots of the exposed surfaces.

Charge as determined by Intensity.—Ten circular inches deposited upon the sphere, evinced an intensity of 20°. The same 10 circular inches upon the plate evinced an intensity of 10°, being inversely as the exposed surfaces, according to the formula

$$E = \frac{1}{S} \text{ (13)}.$$ An additional 4 circular inches raised the intensity

of the plate to 20°, being as the square of the quantity very nearly, so that the relative quantities or charges under the same intensity = 20°, were 10 and 14 ; being as the square roots of the exposed surfaces.

27. This experiment has immediate reference to the beautiful and ingenious investigation of Coulombe, before referred to, relative to the electrical capacity of a sphere and circular plate of twice its diameter, and from which it has been inferred that the capacity of the plate is twice that of the sphere; Coulombe having shown that when the plate in a neutral state and insulated, was made to touch the sphere, also insulated, but charged with electricity, the plate took away from the sphere two-thirds of the accumulation, leaving only one-third with the sphere. It is to be here again especially observed, that both surfaces of the plate are open to external inductive action, whilst one surface of the sphere is closed, and not so exposed. Consequently, the inductive influence of external bodies upon the two exposed surfaces of the plate, is double that of the same influence upon the one exposed surface of the sphere ; and thus, in the sharing of the charge, the plate is in a condition to take from the sphere a greater quantity under the same electrometer indication. The plate no doubt acts inductively from both its surfaces, as compared with the sphere, but this is not, however, electrical charge (1), as applied to a given surface.

In order to estimate correctly the relative capacities of a globe and a circular plate of twice the diameter of the globe, it is necessary to refer each to a third body, as a standard of comparison ; for example, to another sphere which may be of equal diameter to the given sphere. Let, therefore, this third body, or trial sphere, be carefully insulated; bring the charged sphere, also carefully insulated, into contact with it. Then, because the two spheres are equal and similar, the accumulation (as admitted by Coulombe) will be shared equally between them, and the neutral sphere will have taken away one-half the electricity from the charged sphere, so that the capacity of the charged sphere relative to the trial sphere will be represented by the fraction $\frac{1}{2}$.

Let this experiment be now repeated, substituting the neutral plate, carefully insulated, for the trial sphere. In this case, according to Coulombe, the plate takes from the sphere two-thirds of its electricity. The capacity of the plate, therefore, as referred to the trial sphere, is represented by the fraction $\frac{2}{3}$; so that the relative capacities of the sphere and the plate will be as $\frac{1}{2}$: $\frac{2}{3}$, or as $3:4$; that is, as $1 : \sqrt{2}$ nearly, as just shown, Exp. 16. Hence, if three measures of electricity deposited on the sphere produce an electrometer indication of 10°, then four measures deposited on the plate would give the same intensity of 10°, or very nearly, as is demonstrable by experiment. Now, if the capacity of the plate were double that of the sphere, the relative quantities should have been 5 and 10, instead of 3 and 4, which is not the case, as we have just shown. We see, therefore, that the plate has really not twice the capacity of the sphere; and that it is not possible to place upon the plate twice the quantity of electricity under the same intensity. The conclusion, therefore, arrived at by Coulombe relative to the double action of the plate is only exact for the relative surfaces of indication.

28. The following experiment may be further adduced in support of the preceding.

Exp. 17. A copper plate, which we will call A, 10 inches square, compared with the same plate, rolled up into an open cylinder, U, 10 inches long by 3·2 inches in diameter.—Here, as in the last case, the surfaces are actually the same; but the plate having twice the exposed surface of the cylinder, the surfaces of the plate and of the cylinder, electrically considered, are as $1:2$, as in the case of the globe and the plate. On examining the charges of these surfaces, the charge of the plate was 9·75 circular inches under an intensity of 11°. The charge of the cylinder 7 circular inches, under the same intensity of 11°. The charges, therefore, being as $7:10$ nearly, or as $1:1.4$, whilst the surfaces are as $1:2$, we have the charge of the cylinder to the charge of the plate, into which it may be expanded, as $1 : \sqrt{2}$ (11); that is to say, as the square roots of the surfaces of induction. In determining the charge by intensity, we have seen that 7 circular inches upon the cylinder gave an intensity of 11°; while the same quantity upon the plate gave an intensity of $5\frac{1}{2}$°, being inversely as the exposed surfaces. When about 3 circular inches were added to the plate, the intensity advanced to 11°. The quantities or charges, therefore, under the same intensity, were as 7 : 10 nearly, as before, or as 1 : $\sqrt{2}$, being a general law of charge for closed and open surfaces.

Exp. 18. A hollow copper cube having the edge $= 5\cdot7$ inches, the surface 195 inches, compared with a hollow copper sphere, with diameter $=$ edge of the cube, surface 103 square inches nearly. On examining the charges of these surfaces, the charge of the sphere was 9 circular inches, under an intensity of 10°. The charge of the cube 10 circular inches, under the same intensity of 10°. The relative charges of the sphere and of the cube, therefore, may be considered as approaching each other. We may, therefore, consider the charge of a cube as approaching that of a sphere of the same diameter as the edge of the cube, notwithstanding the difference of the surfaces (11), owing to the six surfaces of the cube not being in a separated or disjointed state (9).

29. I have to observe here, in conclusion, that the numerical results of the foregoing experiments, although not mathematically exact in every instance, yet were, upon the whole, so nearly accordant as to leave no doubt upon the mind as to the law in operation. In very delicate experiments of this kind it is next to impossible to determine forces to within a nearer approximation than that of a degree of the electrometer, or to within quantities less than that of $\cdot25$ of a circular inch. In cases in which the attractive forces are considerable, it will be sufficient if the approximations are within a degree or two, or within one measure of quantity. Nevertheless, I found the numerical results generally exact. If the manipulation be skilfully conducted, and the electrical insulations duly provided for (25), it is quite astonishing to find how rigidly exact the numerical results come out.

Two Lectures

ON

ATMOSPHERIC ELECTRICITY

AND

PROTECTION FROM LIGHTNING.

———————

[These Lectures were prepared expressly for the Artillery and Engineers at Woolwich and Chatham, and were to have been delivered in September, 1861. The apparatus was packed and everything prepared, when just before the author's intended departure, he was seized with an attack of acute inflammation in the eyes, the commencement of more than a year and a half of much suffering.—ED.]

FIRST LECTURE.

Gentlemen,

I have the honour of addressing you under the sanction of the Right Hon. the Secretary of State for War on a scientific subject of great physical and practical public interest; the question of the security of ships and buildings and other elevated structures on the earth's surface, against the terrible agency of lightning; a question which, as you may possibly have already learned, has been, ever since the middle of the last century, when the American and French philosophers identified the agency of lightning with that of our ordinary electrical apparatus, a very vexed question—and remains even now at this time not altogether set at rest. The immediate practical result of the new discovery of 1750 was a proposal by the celebrated American philosopher, Franklin, to lead metallic rods or wires along the sides of towers and down the rigging of ships, which being found to transmit easily the matter of lightning, he assumed would parry its appalling thrust, and so avoid damage to the structure to which they were applied. These lightning rods of Franklin excited, however, very severe, and, as you will presently see, very absurd controversy. The laws of electrical force and discharge not being at that time well understood; moreover, the tendency of lightning to settle on pointed metallic bodies, as on spires and ships' masts, on the spears of soldiers, and such like, presenting to the spectator so many luminous points, led to the notion that these bodies attracted the matter of lightning, and that hence the lightning rods of Franklin were to be rather dreaded as a source of danger, than hailed as a means of protection. On the other hand, Franklin's hypothesis assumed that the influence of the rod was such as to charm to rest the otherwise unruly spirit of the thunderstorm, and by attracting to itself the matter of lightning, discharge it silently or in harmless corruscations into the earth. Upon this assumption there arose much difference of opinion relative to the amount of security the lightning rod could ensure, as compared with the quantity of lightning which its active influence might draw down upon the structure to which it was applied.

You will, of course, see, that in an early stage of electrical discovery, when a large induction of facts was wanting, either to settle the matter in favour of, or against, the new invention, such grounds of dissent were by no means either unreasonable or untenable. No great wonder then that men's minds, especially those of mariners, who of all others were frequently exposed to terrific thunderstorms, should have become confused and often superstitiously unsettled upon the question. We accordingly find very extraordinary means resorted to both at sea and on shore, with a view of keeping off lightning; means which forcibly point out the extremely unenlightened state of mankind upon this great and important question. In many countries, not excluding this, it was customary on the approach of a thunderstorm to agitate the air by ringing the bells of church towers, more especially on the Continent, where the bells had been commonly blessed by a religious ceremony. Occasionally the practice of a discharge of cannon or artillery was resorted to. In ships it was customary to apply wet swabs over the pumps and doors of the magazines, with a view to prevent the entrance of lightning. In houses, during a thunderstorm, it was customary, up to a recent period, to throw up the sashes and enclose glass articles, especially looking-glasses, in blankets. Balls of glass were placed on the mast-heads of ships and on steeples, as having the power to repel lightning. This superstition, so grossly absurd, has been revived even in the present day, when science has made such great advances as would lead one to imagine so ridiculous an assumption impossible.

It will be my object on the present occasion to bring under your notice some striking practical deductions and experimental illustrations bearing on the question of security against lightning, and by the assistance of three great lights of Baconian Philosophy—induction, observation, and experiment—so to simplify and clear the subject of irrelevant and sophistical argument, as to leave no doubt on your minds as to the course we should pursue in placing buildings, magazines, ships, and other structures, beyond the reach of danger from one of the most terrible elements of nature.

Before, however, entering more especially upon the laws and mode of operation of the peculiar physical force active in atmospheric electrical discharge, it may not, perhaps, be amiss to call your attention to some of its distinctive effects.

A writer in "Nicholson's Journal of the Progress of Science" calculates the loss by lightning, in Great Britain alone, at £50,000 sterling annually. I believe this to be under the mark. In the Royal Navy, during the late war, the country was losing at least £10,000 a year upon cases of damage found recorded in the journals of H.M.'s ships. We had, between the years 1810 and 1815, that is, within about five years, no less than forty sail of the line, twenty frigates and twelve sloops and corvettes, placed *hors de combat* by lightning. In two hundred and fifty such cases one hundred seamen were killed, and two hundred and fifty, at least, severely hurt. In the Merchant Navy, within a comparatively small number of years, no less than thirty-four ships, most of them large vessels with rich cargoes, have been *totally destroyed*—being either burned or sunk—to say nothing of a host of vessels partially destroyed, or severely damaged. Damage to H.M.'s ships by lightning has happily ceased : it is now not known in the British Navy. In the Merchant Navy, however, it unfortunately continues, no adequate means being taken, as in the Royal Navy, to check it.

If we turn to buildings on the land, here also we find records of destruction of the most frightful character. Fuller, in his "Church History," says, that "scarcely a great abbey exists in England which once, at least, has not been burned by fire from heaven." He quotes, as examples, the abbey of Croyland twice burned, the monastery of Canterbury twice, the abbey of Peterborough twice, the abbey of St. Mary's, in Yorkshire, the abbey of Norwich, and several others. The number of churches and church spires wholly or partially destroyed by lightning is beyond all belief, and would be too tedious a detail to enter upon. Within a comparatively few years, in 1822 for instance, we find the magnificent cathedral of Rouen burned, and so lately as 1850, the beautiful cathedral of Saragossa, in Spain, struck by lightning during divine service and set on fire. So lately as March of the last year, a despatch from our Minister at Brussels, Lord Howard de Walden, was forwarded by Lord John Russell to the Royal Society, dated 24th February, stating that on the Sunday preceding a violent thunderstorm had spread over Belgium, that twelve churches had been struck by lightning, and that three of these fine old buildings had been totally destroyed.

Explosions of gunpowder magazines by lightning have been frequent. Lately, at Sondpore in India—in 1857—lightning fell on the magazine, fired it, and nearly a thousand people lost their lives. In the preceding year, 1856, lightning struck the church of

St. John's in the Island of Rhodes ; the electrical discharge exploded a large deposit of gunpowder in the vaults beneath : the whole building and part of the town were laid in ruins. In 1829 lightning blew up the magazine at Navarre, and destroyed everything around. Admiral Rosimel reports this case to his Minister of Marine at Paris. A lamentable occurrence of this kind in 1769 laid a fourth part of the beautiful city of Brescia in ruins. About twenty such cases may be quoted.

Without going more at length into instances of destruction by the terrible agency of lightning through a long series of years, I will merely mention a few cases that have occurred within the last two or three years.

In 1859 the General Hospital at Jersey was struck and set on fire ; all the centre of the building was destroyed. In 1857 a goods shed on the South Eastern Railway was struck ; 350 feet of the roof fell in. In 1860 several churches were struck and severely damaged.

I cite these few cases merely with a view of showing how much more frequent and ruinous are the effects of lightning in this country than is commonly supposed ; and now having, as I think, sufficiently adverted to this, I will, in conclusion, ask your attention to two instances of the actual effects of this agency, as exemplified on sea and on land. These will better enable you to understand what sort of agency we have to deal with.

In this Diagram, No. 1, you have a faithful representation of one of our frigates, the *Thisbe*, of 36 guns, struck by lightning off Scilly, in January, 1786. I will read you the extract from the ship's log, and I think you will say that nothing can be more frightful :

" Four a.m. Strong gales ; handed main-sail and main top-sail ; hove to with storm stay-sails. Blowing very heavy, S.E. 4.15, a flash of lightning with tremendous thunder disabled some of our people. A second flash set the main-sail, main-top, and mizen stay-sails on fire. Obliged to cut away the main-mast ; this carried away mizen top-mast and fore top-sail yard. Found fore-mast also shivered by the lightning. Fore top-mast went over the side about 9 a.m. Set the foresail."

Well you see here was a ship reduced to a mere wreck. I put this case before you as a powerful practical illustration of what lightning may effect ; and I beg you not to take it as an isolated case, happening once in a hundred years ; such cases are numerous. Here is another similar case in a sister ship—the *Lowestoffe*, struck in the Mediterranean, in March, 1796, some few years after. I will read you an extract from the log of this ship. "North end of Minorca, S. 46 W.; heavy squalls, E. by S.; hail, rain, thunder, and lightning. 12.15, ship struck by lightning, which knocked three men from the mast-head, one killed. 12.30, ship again struck. Main top-mast shivered in pieces; many men struck senseless on the decks. Ship again struck and set on fire in the masts and rigging. Main-mast shivered in pieces, fore top-mast shivered; men benumbed on the decks and knocked out of the top, one man killed on the spot. 1.30, cut away the main-mast; employed clearing wreck. 4, moderate; set the foresail." Here is another ship left a wreck on the sea by lightning.

The Diagram No 2 is a faithful representation of the remains of the beautiful spire of St. Michael's Church, at the Black Rock, Cork, as it appeared on the morning of the 30th January, 1836, after being struck by lightning on the preceding night. Imagination cannot form a picture of a more awful ruin. The side of the steeple was fairly torn out, the top of it swept away, and the stones scattered in all directions. You will, I trust, excuse my troubling you with these introductory remarks; but I am desirous you should fully comprehend the kind of agency with which we are dealing, and the vast importance of the question of security against its effects, under the form of lightning, to the national interests and to the public service generally.

Well, then, having by a sort of *Discours Préliminaire* so far called your attention to the nature of the subject of which we are about to treat, I shall proceed with a few remarks on electrical force generally, at least so far as involving atmospheric electricity and the action of lightning, and electrical discharge. In doing this, I shall not pretend to develop what is usually termed a course of elementary electricity. I have no doubt but that you are fully aware of the ordinary elementary phenomena of electrical action, so that what I shall lay before you will be easily comprehended. We will, therefore, step on to great leading facts and principles of a practical kind, bearing, more especially, on the great question we have in hand.

In the first place, I must request you to dismiss from your minds all notion of an electric fluid or fluids, as the immediate or occult power in which electrical action originates. It is a very common and a very learned error to refer the tremendous power of which I have given you such striking examples to an immaterial fluid of high elasticity; an assumption involving such manifold absurdities

T 2

as to be utterly untenable in treating a practical question such as we have now before us.

When Newton by a sublime geometry and a profound analysis treated the movements of the planetary system, he was content to recognise gravity, in his doctrine of a universal gravitation, as a physical force taken in the abstract ; and in order to study its laws and mode of operation, he did not for a moment, at least in his more immediate researches, seek to determine, or pretend to explain, the occult nature of gravity,—considering such an inquiry without the limit of legitimate physical research. Now, this is just the course I propose to you to follow. We shall view the agency which rent open the steeple—already referred to in this diagram—as a physical force of nature, taken as force merely, and we shall study its laws and its mode of operation. Now it is no sound objection to this to say that, being ignorant of the nature of the agency with which we are dealing, we can scarcely treat it with clearness and precision, because the great end of modern philosophy is rather to trace and apply the uniform relations of certain facts, than speculate upon the source of a mysterious causation, most probably by a decree of the Almighty placed beyond the reach of that mental power with which he has endowed us, and which tells us pretty plainly how far we can *not* reach, at all events.

Suppose I bend a spring—a Kentish bow, for example, with which our forefathers shot their arrows, a cloth-yard long—what do I effect ? You will immediately perceive that all the particles on the concave side of the arc are compressed ; all those on the convex side become expanded. I relax the string : the particles on the convex side contract,—which force, for distinction sake, I may call *positive;* those on the concave side expand,—hence another kind of force which I may call *negative.* Or, reciprocally, these forces make up the total power of the bow, and, in combining, operate so forcibly on the arrow, through the bow-string, as to shoot away the arrow with enormous power and great velocity. Now we are accustomed to designate this power by the term elastic force, or elasticity. And really the fact itself is all we know about the matter. No one can say in what the elasticity consists, or really how the motion is induced; and I think you will see how little we shall be likely to gain by assuming the existence of an elastic fluid of high tension as the source of it. Similar reasoning applies to the assumption of an electric fluid as a source of electric power. For any good, sound, practical purpose we had better come at once to consider electricity as a physical force, the nature of which is quite unknown to us, but the laws and modifications of which are within our reach.

Observing then a peculiar series of phenomena which lead us directly to recognise the existence of a peculiar unknown power of nature which we term electricity, the first important fact we observe is the dual nature of this force: it is made up as it were of two distinct and inseparable modes of force—these have been for the sake of clearness termed *positive* and *negative*—as we named the forces of the bent bow. You cannot have one of these alone,—the other is always present. Hence the analogy of the bent bow helps us in some degree, although it may not help us in other cases. In the bent spring we have, as already observed, compression of particles upon the concave surface, extension upon the convex, from which results a force of expansion on the one side, a force of contraction on the other; both forces are operative in bringing the bent bow to its normal state. You cannot in the bent bow have one force without the other. Now such is the case in electrical action. Electricity, therefore, like elasticity, you will regard as a peculiar property of common matter, or a peculiar force impressed on it. The immediate sources of the phenomenon that we term electrical, or the means of bending, as it were, our electrical bow, are numerous; for instance—changes of temperature; changes of form; mechanical actions, such as friction, pressure, percussion, and the like; to which may be certainly added the contact of bodies; any of these may superinduce upon common matter an apparently attractive force, as you see in all common electrical experiments. In this case the bodies are said to be electrified, or charged with electricity.

A most wonderful and striking fact, characteristic of electrical force, is its apparently sympathetic action or foresight, as it were. It pioneers, in fact, for its subsequent operation; it finds out, in advance and beforehand, what it can effect, what course it may best follow, and it impresses upon common matter a sort of physical preparation for its reception, in the shape of an active power. This species of electrical action, or, to all appearance, action at a distance, has been commonly termed *induction;* and by our scientific neighbours, the French, *électricité par influence.* I must beg of you before entering upon a few elementary experimental illustrations of these and other electrical phenomena, to take the electrical apparatus before you to a certain extent upon trust, or rather upon the faith of your general knowledge of ordinary electricity, or upon the ground of a broad generalisation, without further elaborate explanation and detail. You will be so good, for example, as to consider what is termed the *electrical machine* as an instrument which enables us to call up, through the aid of friction, the force we term *electricity*, that is, the true electrical powers of which we have been speaking, and apply either of them or both to the purposes of experiment. You will look at the Leyden or electric jar as a means of accumulating one or both of these forces, and causing them to assume a dense explosive form, imitating a discharge of lightning. Other kinds of apparatus or auxiliary instruments you will view as instruments for facilitating experiments on electrical accumulation and discharge, or other electrical investigation. We shall occasionally offer a few explanatory remarks on all these instruments as we proceed, many of them not being generally known.

It may, perhaps, be as well to observe in passing, that the physical agency or power we term electricity, is subservient in a certain sense to the kind of common matter with which it is associated ; some bodies, for example, propagate or transmit it with rapidity from one substance to another. Such bodies have been since termed electrical *conductors*. The metals, acids, carbon, &c., do this ; other kinds of matter, such as vitreous and resinous bodies, arrest its progress, and stop its course ; they, in fact, imprison it, as it were ; these have been termed *insulators* of electricity.

We will give a few practical illustrations of this.

Exp. 1. Electrical machine-conductors and non-conductors.

All bodies have conducting power; none are perfect insulators. Division into most perfect, less perfect, and imperfect. See Table.

We have further to observe that an electrified body is any insulated substance having either of the electrical forces dominant in it, or in different parts of it; and it is said to be charged positively or negatively, according to the particular modification of force with which it is thus electrified. In common parlance, the two forces have been designated positive and negative electricities; positive electricity being arbitrarily associated with the electrical development induced by the friction of glass with silk; whilst negative electricity is identified with the development induced by the friction of resinous bodies with woollen ; hence, as you know, the terms, *vitreous* and *resinous* electricity, often called the *vitreous fluid* and the *resinous fluid.* We may certainly employ this figurative language as a matter of convenience, but we must take very especial care not to build on it as fact. As already noticed in the case of the bent bow, we may for convenience call the force of the convex surface *positive* elasticity, and the force of the concave surface *negative* elasticity; but by this we are after all only designating force, or kind of force.

Thus premised, we will first take a case of what is termed *electrical attraction*, and I propose to show you that in this the previous induction or foresight of the electrified body is present. The phenomenon we term attraction, you will observe, is the first and most palpable evidence of the presence of electrical force. It was in fact from this apparent power developed by the friction of amber, that the Greek philosopher, Thales, inferred that mineral to possess a species of animation; hence, as you are aware, the term electricity, from the Greek ηλεκτρον, signifying "amber."

Exp. 2. Suspended movable arm and disc with gold leaf on opposed electrified bodies; the leaf diverges, and attraction ensues. Explain the experiment.—The quantity jar, carrier-ball, &c., &c.

We see here a striking instance of what is termed electrical induction—apparently action at a distance. We will take it as such, without entering upon theoretical explanations. So that you see before a body is attracted electrically, it is first rendered attractable; the twin power or force is first developed in it by the influence of the conjugate or attracting force; and this is a universal law of electrical action. But we have apparently a second species of electrical force to bring under your notice, evident in the divergent leaves, also important to illustrate. In the previous experiment we see the immediate result of a development of the two electricities, viz., attractive power. But let us see what would happen, supposing each of the bodies were electrified alike, either positively or negatively. In this case the tendency to a final attractive force is still present, and an effort is made by each body, as in the former case, to develop an opposite power in the other, without which it may be shown that no attraction could ensue. Before this attraction can be effected, the electrified state of each must be more or less changed, their states must be in some considerable degree not merely reduced to zero, but positively reversed. The consequence of this effort is an apparent repulsion of the two bodies, if free to move. The already existing electrical states refuse, as it were, under existing circumstances, to be coerced in this way; they resist, and the bodies separate. Still here, as before, the advanced influence is first operative, before divergence or separation of the two electrified bodies ensues.

Exp. 3. Show the increased divergence of the electrified leaf on the movable arm at the instant of repulsion.

It is from this and other phenomena that we have the well-known electrical formula—dissimilar electricities attract—similar electricities repel. That is to say, two positive or two negative forces repel; a positive and a negative force attract. In fact, attractive force cannot exist except in the case of these opposite forces.

Now whatever be the source or occult cause of electrical force, certain it is, the two forces it develops are singularly and peculiarly associated with common matter. Different kinds of matter have a disposition to develop positive electricity; others negative electricity. Thus sulphur in falling through the air becomes negatively electrified ; red oxide of lead positively electrified. The following experiment, whilst it shows the attraction of opposite forces, illustrates in a remarkable way this disposition of certain bodies to assume a positive or a negative electrical force.

> *Exp.* 4. Sift red lead and sulphur on a resinous plate, and explain.

We will here stop to illustrate two important laws of electrical force immediately bearing on our present subject. We have already shown the tendency of two bodies charged with opposite electricities to approach each other. Now such is precisely the case with electrified clouds—such, more especially, as those we term thunder-clouds—and the opposed surface of the earth. It is well we should understand, therefore, the law according to which the force increases, as we increase the *quantity,* as we may term it, of the electrical charge, or as we diminish the distance.

Now it may be shown that the force between a mass of electrified clouds and the earth increases with the square of the quantity of electrical charge directly, and decreases with the square of the distance inversely. Very much intricate discussion and experimental and analytical calculation has been expended upon these laws of electrical force ; but I believe I shall be enabled to show you, by the balance apparatus before us, that, as mere experimental facts, the demonstration of this is simple and easy.

> *Exp.* 5. Discs and electrical balance. Accumulate a certain number of measures of charge on discs. Constant distance. Note weights for these measures. Double the measures, and again note the force. F $\propto 2^2$.
>
> *Exp.* 6. Double the distance for the last noted quadruple

force. The force will be now one-fourth and same as at the unit of distance.

It would be doubtless interesting to pursue this question of electrical attractive force, and unfold all its very wonderful phases and physical bearings on bodies of various forms,—as, for example, on spheres, planes, and other surfaces; but, as I have already remarked, we cannot possibly on this occasion treat the whole science of electricity. I must, however, call your attention to an extremely important phenomenon of electrical force as thus exemplified, viz.—wherever induction can go on, attractive force follows. If it cannot, or at least, to any extent, we have little or no attractive force. Be so good as to fix attention on this, because it bears immediately on the question of protection from lightning. We have all heard of the notion entertained by the generality of persons of the danger incident to metallic bodies on the occurrence of lightning storms, because of their superior attractive force,—vanes, spindles, knives, and spears, to say nothing of metallic substances generally, have all been commonly deemed dangerous in the extreme. This, however, is an unlearned and a very vulgar error, as you will presently find. Pointed metallic bodies, or other metallic bodies, have no more power of drawing or attracting lightning than non-metallic substances, or even insulators of electricity, taken in the sense implied,—that is to say, as having a specific affinity or inherent attractive force for the matter of lightning. Both Cavendish and Coulombe found electrical charge to be quite independent of the nature of the substance charged. An equal division of electricity always obtains between bodies sooner or later without any regard to the kind of substance electrified. I proceed to one or two illustrations of this.

Exp. 7. Electrical balance—wood disc—metal disc equal. To show attractive forces equal, or spheres.

In these and the preceding similar cases you perceive, as indicated in the third experiment with the movable disc, a free induction upon the earth, or upon the source of charge, is open to the attracting and attracted bodies. But let us arrest this induction in some considerable degree, by insulating the attracted body,—that is, cutting off its free induction upon the ground,— then the opposite electrical force is no longer developed to an extent sufficient for attractive effect,—that is, in the especial case I am about to show you; for you are to recollect that it is upon the opposite electrical forces the action depends—not upon the kind of matter absolutely with which they are associated.

Exp. 8.—Suspend the thin disc by silk or gutta-percha or fine silk gut; compare, first, the attractive force of a given number of measures, when uninsulated by a fine wire. Now remove wire and weight of wire. Little or no attraction.

You perceive in these instances how ill supported by experiment, and, as you will presently find, by the great physical operations of nature, are the common notions of lightning as being especially attracted by certain kinds of matter in preference to other kinds of matter. There is, however, one peculiar kind of action which appears to favour the conclusion that attractive force is resident in pointed metallic bodies, seeing that lightning falls on the metallic-pointed terminations of church spires, the vane spindles of ships, and such like, and, as it is assumed, with a disregard of other bodies. In treating this question, which is of a somewhat vexed character, you will have to draw a very nice line between absolute attractive force and apparent attractive force, and limit by clear definition, what you really understand by attraction. We will first proceed to one or two experimental illustrations.

Exp. 9.—Let a current of sparks pass between two balls, as in Exp. 1; present a pointed uninsulated metallic conductor at a distance; sparks stop; take away, they return, &c. The point becomes luminous at a distance.

It may be here inferred that this pointed body has actually attracted the matter of lightning towards itself; well, let it be so ; we will by way of a *reductio ad absurdum*, grant it. Then comes the question, is this result peculiar to pointed metallic bodies ? Is it not also an effect producible by other bodies not metallic ? Let us see :—

Exp. 10.—Repeat former experiment, and present a wooden point : same effect ensues.

You see common wood has the same property; it is just as attractive of the matter of lightning as metal. Well, now comes the question (for you see I am treating this matter pretty plainly and practically, and with an eye to our subject), will the metal point always, under all circumstances, act in the way just exemplified ? No, it will not.

Exp. 11.—Repeat stream of sparks as before; present an insulated point; sparks do not stop.

We may fairly question the assumption, upon the faith of these facts, of the peculiar affinity or attractive influence of metallic bodies for the matter of lightning. This leads us to inquire further into the immediate source of the phenomenon in question, viz., this peculiar influence of pointed bodies.

On a careful investigation of this phenomenon, we find it to depend on a peculiar adaptation of acute terminations to the purposes of electrical induction. Acute terminations in free conducting bodies so dispose and prepare the particles of any resisting medium, intermediate between attracting bodies in opposite electrical states, as to greatly facilitate electrical action in the direction or position in which they happen to be placed; when they are not enabled to do this, then, as you have just seen, they are inert. Thus, in Exp. 10, as we have just witnessed, the force is originally exerted between the discharging balls. The two forces are developed between these; their union is attended by a breaking down of the particles of the resisting atmosphere intermediate between them. Light and heat are both developed in this rupture, and a succession of explosive sparks ensues, as we continually evolve electricity by the electrical machine. But now, when I bring a distant free point into competition with the receiving ball of this system leading to the earth, then by its peculiar action on the arrangement of the particles of the intermediate resisting air, it facilitates the inductive force on the direction of its position to so great an extent, as to supersede the explosive discharge between the discharging balls altogether, and thus cause what we may term the current of discharge to pass in a different direction, and in a thin attenuated stream upon the point, without explosion. This is an important phase of electric force. You may call it *attraction* if you like; but it is not that species of attraction as commonly understood. If we place a large hollow pipe in communication with the gutter leads of a building, which, by its facilitating the discharge of heavy rain, prevents its accumulation and discharge in some other direction, we may, in a certain sense, and if we choose to indulge in such a figure, attribute the operation of this pipe to a species of attractive force upon the water, by causing the water to gravitate in a certain direction; but we surely should not dream of attributing to such a pipe a power of attracting the rain from the clouds, and so drawing down the water upon the building!

Let me for an instant, by way of concluding this part of our subject, ask the favour of your momentary attention to a final experimental illustration bearing on this question. We have seen that without the presence of the two electrical forces, without this

pioneering influence of induction, we have little or no result; let
us now see how this may be promoted by a pointed free body—I
mean a conducting body in connection with the earth, and acting
at a distance; and also, when this induction is perfected, how
immediately the attraction ensues.

> *Exp.* 12.—Repeat Exp. 9. Collect a certain number of measures
> sufficient to induce attraction in the free state ; then present
> a point over the suspended disc; attraction immediately
> ensues by completing the induction; so it does with wood.

We see here a case of an apparent repulsion of a point.

We shall conclude this First or Introductory Lecture here, our
time being pretty well expended. Before I do so, however, it
may be well to pass over again, by a brief recapitulation, the path
we have been following.

RECAPITULATION.

We have, first, in the way of introductory matter, called attention to Franklin's invention of the lightning-rod, and the objections to it ; we have next offered remarkable instances of the
effects of lightning and its ruinous consequences to public and
private interests ; we have taken a general view of electrical force,
on which the destructive effects we have called attention to
depend: we find it a sort of dual power, and made up of two
modes of force ; we have seen that in every case of what is termed
electrical action, both these kinds of force, termed, for distinction's
sake, positive and negative, are called up by a peculiar operation,
termed induction, apparently action at a distance; and we have
seen this to be a sort of pioneering action, and precedes the
electrical phenomena of attraction and repulsion.

Considering attraction as the immediate consequence of electrical
development, we have called attention to two important laws ; we
have found the force to be directly as the square of the development, or what for convenience we may term the quantity of
electricity accumulated, and inversely as the square of the distance
between the attracting bodies. We have shown these laws to
depend on a free induction : if either body be limited in this
respect and the induction arrested, we obtain little or no attractive
force. We have called attention to the peculiar action of pointed
bodies, considered as sources of attraction, and shown upon what
their operation depends.

SECOND LECTURE.

HAVING in my last lecture considered certain elementary principles of ordinary electrical action, bearing on the phenomena of thunder-storms, I now proceed to the more immediate object of these lectures —the laws and operation of electrical discharge under the form of lightning, and the means of protection from its destructive effects. With a view to continuity, however, you will, I hope, excuse a brief recapitulation of my former observations.

It will be convenient and perhaps judicious, in continuing these remarks, to commence with a general view of Atmospheric Electricity and the nature of thunder-storms. By certain natural actions, such as changes of temperature, chemical or other operations, the agency which we term *Electricity* becomes developed in our atmosphere, and some thousands of acres, perhaps, of condensed vapour, under the form of cloud, become charged with electricity. The air intermediate between the plane of this area of cloud and the earth's surface participates in the inductive force upon the plane of the opposed surface of the earth beneath. The two electrical forces thus disjointed and brought into play seek, by a law of nature, to reunite and recover their normal state of repose, just as water, by a law of gravity, seeks its level of rest. But as you saw in the preceding lecture, air is an insulator, and will not admit of this immediate union of the two forces, until at length the attractive force between the plane of the area of cloud and that of the earth's surface beneath is so powerful as to break down the intervening air with a terrific and dense explosion. Now it has been a very common misapprehension, in ordinary views of lightning discharge, to consider strokes of lightning on buildings as a species of force or action set up between the building, or some part of it, and the thunder-sky which sends forth its terrific flash, and which alarms us by the power of its dreadful artillery. This is a great mistake : the action is not between the clouds and the building, but

between the clouds and a large area of the earth's surface; the
building is only acted on and struck by the electrical discharge, as
being a point in the earth's surface, and at the moment placed in
a given electrical position.

Well, then, in this point of view, we have to take into account
the area of watery vapour, the area of the earth's surface opposed
to this, and the intermediate air; and we have to consider the
peculiar conditions of these three elements. First, we have a
development of force by induction between the clouds and the
earth, by which opposite forces, as already observed, are called into
play. These may be taken as the terminating planes of the action.
Now the condition of the intermediate air, termed by Faraday the
Dialectric Medium, or medium through which the forces operate,
is an extremely interesting question.   It is really difficult to
realise action at a distance, as if by influence or sympathy; and yet
we can scarcely tell how to avoid it.   According to Faraday, the
intermediate particles of the atmosphere—that is, the dialectric—
become *polarized*. Electrical forces in these particles developed by
influence become disposed as in the diagram. The positive and
negative forces, which originally existed in a state of union or
repose, are separated and opposed to each other in a series, and so the
force becomes propagated to or from the earth's surface. So long
as the particles can sustain this forced state, so long *insulation* is
the result, and the forces upon the terminating planes of the action
—that is, the area of cloud and the earth's surface—are maintained;
but if they break down, as it were, or give way to reversion of the
polarized state, the whole system restores itself to its normal condition by what is termed *Disruptive Explosive Discharge*. Light
and heat are evolved, attended by violent expansive power. Some
little illustration of this by means of experiments may perhaps be
useful here.

Exp. 13, with illustrations.   Two circular discs, 18 inches and
1 foot, fitted with electroscopes, show induction, insulation,
and discharge, and condition of air between the plates, &c.

Conceiving the fixed large circular disc to be a portion of the
earth's surface, and the lesser disc an area of cloud over it, let us
see first what is the immediate result of an electrical development
on this upper area. I may be supposed to hold in my hand here
some thousand acres of charged cloud; the divergent electroscope
is evidence of the presence of this charge; I bring it immediately
over the lower fixed plane. You see its inductive effect: all the intervening air is now in a state of great excitation, and if the force

induced between the two planes were sufficiently great, it would break down with an explosive spark, termed by Faraday, as already noticed, the disruptive discharge. This explosive disruption would occur in the points of least resistance—that is to say, in points in which, from a great many incidental circumstances, such as the presence of foreign matters, &c., the particles of the polarized atmosphere would most readily yield. Imagine the small body I place upon the lower disc to represent a building : this would be a mere point in the great surface. It would not produce discharge as upon itself. It would be nothing more nor less than the door-way, as it were, through which the electrical forces operated between the opposed planes. An easier course, not far distant from the building, might arise from many causes, in consequence of the easier yielding of the air in that line. We have represented in this diagram a mass of clouds, c, a portion of the earth's surface, 2 N, with an intermediate atmosphere; R, the position of a building or ship; and suppose a thunder-stroke to originate at c, we at once perceive that resistance from many causes, through c N or c o, might be much less than through the course c R, so that the thunder-stroke is in no sense determined by attractive force in the building; it is the total force of the whole surface which deter-mines it. And when we consider how small comparatively is the magnitude and elevation of any building or other artificial structure, when taken in relation to a vast mass of clouds and electrified air, we see that it vanishes in the general surface. We find accord-ingly that lightning often strikes near a building or ship directly through the air without touching it ; and so definite and determined is its course, that instances have occurred, especially in the navy, in which the sleeve of a coat has been rent from the arm of the wearer without damage to his person, so that this is purely a question of resistance in this or that direction, as we shall presently see.

Before I proceed to illustrate the laws of destructive electrical discharge, and the means of palliating or avoiding its terrible effects by an efficient and systematic application of metallic con-ductors, it may be interesting to exhibit, by very conclusive and pretty experimental illustrations, the operation of a charged atmo-sphere such as we have alluded to. I have little doubt but that highly charged clouds and a polarized air are the sources of many of the phenomena classed as *whirlwinds*, *waterspouts*, and such-like, and certainly of the luminous phenomena frequently seen on the vane-spindles of ships and towers, termed by sailors *comazants*, and noticed on soldiers' spears by ancient writers. Thus Cæsar, " On the

African War," *De Bello Africano*, says, "That same night the points of the spears of the fifth legion burned with fire," &c. Seneca, again, in his "Natural Questions," remarks, "The spears in the Roman camp seemed to be on fire."

Franklin, after his great experiment with the kite, in which he obtained electricity from the atmosphere as we do now artificially by the conductor of the electrical machine, resorted to an extremely elegant and simple contrivance, by which he obtained notice of the passage of charged clouds over his house. I have in my last discourse shown and explained the operation of pointed free conductors. I have here so modified Franklin's beautiful contrivance as to bring it within the limits of ordinary experiment.

Exp. 14. The Thunder-cloud and Bells.

The source of the electrical accumulation is here our electrical machine, the means of accumulation the electrical jar, a swinging arm poised on the charging rod carries an artificial cloud, which, being composed of light cotton wool, freely receives the concentrated electricity, and spreads its fibres in the way of the electrified vapours of the atmosphere. The air under this charged substance becomes affected by induction, which reaches to the bodies beneath, just as in the case of a charged atmosphere, and will hence influence a building with a pointed conductor when passing over it. If the conductor be insulated as in the model, and be connected with an insulated bell, the charge it receives will accumulate on the bell, and will readily attract a small insulated ball of metal freely suspended near it by a silk thread. But when this ball becomes, by contact with the electrified bell, also electrified, and with the same kind of electricity, then, as we have seen, it tends to separate from it by a species of repulsion; so that if a second bell uninsulated—that is, in connection with the earth—be placed on the other side of the suspended ball, it will pass to that, and discharge its electricity upon it, in which case it will return again to the electrified bell, and so continuously as long as the artificial cloud throws off electricity and charges the intervening air. We have in this way a continued sounding of the bells, a constant ringing. Such was Franklin's arrangement through his house. The passage of a cloud charged with electricity over his conductor invariably gave notice of its presence.

The phenomena of whirlwinds are easily imitated in this way, as in the whirling columns of sand in a desert, and such like.

Exp. 15. Some light bran is to be strewed on a conducting surface, and an artificial cloud made to swing over it. The bran will be caught up in a sort of whirlwind.

These experiments apply more especially to certain phenomena of a charged atmosphere unaccompanied by condensed explosion or disruptive discharge. This we will now proceed to consider.

When the disjointed electrical forces are so powerfully attractive of each other as to break down all insulating resistance, and unite with spasmodic violence, then it is that the phenomena termed *Lightning and Thunder* present themselves. Lightning is the extrication of brilliant light, accompanied by heat, from the air or dialectric medium, whose particles are violently broken through ; that is, *depolarized*. Thunder is the *crash* of this effort, prolonged by repeated echoes and reverberations from the surface of clouds or other bodies, just as the sound of artillery reverberates amongst distant hills. The crash itself, heard near at hand and apart from reverberation, is something very horrid : it is as if an enormous number of earthenware jars of great magnitude, suppose ten thousand in number, were smashed by falling from a great altitude upon a reverberating stone pavement. When the thunder-stroke is close, a vivid gleam of light immediately precedes this. If distant, the crash is modified and lost in a roar of reverberation, and is not heard until some seconds, more or less, after the lightning flash ; so that by noting the time which elapses between the flash and the sound, we may, as in the case of cannon, estimate the distance : the more immediate is the crash upon the light, the nearer is the discharge.

The distance of air through which disruptive discharge can break is directly as the electrical disturbance ; or, to use other but more figurative language, directly as the quantity of electricity accumulated between the terminating planes of the action— viz., the clouds and earth—or artificially between the coatings of a fulminating square of glass, or the Leyden jar. The following experiment will pretty fairly demonstrate this.

Exp. 16. If two smooth balls be opposed to each other, one of them connected with the rod of a charged jar, the other with its outer coating, and measured quantities of electricity be accumulated, the jar will discharge by breaking through the air intervening between the balls ; and the distances between the two balls will measure the accumulated quantity, or very nearly.

υ

We saw in a former experiment that the attractive force of the accumulation, as measured by a delicate balance, was as the square of the quantity of electricity; and we have to reconcile these apparently opposite results, and show how it is that whilst the attractive force is as the quantity squared, the explosive force or distance is as the quantity simply. A very little examination will suffice to elucidate this. You will immediately perceive, by reference to this diagram, that since the force increases with the quantity squared directly, and decreases with the square of the distance inversely, the force for explosion must remain the same at all distances, and must be as great as is requisite to overcome the atmospheric resistance, which is a constant quantity.

Imagine discharge to occur at distance D, between two balls, P, N, with a given quantity of electricity, which we will call 1 or unity; then whilst undischarged, or before discharge takes place, let the receiving ball N be removed to o, or at distance D^2 from P. In this case the force between P o will be only one-fourth of what it was before, since the force is as $\dfrac{1}{D^2}$. Discharge could not, therefore, take place at this distance, P o, with the unit of quantity. But now suppose we double the quantity and make it 2, or twice as great; then, because the force increases as the square of the quantity directly, the force in P o becomes four times as great, and is hence from its reduced state of one-fourth raised up to one—that is, to what it was before—and now the explosive discharge again ensues, so that what we lose by distance we gain by increased quantity.

I have now to call attention to certain laws of explosive electrical discharge very important to keep in view in all our arrangements for obviating the destructive effects of lightning. On examining numerous instances in which ships and buildings have been struck and damaged by lightning, we find the electrical discharge always pursuing a determinate line or lines, which upon the whole present the least resistance to the neutralisation of the two opposite electrical forces. Both space and time appear to be economised as it were in its progress. For however small we assume the duration of the thunder-stroke, or however limited the distance through which the lightning strikes, both these quantities would become still less were other lines of transit provided opposing still less resistance. This is the leading character of electrical disruptive discharge. Hence a stroke of lightning seizes upon bodies which lie convenient and ready for its progress, absolutely avoiding other bodies, however near, from which it can receive no assistance, as I have already remarked; and I must again here observe, as a most marvellous and interesting fact, that at the instant before electrical explosive discharge occurs, the explosive stream, in the act of moving between the opposed planes of action to a condition of neutrality, feels its way as it were in advance, and marks out by a species of foresight the course it is about to take. The course of lightning is not, as many imagine, left to the chances of the instant, to be drawn aside this way or that by adjacent metallic or other matter. On the contrary, the course of a thunder-stroke is already fixed and absolutely settled before the lightning appears and the actual discharge takes place. The following experiments will serve to illustrate this selection of a least resisting course, and how electrical discharge seizes upon some bodies and avoids others, according as they lie in positions favourable or unfavourable to its progress.

Exp. 16 and *Exp.* 17. Disjointed gold on papers, and also illustrating ships' conductors.

These separated fragments of gold leaf may be taken to represent detached perfect or imperfect conducting masses anyhow placed in a building or ship: the course of a discharge of lightning through them will be found to obtain on precisely the same principles.

THE END.

VIRTUE AND CO., PRINTERS, CITY ROAD, LONDON.

www.ingramcontent.com/pod-product-compliance
Lightning Source LLC
Chambersburg PA
CBHW021502210326
41599CB00012B/1097